THEMATIC MAPPING FROM SATELLITE IMAGERY
a guidebook

CARTOGRAPHIE THÉMATIQUE DÉRIVÉE DES IMAGES SATELLITAIRES
un guide

THEMATIC MAPPING FROM SATELLITE IMAGERY
a guidebook

CARTOGRAPHIE THÉMATIQUE DÉRIVÉE DES IMAGES SATELLITAIRES
un guide

Edited by J. Denègre

Institut Geographique National, 136 bis, rue de Grenelle, 75700 Paris, France

INTERNATIONAL CARTOGRAPHIC ASSOCIATION

ASSOCIATION CARTOGRAPHIQUE INTERNATIONALE

Pergamon

Published on behalf of the International Cartographic Association by Elsevier Science

U.K.   Elsevier Science Ltd, The Boulevard, Langford Lane,
            Kidlington, Oxford, OX5 1GB, U.K.

U.S.A.   Elsevier Science Inc., 660 White Plains Road,
            Tarrytown, New York 10591-5153, U.S.A.

JAPAN   Elsevier Science Japan, Tsunashima Building Annex,
            3-20-12 Yushima, Bunkyo-ku, Tokyo 113, Japan

First edition 1994

Library of Congress Cataloging in Publication Data

Thematic mapping from satellite imagery: a guidebook = Cartographie
thématique dérivée des images satellitaires: un guide/edited by J. Denègre
-- 1st ed.
p. cm.
English and French.
Includes bibliographical references and indexes.
1. Cartography -- Remote sensing. I. Denègre, Jean. II. Title: Cartographie
thématique dérivée des images satellitaires.
GA102.4.R44T47   1994
526.9'82--dc20   94-19073

British Library Cataloguing in Publication Data

A catalogue record for this book is available from
the British Library.

ISBN 0 08 042351 5

Printed in Great Britain by BPC Wheatons Ltd, Exeter

Contents/Table des matières

Preface

The last decade of the twentieth century is seeing a marked increase in the number of programmes for Earth observation from space. This expansion reflects a growing concern for monitoring development and managing natural resources, taking into account environmental priorities. The programmes also respond to the needs of scientific researchers in understanding the mechanisms and evolution of global change.

It is likely that very soon the bulk of geographical data, at both global and national scales, will be derived from Earth observation from space. The conversion of this data into information which is portrayed in cartographic form is of key importance. The visual cartographic image will facilitate communication, analysis, synthesis, modelling and forecasting of both natural and human phenomena, as well as the important interactions between the physical and cultural domains.

In 1988, the Commission on Thematic Mapping from Satellite Imagery proposed that a guide-book on the topic be produced, given the increasing significance of these techniques. I am pleased to write the preface of that volume today. The *Guide-Book* is the result of a major international co-operative effort of specialists from many nations and its purpose is to draw on practical application experience to formulate general rules for cartographic production from satellite imagery. The volume is, therefore, useful both for educational purposes and applied production as it attempts to put the experience and knowledge already acquired within the cartographic community at the disposal of future practitioners.

I am very pleased to see this volume appear as it meets a significant and growing need and will be of great benefit to cartographers throughout the world.

D. R. F. TAYLOR
President
International Cartographic Association

Introduction

Satellite data contribution to cartography sets cartographers quite a number of specific technical problems which must be overcome should they want to exploit to their utmost the considerable opportunities (hardly seized as yet) offered by Earth observation satellites. Actually the latter appear as a crucial stage, not only in the matter of generalized remote sensing, but also of data-collection automation, since they cover the whole globe, and this several times a year, a month, or even a day in some cases. This automation converges with the one developed in computer-aided cartography and geographic information systems (GIS). The so formed synergy leads potentially to high power tools—the sole ones, to tell the truth, measuring up to present and future planetary challenges.

The purpose of this guide-book is to present as pedagogically as possible map (most often thematic ones) producing methods from satellite images. Such a presentation is meant for inexperienced satellite imagery users desiring practical guidance on methods employed and their expected results. For so doing, this guide-book is broken down into five chapters, each one representing one process stage and setting out its particular problems.

The first technical problem to be met is, of course, how to interpret the collected satellite data, in other words, how to convert these "data" into "information". This problem is itself subdivided into two parts:

(1) Selection of image types fit for highlighting information which is wanted to be particularly identified, such image types being characterized by their geometrical ground resolution, their spectral band(s), their repetitiveness, etc.; therefore, image types available nowadays and space programmes developed over the ongoing decade are described in Chapter 1.

(2) Data interpretation, according to characteristics of images and information looked for: it uses a number of methods belonging either to visual photo-interpretation or digital image processing or again to a combination of them both; these methods are explained in Chapter 2.

The second technical problem to be encountered is integration of information acquired from conventional cartography. Indeed, merely satellite-derived information is rarely sufficient for producing maps in the full meaning of the word: it must be combined with external conventional information, to a greater or lesser extent, which will come out as complementary in various fields: geographic reference systems, toponymy, urban centres outlines, road networks, etc. Such integration also takes on two aspects:

(1) "Digital" combination of both satellite and conventional information: integration into a same reference system (image rectification), adaptation of contrasts and colours, conversion to vector or raster mode, selection of information to be represented, screen or plotter display; these procedures which result in the GIS general functions are described in Chapter 3.

(2) Final graphic representation of combinations so obtained, in terms of semiology (choice of symbols, colours, letters sizes, etc.), scale and accuracy, leads to several types being distinguished ("iconomaps"); these drafting methods and resulting documented typology are described in Chapter 4.

To end with, a special chapter presents applications and samples of produced maps. Samples have been deliberately selected among various thematic categories in several countries, thus providing a panorama of world production of thematic iconomaps, themselves classified per theme (general cartography, geology, hydrography, oceanography, glaciology, agriculture, etc.). This forms Chapter 5.

Owing to this guide-book's bilingual nature, and also for some practical reasons, the following frame has been adopted:

English version: Chapters 1, 2, 3 and 4.
Bilingual version: Chapter 5 (applications and colour illustrations only).
French version: Chapters 1, 2, 3 and 4.

JEAN DENÈGRE
President of the ICA Commission
"Thematic mapping from satellite imagery" (1984–91)

Acknowledgements

The present guide-book is the result of international co-operation developed within the framework of ICA (Commission on Thematic Mapping from Satellite Imagery) during the 1987–92 period. Among the specialists from various countries who brought their contributions, six of them are the main contributors:

Prof. Andrzej Ciolkosz (Poland) for Chapter 1.
Jean Denègre (France) for Chapters 3 and 4.
Sten Folving (Italy) for Chapter 3.
Prof. Andrzej B. Kesik (Canada) for Chapter 1.
Donald T. Lauer (USA) for Chapter 2.
Dr Janos Lerner (Hungary) for Chapter 4.

In addition, Chapter 5 which presents some twenty practical applications is the result of numerous contributions, from Austria, Brazil, Czechoslovakia, Germany, Tunisia, etc. The names of the contributors have been mentioned each time in the title of the relevant applications. Printing of the colour illustrations has been carried out through the courtesy of ITC (International Training Center for Earth Sciences, The Netherlands) with the help of Mr Koert Sijmons.

Finally, as a symbol of international co-operation, this book is bilingual, using both official languages of ICA: English and French. This could not have been possible without the assistance of a number of benevolent translators from the Institut Géographique National de France: Jean-Claude Barret, Viviane Bois, Laurent Breton, Patrick Dayan, Jean-Pierre Delmas, Alain Jacqmin, Philippe Houssay, Allan Moore. In addition, let us mention the names of the typists: Annick Clavaud, Françoise Coulon, and Simone Chavaroche.

To all we express our best thanks.

Jean Denègre
Editor

Index of countries covered

Index of satellite images

Avant-propos

La dernière décennie du 20ème siècle voit se multiplier les programmes spatiaux d'observation de la Terre. Cette expansion correspond aux préoccupations croissantes en matière de maîtrise du développement et de gestion des ressources naturelles, tout en tenant compte des priorités de l'environnement. Elle répond aussi aux besoins de la recherche scientifique pour comprendre les mécanismes du "changement global" de la planète et prévoir son évolution.

Aussi est-il vraisemblable que les données d'origine satellitaire fourniront demain la majeure partie de l'information géographique à l'échelle du globe, aussi bien qu'à celle d'un continent ou d'un pays. Leur traduction sous forme cartographique demeure le point de passage obligé pour communiquer, sous forme visuelle et intelligible au plus grand nombre, les constats, analyses, modèles, synthèses et prévisions concernant les phénomènes naturels ou humains, et leurs interactions.

C'est pourquoi, après avoir présenté, en 1988, un aperçu de l'état de l'art en "spatiocartographie" thématique, la Commission spécialisée de l'Association Cartographique Internationale s'est proposé d'établir un guide des techniques spatiocartographiques. Je suis heureux d'en écrire aujourd'hui la préface. Le Guide est le résultat d'un large effort de coopération internationale de la part de spécialistes de nombreux pays. Il s'agit, à partir, des nombreuses applications apparues au cours des dernières années, d'énoncer les règles pratiques qui sont à la base des processus de production cartographique. Ouvrage pédagogique autant que pratique, ce guide s'efforce de mettre à la disposition des futurs praticiens l'expérience et le savoir-faire déjà acquis au sein de la communauté des cartographes.

J'ai le plaisir de saluer l'édition de cet ouvrage, qui répond à un besoin significatif et croissant, et qui sera d'une grande utilité pour les cartographes du monde entier.

D. R. F. TAYLOR
Président
Association Cartographique Internationale

Introduction

L'apport des données satellitaires à la cartographie pose aux cartographes un certain nombre de problèmes techniques spécifiques. Il importe de maîtriser ces problèmes si l'on veut exploiter pleinement les possibilités considérables (mais peu utilisées jusqu'ici) apportées par les satellites d'observation de la Terre. Ces derniers représentent en effet une étape décisive, non seulement en matière de télédétection généralisée, mais également d'automatisation de la collecte des données, puisqu'ils couvrent la totalité du globe, et cela plusieurs fois par an, par mois, ou même par jour dans certains cas. Cette automatisation converge avec celle développée en matière de cartographie assistée par ordinateur et de systèmes d'information géographique (SIG). La synergie ainsi formée conduit potentiellement à des outils d'une très grande puissance—les seuls, à dire vrai, à la mesure des enjeux planétaires d'aujourd'hui et de demain.

L'objectif du présent manuel est de présenter de manière aussi pédagogique que possible, les méthodes développées pour établir des cartes (thématiques le plus souvent) à partir des images-satellites. Cette présentation s'adresse donc à tous ceux qui abordent l'emploi des images-satellites et souhaitent disposer d'indications pratiques sur les méthodes utilisées et les résultats que l'on peut en attendre. Aussi le manuel est-il articulé en cinq chapitres , chacune représentant une étape du processus et comportant ses problèmes propres.

Le premier problème technique rencontré est évidemment celui de savoir interpréter les données satellitaires collectées, autrement dit, de transformer ces "données" en "informations". Ce problème se subdivise lui-même en deux volets:

(1) Le choix des types d'images les plus aptes à mettre en évidence les informations que l'on veut pouvoir identifier, les types d'images se caractérisant par leur résolution géométrique au sol, leur(s) bande(s) spectrale(s), leur répétitivité, etc.; c'est pourquoi la description des types d'images aujourd'hui disponibles et celle des programmes spatiaux de la décennie en cours font l'objet du chapitre 1.

(2) L'interprétation des données, en fonction des caractéristiques des images et des informations recherchées: elle met en oeuvre un certain nombre de méthodes, appartenant soit à la photo-interprétation visuelle, soit au traitement numérique d'image, soit à une combinaison des deux; ces méthodes font l'objet du chapitre 2.

Le deuxième problème technique rencontré est l'intégration des informations obtenues dans la cartographie classique. En effet les informations d'origine purement satellitaire sont rarement suffisantes pour produire à elles seules des cartes au plein sens du terme: elles doivent être combinées avec des informations extérieures de type classique, plus ou moins nombreuses, qui les complètent dans divers domaines: référentiel géographique, toponymie, contours d'agglomérations, réseau routier etc. Cette intégration revêt, elle aussi, deux aspects:

(1) La combinaison "numérique" des informations satellitaires et des informations classiques: intégration dans un même référentiel (rectification des images), adaptation des contrastes et des couleurs, passage en mode vecteur ou en mode trame, sélection des informations à représenter, affichage sur écran ou sur traceur; ces procédures, qui débouchent sur les fonctions générales des SIG, font l'objet du chapitre 3.

(2) La représentation graphique finale des combinaisons obtenues, en termes de sémiologie (choix des symboles, des couleurs, des tailles de caractères, etc.), d'échelle, de précision, conduit à distinguer plusieurs types ("iconocartes"); la description de ces méthodes de rédaction cartographique et la typologie des documents obtenus font l'objet du chapitre 4.

Enfin, un chapitre spécial est consacré à la présentation d'applications et d'échantillons des cartes produites. Les échantillons ont été choisis délibérément dans différentes catégories thématiques et dans divers pays. Ils donnent ainsi un panorama de la production mondiale d'iconocartes thématiques, classées elles-mêmes par thèmes (cartographie générale, géologie, hydrographie, océanographie, glaciologie, agriculture, etc.). C'est l'objet du chapitre 5.

En raison du bilinguisme du manuel, et pour certains motifs pratiques, la composition adoptée est la suivante:

Version anglaise: chapitres 1, 2, 3 et 4.
Version bilingue: chapitre 5 (applications et illustrations couleur uniquement).
Version française: chapitres 1, 2, 3 et 4.

JEAN DENÈGRE
Président de la Commission ACI
"Cartographie thématique dérivée des images satellitaires"(1984–91)

Remerciements

Le présent ouvrage résulte de la coopération internationale instaurée dans le cadre de l'ACI (Commission de cartographie thématique dérivée des images satellitaires), au cours de la période 1987–92. Il a bénéficié du concours de spécialistes de divers pays, parmi lesquels les rédacteurs principaux sont au nombre de 5:

> Dr. Andrzej Ciolkosz (Pologne) pour le chapitre 1.
> Jean Denègre (France) pour les chapitres 3 et 4.
> Sten Folving (Danemark) pour le chapitre 3.
> Prof. Andrzej B. Kesik (Canada) pour le chapitre 1.
> Donald T. Lauer (USA) pour le chapitre 2.
> Dr. Janos Lerner (Hongrie) pour le chapitre 4.

En outre, le chapitre 5, qui présente une vingtaine d'applications pratiques, est le fruit de contributions ancore plus nombreuses, venant de pays comme l'Allemagne, l'Autriche, le Brésil, la Tchécoslovaquie, la Tunisie, etc. Les noms des contributeurs ont été indiqués chaque fois dans l'intitulé de l'application correspondante. L'impression couleur des illustrations a pu être réalisée grâce au concours gracieux de l'ITC (International Institute for Aerospace Survey and Earth Sciences) des Pays-Bas et à la diligence de M. Koert Sijmons.

Enfin, symbole de coopération internationale, cet ouvrage s'est voulu bilingue, dans les deux langues officielles de l'ACI: anglais et français. Il n'aurait donc pu voir le jour sans le concours de nombreux traducteurs bénévoles de l'Institut Géographique National France, dont nous souhaitons citer ici les noms: Jean-Claude Barret, Viviane Bois, Laurent Breton, Patrick Dayan, Jean-Pierre Delmas, Alain Jacqmin, Philippe Houssay, Allan Moore. A ces noms s'ajoutent ceux des dactylographes qui ont assuré la saisie partielle des textes: Annick Clavaud, Françoise Coulon, et Simone Chavaroche.

A tous nous exprimons nos remerciements les plus chaleureux.

JEAN DENÈGRE
Editeur

Index des pays couverts

Index des satellites utilisés

Chapter 1

SATELLITE REMOTE SENSING IMAGING AND ITS CARTOGRAPHIC SIGNIFICANCE

Compiled by
Andrzej Ciolkosz* and Andrzej B. Kesik†

*Institute of Geodesy and Cartography, Warsaw, Poland
†Department of Geography, University of Waterloo, Waterloo, Ontario, Canada

CONTENTS

1.1 Introduction

The last three decades of development in cartography, remote sensing and geographical information system (GIS) have been characterized by the emergence of a new interface between established technical disciplines. Many attempts have been made for better integration of different types of cartographical data and for more successful implementation of remote sensing data into the global and regional environmental data banks. The importance of

1

remote sensing, as a subsystem of environmental information, increased substantially due to the addition of several new satellites with photographic and electronic sensors capable of providing data and images related to the physical and human elements of the geosphere.

Cartographic presentation of the spatial and temporal variations of elements of the geosphere represents the main objective of thematic cartography. This objective is realized in many ways, by different mapping programmes carried out on the global, regional and local levels. Thematic mapping has been recognized by many societies as an important activity, vital for resource inventory, environmental management and planning.

Rapid development of satellite systems caused an increase of remote sensing data acquisition and application into cartographic operations and map compilation. The interface of cartography/remote sensing/and GIS could be presented in different models (Fisher and Lindenberg, 1989). The three-way interaction model (Fig. 1.1) seems to reflect present relationships between the three disciplines without any particular discipline dominance. Overlapping areas represent zones of interdisciplinary research with potential for the development of new tools, methodologies and strategies for a better understanding of the Earth system (NASA, 1986).

The use of satellite images as an input into thematic mapping occurs when one of the three following situations develops:

—Input from the remotely sensed satellite data is needed because it represents an exclusive source of desired information.
—Satellite remote sensing data is needed as the supplementary source of information.
—Remotely sensed satellite data is desired as the substitute for other sources of information which may be evaluated as less

reliable, less economical and unsatisfactory due to time requirements.

Satellite images or digital data are acquired, processed and reproduced for practical use by international or national organizations responsible for the management of the experimental or operational satellite systems. Dissemination and application of the satellite images for cartographic objectives on the global, regional and local levels depend upon the remote sensing data availability, access to data, budgeting of the undertaking projects and the methodology of mapping.

Satellite remotely sensed information is presented and used in two different formats:

—*Images:* generated by the photographic and electronic systems of remote sensing and reproduced by photoprinting and image reproduction techniques generating hardcopy output (Table 1.1).
—*Data:* collection of digital values representing electrical signals produced by sensors. Data are normally recorded on computer compatible tapes (CCTs) and are ready for the computer-assisted analysis, which may involve image enhancement and image classification.

Satellite images reproduced as black and white copies or colour composites are suitable for the direct visual/manual interpretation and analysis. This process usually involves identification and delineation of features and feature associations, using primarily the cerebral, eye/brain system of image interpretation, supported by optical/mechanical devices. Classification and categorization of analysed features is done on the basis of the subjective interpretative process which is guided by the analyst's

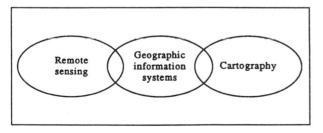

Fig. 1.1 The model of three-way interaction between remote sensing, cartography and GIS (after Fisher and Lindenberg, 1989).

Table 1.1 Hard-copy Output Devices (after Ferns, 1984)

Hard-copy device	Quality	Cost
Colour film writer/plotter (e.g. Optronics, Dicomcd)	****	*****
Black and white film writer/ plotter	*****	****
Colour camera (e.g. Matrix, Dunn, Modgraph)	***	***
Ink-jet printers	**	***
Colour dot matrix printers	*	**
Camera (35 mm) screen photography	**	*

background and his knowledge of the subject of study. Taxonomic classifications of analysed and mapped features represent the derivative products of image interpretation. They frequently show dependency on existing systems of feature classification elaborated by particular branches of the geosciences. An excellent example of this technique is documented in the *Atlas of Interpretation of Multispectral Aerospace Photographs; Methods and Results* (Sagdeyew *et al.*, 1982).

The visual/manual image analysis is frequently assisted by the use of image enhancement techniques, which, when properly used, may improve and speed up the visual image analysis by enhancing the tonal or colour representation of different groups of features and by making them more distinctive and easier to detect and delineate. Enhancement techniques are usually performed through the computer-assisted processing of data. After implementation of selected enhancement algorithms, derivative data are used for production of hard copies: black and white or colour composites reproduced at the desired scale.

Digital, computer-assisted image analysis techniques have an advantage of relatively easy data manipulation by application of algorithms for image preprocessing (radiometric and geometric corrections), enhancement and classification. Image display and production of hard copies could be done at different levels of data manipulation. Digital image analysis requires proper hardware and software for data processing and display.

The advances in computer technology led to the development of two different levels at which digital image analysis is performed: at the high level of computer technology there are specially designed and built-up image analysis systems based on parallel computer architecture. Such systems are capable of handling large data volumes and demanding data processing, enhancing and classifying algorithms (Fig. 1.2).

On the low level, which is gaining fast popularity with users, are the microcomputers with available packages of software. They are suitable for consulting companies, educational institutions and professional experts. The new generation of micros can exceed storage capacity and computing capabilities of the minicomputers of a few years ago. Technological trends indicates that microcomputers with packages of remote sensing processing software and accessories for printing, plotting and video display will further stimulate the digital image analysis, making it a common tool in the cartographic operations requiring an input from the remote sensing data.

Production processes for extracting information from satellite data are covered in much greater detail in Chapter 2 of this book.

1.1.1 Spectral, spatial and temporal resolution of satellite images

Cartographic applications of satellite images depend upon the ability of images generated by a particular sensing system to render a sharply defined image of the investigated features. This is frequently described, with limited precision,

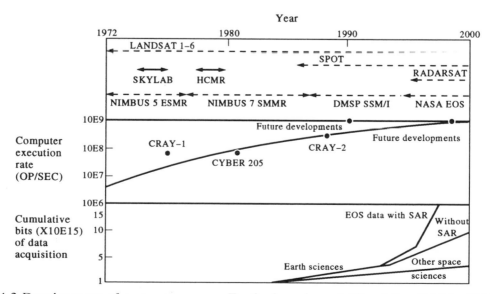

Fig. 1.2 Development of computer power, Earth observing satellites (after NASA, 1988).

as an image resolution, or spatial resolution, expressed by the size of objects or distances and described as the ground resolution.

The term image resolution is complex. Firstly, it applies to the image properties which result from the interaction of sensor characteristics, processing procedures, display characteristics and image analysis conditions. Secondly, the term may also refer to the spectral and temporal resolution. Spectral resolution refers to the dimension and number of specific wavelength intervals in the electromagnetic spectrum used by a particular sensor.

Temporal resolution refers to the frequency of imaging of a particular area by a given sensing system. Spectral and temporal resolutions of remotely operating sensors are essential for the characteristic dimension of recorded data, and for data discrimination capability. They are also important for the assessment of sensor ability for the change detection studies. The first application of Chapter V illustrates differences in spatial resolution of the Toronto Islands scene recorded by four satellite systems: Landsat MSS, and TM, and SPOT in panchromatic and multispectral mode.

1.1.2   *Classification of satellite sensors*

Classifications of satellite sensors are based on different criteria. Most commonly used are:

—sensor characteristics,
—platforms and satellite orbits characteristics,
—application domains.

A great variety of satellite sensors have been launched since 1960 and many sensors have been abandoned, improved or modified. Table 1.2 lists the main remote sensing systems in descending order of their spatial resolution. Temporal resolution, swath width of the image path and the main applications of imagery are listed due to their relevance for cartographic considerations.

1.2   Geostationary and near-polar meteorological satellites

Meteorological satellites operate either from the geostationary, high orbits (called GO), located above the equator at an altitude of 35 900 km, or from the low, near-polar orbits (called LEO), at altitudes of 200–1000 km above the Earth (Fig. 1.3, Table 1.3).

Geostationary orbits of satellites allow the synchronization of the satellite movement with the rotation of the Earth. The result is an apparent fixed position of the satellite above the same point of the surface of the Earth. The angle of view of the sensor located on a geostationary satellite covers the whole Earth's disc of 80 degrees (about 6200 km radius), so that the five satellites properly located above the equator at different longitudes provide the complete coverage for the globe within the range of 80 degrees North to 80 degrees South latitudes. However, due to the curvature of the Earth, surface polar regions are presented on image with great distortion.

Geostationary satellites perform the following operations:

—*Imagery missions.* Images are produced from data generated by the spin/scan radiometers operating in the visible and infrared portions of the spectrum. The resolution of the image data at the sub-satellite point is 0.9–2.5 km, (depending on the type of spacecraft), in the visible spectrum and about 5–7 km in the infrared spectrum.
—*Data dissemination missions.* Satellites carry one or two S-band transponders to relay preprocessed analogue or digital image data to the user receiving stations within the radio range of the satellite. The APT (Automatic Picture Transmission) transmission, called Weather Facsimile Service (WEFAX), allows the transmission of sectorized image data to the large group of users and image data implementation into the analysis and weather forecasting.
—*Data collection missions.* Satellites are capable of collecting meteorological, hydrographical and oceanographical data from large numbers of stationary or movable Data Collection Platforms (DCPs), and of relaying these data to central ground station for processing and application.

Near-polar meteorological satellites operate from low orbits, ranging from 600 to 1500 km. Satellite movement is Sun-synchronous, that is the orbit remains in a constant plane relative to the Sun, while the Earth spins below. The global coverage is obtained by successive orbits. At an altitude of 900 km, the satellite has a period of 103 min and consecutive orbits are 2860 km apart at the equator. Some sensors have the

Table 1.2 Main Satellite Remote Sensing Systems

Satellite System(s) Country	Operational period	Image repeat possibility	Wavelengths frequency	No. of bands	Spatial resolution	Swath width	Applications
GOES Visible and IR Spin Scan Radiometer (VISSR) USA	From October 1975	19 min	Visible Thermal IR	2	0.8 km 9.9 km	Full Earth disc or quarter disc	Meteorology
METEOSAT Radiometer ESA	From November 1977	30 min	Visible Middle IR Thermal IR	3	2.4 km 5 km 5 km	Full Earth disc or sectors	Meteorology Environmental Studies
GMS Visible and IR Spin Scan Radiometer Japan	From July 1977	30 min	Visible Thermal IR	2	1.25/4 km 5/7 km	Full Earth disc or quarter disc	Meteorology
INSAT I Very High Resolution Radiometer India	From April 1982	30 min	Visible Thermal IR	2	2.75 km 11 km	Full Earth disc	Meteorology
TIROS-N/NOAA Advanced Very High Resolution Radiometer (AVHRR) USA	From October 1978	12 hours	Visible Near IR Middle IR Two thermal bands	5	1.1 km	2400 km	Meteorology Oceanography Hydrology Vegetation
DMSP Defence Meteorological Satellite Programme Operational Linescan System (OLS) USA	Block 5D from September 1978	12 hours	Visible and near IR Thermal IR	2	0.6 km	620 km	Meteorology
METEOR 1-PRIRODA Multispectral Scanners USSR	From 1974 to 1980						
MSU-M			Visible Near IR	4	1.7 km	1930 km	Environmental Studies
MSU-S			Visible	2	0.24 km	1380 km	Environmental Studies

(*Continued overleaf*)

Table 1.2—(*Continued*)

Satellite System(s) Country	Operational period	Image repeat possibility	Wavelengths frequency	No. of bands	Spatial resolution	Swath width	Applications
METEOR-2 PRIRODA	From 1980						
MSU-SK			Visible Near IR	4	0.24 km	600 km	Environmental Studies
MSU-E			Visible Near IR	3	28 m	28 km	Environmental Studies
Fragment Resource—O	Operating		Visible Near IR Middle IR Thermal IR	6	0.8 km	85 km	Environmental Studies
LANDSAT USA Multispectral Scanner (MSS)	From 1972						
Landsat 1–3		18 days	Four bands in visible and near IR Thermal IR (Landsat 3 only)	4	80 m 237 m	185 km	Land use Vegetation geology Geomorphology Hydrology
Landsat 4–5	From 1982	16 days	(As above)	4	80 m		
Thematic Mapper (TM) Landsat 4–5	From 1982	16 days	Visible Near Middle IR Thermal IR	6	30 m 120 m	185 km	Land use Vegetation Geology Geomorphology Cartography
SEASAT Synthetic Apperture Radar (SAR) USA	In 1978	Limited cover	23.5 cm L-bank	1	25 m	100 km	Oceanography
SPOT High Resolution Visible (HRV) France	From 1986	2.5 days off nadir	Visible Near IR Panchromatic	3 1	20 m 10 m	60 km	Land use Agriculture Cartography

	Date	Coverage	Spectral band	No.	Resolution	Swath	Applications
MOS-1 Japan	From 1987						
MESSR			Visible Near IR	4	50 m	100 km	Oceanography
VTIR			Visible	1	0.9 km 32 km 23 km	1500 km	Oceanography
MSR			23.8 and 31.4 GHz			320 km	Oceanography
IRS-1A LISS India	From 1983	22 days	Visible Near IR	4	36.5 m	148 km	Environmental Studies
SPACE SHUTTLE	From 1981						
SIR-A	1981	Limited Cover	23.5 cm L-band		40 m	50 km	Geology Geomorphology Soils
SIR-B USA	1984	Limited Cover	23.5 cm L-band		30 m	20–50 km	Land Use Oceanography
ERS-1	From 1991	Global Coverage	C-band 5.3 GHz		20–30 m	80.4 km (full performance)	Cartography Oceanography Ice Studies
Metric Camera ESA	1983	Limited Cover	Panchromatic		20 m		Cartography
Large Format Camera (LFC) USA	1984	Limited Cover	Panchromatic Colour IR	2			
MOMS West Germany	1984	Limited Cover	Visible Near IR Thermal IR		20 m 10–20 m	140 km	Cartography Geology Soils Vegetation
COSMOS Space Photography USSR							
Resource F-1 KATE-200	Operating	Limited Cover	Multispectral (4-lenses)		15–30 m		Multidisciplinary Applications
KFA-1000			Panchromatic		5–7 m		Cartography
Resource F-2 MK-4			Multispectral (4 lenses)				
Soyuz-22, Salyut 6 MIR-Mission MKF-6		Limited Cover	Multispectral (6 lenses)		9–13 m		

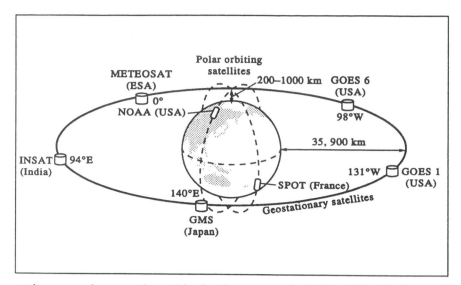

Fig. 1.3 Geostationary and near-polar orbits for the meteorological and Earth observation satellites.

Table 1.3 Orbital Altitudes of Selected Earth Satellites

	Miles	Km
Communication satellite (Westar)	22 300	35 680
Space shuttle	115–690	184–1104
Landsat 1, 2 and 3	570	912
Landsat 4 and 5	423	705
Seasat	480	800
Tiros 1 (weather satellite)	480	800
Nimbus (weather satellite)	594	950
GOES (weather satellite)	21 480	35 800

swath width less than 2860 km and hence provide incomplete coverage in the equatorial zone.

Near-polar satellites have different sensors and payload which generally are capable of performing the following missions:

—*Imagery missions.* Imagery data are generated by radiometers operating in the visible and infrared portions of the spectrum. The ground resolution varies from 1–4 km in the visible, to 1–7 km in the infrared along the sub-satellite track.

—*Sounding missions.* Sounding of the atmosphere is performed by vertical radiometers which operate in the infrared and in the microwave spectral range and provide information regarding temperature and humidity at different altitudes.

—*Readout missions.* Satellites provide a direct readout service which supports reception of satellite data in a real time by the ground stations located within the radio range of

reception. The medium resolution imagery data is transmitted by the APT (Automatic Picture Transmission), while the high resolution imagery data is transmitted by the HRPT (High Resolution Picture Transmission).

—*Data collection missions.* Near-polar meteorological satellites are equipped for collecting data from the Data Collection Platforms (DCPs).

1.2.1 Geostationary meteorological satellites

1.2.1.1 Geostationary operational environmental satellite (GOES)

American meteorological satellites (GOES) began their operation in 1975. Satellites are owned and operated by the US National Oceanic and Atmospheric Administration (NOAA). GOES-7 was launched in 1987 and the service of the following satellites is expected until 2000. Instruments on board the satellite include the Visible and Infrared Spin-Scan Radiometer (VISSR) to record images of the Earth's disc in the visible (0.66–0.7 μm) and the thermal infrared (10.5–12.6 μm) wavelengths with a spatial resolution of 0.8 km for the visible and 6.9 km for the thermal band. Image repeat possibility is 19 min and the coverage is a full Earth disc or a quarter disc. Since GOES-3, satellites have been also equipped with the atmospheric sounder for gathering vertical atmospheric profile information.

An operational GOES system consists of the three satellites GOES-E (Western Atlantic),

GOES-W (Eastern Pacific) and GOES IO (Indian Ocean), which performed with several technical difficulties. Data from GOES satellites are used together with data from other geostationary satellites operated by ESA, Japan, India and the former USSR to gather meteorological information on a global scale.

1.2.1.2 Meteosat operational programme

The Meteosat programme was initiated by the French space and meteorological authorities in the 1970s. In 1972, the programme was transferred to the European Space Agency (ESA) which in 1973 agreed to the development of two Meteosat satellites. Meteosat-1 failed in 1979 but was replaced in 1981 by Meteosat-2, which remains in orbit today. In June 1988 Meteosat-P2, now called Meteosat-3, took over day-to-day operations, initiating the Meteosat Operational Programme (MOP), executed by ESA on behalf of Eumetsat, the European Meteorological Satellite Organization. Eumetsat is planning further Meteosat launches.

Meteosat carries a radiometer operating in three bands: 0.4–1.1 μm (visible/IR), 5.7–7.1 μm (water vapour) and 10.5–12.5 μm (thermal infrared). Imagery is provided for the zone extending to 55 degrees North and South. Resolution at the equator and 0 degrees longitude is 2.5 km for the visible/IR band and 5.0 km for the two remaining bands. Image repeat occurs once every 30 min.

For the Meteosat mission in the mid-1990s, called Meteosat Second Generation (MSG), Eumetsat is planning to add new instruments:

— a radiometer with very high resolution of 500 m operating in the visible portion of the spectrum;
— an advanced radiometer operating in the visible and IR bands with sounding elements;
— a broadband radiometer capable of measuring solar and terrestrial radiation with a spatial resolution of about 200 km;
— an infrared sounder with very high spectral resolution.

1.2.1.3 GMS, Japan geostationary satellites

Japan's first geostationary satellite, GMS-1, also called Hamavari 1, was launched in June 1977, the second one, GMS-2, in August 1981, which after failure in January 1984 was replaced in August 1984 by GMS-3 (Fig. 1.4).

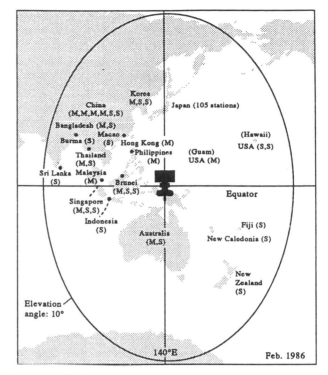

Fig. 1.4 GMS-3 area coverage and data utilization stations. M, medium-scale data utilization station; S, small-scale data utilization station.

GMS-4 was planned for 1989, GMS-5 for 1993 and GMS-6 for 1994. All GMS satellites are positioned over the equator and west Pacific at 140 degrees East longitude (Matthews, 1988).

1.2.1.4 India's INSAT geostationary satellites

The first Indian geostationary satellite, INSAT (IA), was launched in April 1982. INSAT IA used the Very High Resolution Radiometer operating in the visible and infrared bands. Spatial resolution was 2.75 km for the visible band and 11 km for the infrared. The first INSAT was placed over the equator at the geographical longitude of 74 degrees East. The operation lasted until September 1982. In August 1982 the replacement by INSAT 1B was installed with the assistance of the NASA Space Shuttle. It was positioned over the equator at the geographical longitude of 94 degrees East.

1.2.2 Near-polar meteorological satellites of the eighties

In the United States the third generation of near-polar meteorological satellites is represented by TIROS-N satellites.

The TIROS-N series began its operation in 1978. Satellites were named NOAA 6 and 7,

Table 1.4 Characteristics of the NOAA 6 to 10 Satellite Missions

Parameter	NOAA-6, 8 and 10	NOAA 7 and 9
Launch	6/27/79, 3/28/83, 9/17/86	6/23/81, 21/12/84
Altitude (km)	833	833
Period of orbit (min)	102	102
Orbit inclination	98.9 degrees	98.9 degrees
Orbits per day	14.1	14.1
Distance between orbits	25.5 degrees	25.5 degrees
Day-to-day orbital shift	5.5 degrees East	3.0 degrees East
Orbit repeat period (days)	4–5	8–9
Scan angle from nadir	55.4 degrees	55.4 degrees
Optical field of view (mr)	1.3	1.3
IFOV, at nadir (km)	1.1	1.1
IFVO, off-nadir, max. (km)		
along track	2.4	2.4
across track	6.9	6.9
Swath width	2400 km	2400 km
Coverage	Every 12 hr	Every 12 hr
Northbound equatorial crossing (pm)	7:30	2:30
Southbound equatorial crossing (am)	7:30	7:30
AVHRR spectral channels (μm)		
1	0.58–0.68	0.68–0.68
2	0.72–1.10	0.72–1.10
3	3.55–3.93	3.55–3.93
4	10.50–11.50	10.30–11.30
5	channel 4 repeat	11.50–12.50

followed by Advanced TIROS-N satellites flown as NOAA 8, 9 and 10. NOAA satellites have the Advanced Very High Resolution Radiometers (AVHRR) working in a scanning mode. They were built as four- or five-channel instruments (Table 1.4), operating in the visible, near, middle and far infrared bands with a spatial resolution of 1.1 km.

The TIROS-N satellites are placed on Sun-synchronous orbits and provide data at the same local time every day. TIROS-N uses orbits at about 850 km altitude. The system of TIROS-N satellites consists of two operational satellites, one crossing the equator southbound at 07:30 and the other northbound at 15:30 local time.

An important contribution to the knowledge of oceanography was obtained from the data collected by the Costal Zone Colour Scanner (CZCS), which operated in the period 1978–86 on board of the NIMBUS-7 satellite. CZCS acquired over 66 000 images, each of which covers approximately 2 million square kilometres of ocean surface. Specification of the Costal Zone Colour Scanner is provided in Table 1.5.

Table 1.5 NIMBUS-5 and Coastal Zone Colour Scanner (CZCS) Characteristics

Orbital altitude	955 km
Equatorial crossing	12:00
Nadir ground resolution	825 m
Swath width	1566 km
Field of view	$\pm$ 39 degrees
Spectral bands (μm)	
(1) 0.43–0.45	Blue (chlorophyll absorption)
(2) 0.51–0.53	Blue/green (chlorophyll absorption)
(3) 0.54–0.56	Green (yellow substance)
(4) 0.66–0.69	Red (chlorophyll concert.)
(5) 0.70–0.80	Near IR (surface vegetation)
(6) 10.50–12.50	Thermal IR (surface temperature)

In the USSR, after 25 missions of the METEOR-1 series of meteorological satellites (1969–76), the new METEOR-2 series began service in 1977. Resolution of the scanner in the visible portion of the spectrum is 2 km. Operational level and real applications began in the 1980s.

Information obtained from geostationary and polar satellite systems provide a major contribution to the recognition and better understanding of the global, continental and regional environments. Information is relevant to the quantitative and qualitative studies in meteorology, hydrology and oceanography.

Satellite images are subjected to processes of analysis and interpretation which generated many cartographic products: thematic maps, which refer to different elements and characteristics of the atmosphere, hydrosphere and biosphere. Experimental maps produced on the basis of satellite data present the following categories of features:

Atmosphere
Mean temperatures of isobaric layers
Total water vapour content and its distribution by layers
Total ozone content and its distribution by layers
Wind speed and direction in the troposphere
Clouds
Spatial distribution and structure of clouds
Height and temperature of cloud tops
Total water content of clouds
Location and intensity of precipitation
Ocean Surface
Temperature of the ocean surface
Location of major ocean surface currents
Roughness of the ocean surface
Ice conditions
Location of polluted areas on the ocean surface
Land Surface
Temperature of land surface
Degree of soil moisture
Distribution of snow cover
Location of areas of melting snow and ice
Distribution of soil and vegetation cover

1.3 Land observation satellite systems

The land observation satellite system, also called Earth Resource Satellites, developed as an offspring of meteorological satellites and as a further advancement of the early systems like Mercury, Gemini, Apollo and Skylab. These early missions involved experimental use of photographic and electronic sensors which provided the first generation of space images of the Earth, stimulating multidisciplinary investigations and research.

Cartographic applications have been investigated and thematic maps presenting morpho-structural and physiographical regionalizations of different parts of the globe have been revised and enriched.

Multispectral space photography, the precursor of multispectral imagery, contributed not only to the better detection of spatial patterns of terrestrial features, but also to the development of the techniques of image enhancement and image analysis which up to the present are essential for the visual/manual analysis of satellite images.

The era of the land observation satellites began at the planning stage in 1968 and at the operational level in 1972 when the first ERTS-1 satellite was launched by NASA. Between 1972 and 1984 five Landsat satellites were launched, with the expected continuity of this programme for at least two more satellites (Table 1.6).

Table 1.6 Landsat 1 to 5 Missions

Satellite	Launched	Retired	Sensors
Landsat 1	23.06.72	1.06.78	MSS, RBV
Landsat 2	22.01.75	30.09.83	MSS, RBV
Landsat 3	5.05.82	30.09.83	MSS, RBV
Landsat 4	16.07.82	—	TM, MSS
Landsat 5	1.03.84	—	TM, MSS

In 1978, satellite SEASAT provided important Synthetic Aperture Radar (SAR) data, proving the usefulness of SAR imagery for environmental studies. In the period 1978–80 the Heat Capacity Mapping Mission generated broad coverage of thermal data. In 1986, France launched the first SPOT satellite with a new innovative type of high resolution sensor.

In the USSR, satellites from the Cosmos series were equipped by four-band scanners and by the photographic multispectral and single-band cameras. Photographic and electronic sensors have been used by USSR on Soyuz and Mir-Kvant space stations (Gatland, 1989).

In the United States, from 1981, a new platform/carrier, the Space Shuttle has been used for several missions during which new sensors have been tested.

A very significant milestone was reached in July 1991, when with the launch of the satellite ERS-1 the European Earth Observation Programme began. The following section provides a concise rundown of the systems mentioned above.

1.3.1 Landsat satellite programme

Landsat represents a significant, operational, global resources monitoring programme which is reaching its 20 years of continuation. The programme initiated by NASA in 1967 resulted in a planned series of six satellites called ERTS (Earth Resources Technology Satellite), based on the modified version of the Nimbus weather satellite. The basic objective was the gathering of data about Earth resources on a systematic, repetitive, multispectral and medium resolution basis, with nondiscriminatory access to data from the world community.

ERTS-1 was launched on 23 July 1972 as a first unmanned satellite, specifically designed for the systematic acquisition of data related to the Earth's resources. Prior to the launch of the second satellite on 22 January 1975, NASA renamed the ERTS programme Landsat. This name remains, including the conversion of ERTS-1 to Landsat 1.

The Landsat programme from the period 1972–90 refers to all operations related to the five launched satellites and to the proposed future satellites; Landsat 6 and 7. Due to the existing differences between satellite orbits, sensor configuration and specification of collected data, the Landsat satellites are grouped into two generations: first, Landsat 1, 2 and 3; second, Landsat 4 and 5.

First generation of Landsat satellites: 1, 2 and 3. The first three Landsat satellites have the same or very similar configuration as two independent sensors: the Multispectral Scanner (MSS) and the Return Beam Vidicon (RBV) cameras. The sensors responded to the reflected electromagnetic energy which was recorded as the multispectral, four-band (MSS) and three-band (RBV) data, having a pixel resolution of 79 m for Landsat 1, 2 and 3 MSS and Landsat 1 and 2 RBV, and resolution of 30 m for Landsat 3 RBV (Table 1.7).

Located on a semicircular, near-polar orbit, satellites provided global coverage for the zone between 81 degrees North and 81 degrees South, with a period of orbiting of 103 min and equatorial crossing at 9:30 a.m. local time. The cycle of global coverage required 18 days, allowing for 20 cycles per year.

Table 1.7 Landsat 1 to 3 Sensor Specifications

Sensor	Band	Spectral sensitivity (μm)
Landsat 1 and 2		
RBV	1	0.475–0.575 (green)
RBV	2	0.58–0.68 (red)
RBV	3	0.69–0.83 (near IR)
MSS	4	0.5–0.6 (green)
MSS	5	0.6–0.7 (red)
MSS	6	0.7–0.8 (near IR)
MSS	7	0.8–1.1 (near IR)
Landsat 3		
RBV		0.5–0.75 (panchromatic)
MSS	4	0.5–0.6 (green)
MSS	5	0.6–0.7 (red)
MSS	6	0.7–0.8 (near IR)
MSS	7	0.8–1.1 (near IR)
MSS	8	10.40–12.60 (thermal IR)

The primary source of data from Landsat 1, 2 and 3 was generated by the four-band optical/mechanical scanner (MSS). Sampling of the analogue signal by A/D converter generated a nominal ground spacing of 56 m between readings and consequently the matrix of 56 × 79 m cells. However, the brightness value for each pixel was derived from the full 79 × 79 m ground resolution cell controlled by the Instantaneous Field of View (IFOV). The continuous MSS data for the swath of terrain of 185 km was framed into scenes covering approximately 185 × 185 km. Each scene consists of 2340 scan lines with about 3240 pixels per line, which results in about 7 581 600 pixels per channel. A multispectral (four bands) representation of a single scene contains over 30 million pixels of digital data. MSS data obtained from the A/D converter on board of the satellite had only a digital number range of 0–63 (six bits) but was subsequently rescaled during ground processing to a range of 0–127 for the bands in the visible spectrum.

The secondary RBV data obtained from the three multispectral (Landsat 1 and 2) and two panchromatic and near IR (Landsat 3) cameras were less significant despite an increase of ground resolution to 30 m for the RBV Landsat 3 cameras.

During early operation of Landsat 1 there were only four ground receiving stations: three in the United States and one in Canada. Data from areas outside of the range of receiving stations were collected by the recorder located on board of the satellite and played back when the satellite passed over one of the US ground receiving stations.

Limitations related to this system were slowly overcome by building several Landsat receiving stations in different countries. These stations also performed data processing and image and data dissemination.

In the United States, products of Landsat 1, 2 and 3 in the form of standard black and white multispectral images, colour composites and computer compatible tapes (CCTs) have been disseminated to user communities by the Earth Resources Observation System (EROS) Data Centre at Sioux Falls, South Dakota. In 1984 the US Congress passed the Land Remote Sensing Commercialization Act to enable the transfer of Landsat from the public (NOAA) administration to the private sector. In 1985, EOSAT, a new joint venture of Hughes and RCA, resumed dissemination of all Landsat data, as well as operational control of Landsat 4 and 5 and preparation for the future Landsat 6 and 7.

Second generation of Landsat satellites: 4 and 5. The second generation of Landsat satellites began in 1982 when the fourth satellite of the series was launched. It was followed by Landsat 5, launched in March 1984. Landsat 4 and 5 remain operational (August 1991).

Landsat 4 and 5 are different in many aspects. The design of satellite platform is different from the original Nimbus-type platform used for Landsat 1–3. The orbits of satellites are lower, approximately 705 km, which results in a shorter revisit cycle, reduced to 16 days. The orbit has an inclination angle of 98.2 degree with respect to the equator. Crossing of the equator along the North–South path, appears at 9:45 a.m. local solar time. Each orbit lasts 99 min with 14.5 orbits per day. Orbits for Landsat 4 and 5 were designed in such a way that during operation of both satellites the repeat coverage cycle is 8 days with alternating coverage by each satellite (Fig. 1.5).

The most significant innovation refers to the sensors. Landsat 4 and 5 still carry the old MSS sensor with the ground pixel resolution of 82×82 m, but on top of MSS the new sensor called Thematic Mapper (TM) was added.

The Thematic Mapper is a high resolution seven-band scanning radiometer. Specification of bands are given in Table 1.8.

Table 1.8 Landsat 4 and 5 Sensor Specifications

Sensor	Band	Spectral sensitivity (μm)
TM	1	0.45–0.52 (blue–green)
TM	2	0.52–0.60 (green)
TM	3	0.63–0.69 (red)
Tm	4	0.76–0.90 (near IR)
TM	5	1.55–1.75 (mid IR)
TM	6	10.40–12.50 (far IR)
TM	7	2.08–2.35 (mid IR)
MSS	1	0.5–0.6 (green)
MSS	2	0.6–0.7 (red)
MSS	3	0.7–0.8 (near IR)
MSS	4	0.8–1.1 (near IR)

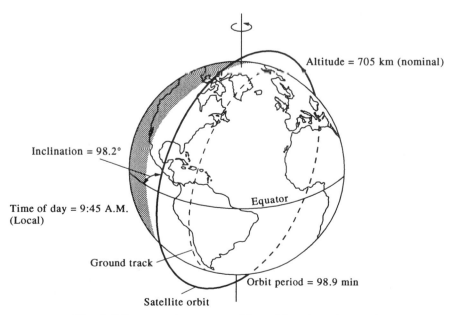

Fig. 1.5 Sun-synchronous orbits of Landsat 4 and 5.

Thematic Mapper data have a spatial resolution of 30 m for bands 1–5 and 7, and 120 m resolution for band 6. The range of quantization of TM data has 255 levels (eight bits) versus 63 levels for MSS. Spectral bands of TM provide opportunity for multidisciplinary applications and mapping (Table 1.9).

Image Interpretation and Thematic Mapping based on Landsat Data. From the beginning of the Landsat era in 1972 scientific communities all over the world took advantage of non-restricted access to the new, environmentally significant data. Numerous investigations exploited the spectral, spatial and temporal characteristics of Landsat data. Images and data oriented investigations and research related to techniques of digital image enhancement and classification flourished. Landsat data as a tool for geoscience research have been accepted and adopted by many disciplines dealing with components of the geosphere. Applications of Landsat data range from studies of bedrock and superficial geology, to geomorphology, hydrography, land use/land cover studies, crop investigation, forest classification and urban change detection. Results of research are quoted in many publications listed in *Geo-Abstracts* or other source books like *Remote Sensing Yearbooks, 1987, 1988/89* and publications by Carter (1986) and Hyatt (1988).

Satellite image maps with descriptive texts have been published in many countries. Some examples are presented in the ICA publication edited by J. Denègre (1988). Collections of maps elaborated on the basis of Landsat data were also published as atlases. The content and methodologies of map compilation vary, ranging from the traditional compilation of line maps based on the visual/manual interpretation of Landsat images (e.g. Chen-Shu-Peng, 1986), to the preparation of thematically related sets of digital maps based on the computer-assisted data analysis and classification (e.g. Adeniyi and Bullock, 1988). European experiences with the Thematic Mapper were summarized by Guyenne and Calabresi (ESA-SP-1102).

The *Landsat Data User's Handbook* (US Geological Survey, 1979) and the *Landsat Tutorial Workbook* (Short, 1982) provide good introductory background information to the Landsat system. Textbooks on remote sensing by Lillesan and Kiefer (1987), Sabins (1987) and Campbell (1987) contain expanded sections on the Landsat system. Introductory information on digital image analysis is presented by Jensen (1986) and Richards (1986).

1.3.2 Heat Capacity Mapping Mission

The Heat Capacity Mapping Mission (HCMM) operated in the period 26 April 1978 to 31 August 1980. The HCMM satellite was the first to test thermal properties of the Earth's surface. The satellite was located on the circular, Sun-synchronous orbit with 97.6 degrees inclination and 620 km altitude above the Earth. The sensor of the HCMM was the Heat Capacity Mapping Radiometer (HCMR), which operated in the visible, near and thermal infrared regions of the spectrum (Table 1.10), providing 500 and 600 m resolution respectively.

HCMM was an experimental mission which did not provide global coverage. The lack of an onboard tape recorder restricted data coverage to North America, Europe and Australia, which

Table 1.9 Thematic Mapper Spectral Bands and Their Applications

Band	Wavelength (μm)	Nominal spectral location	Main applications
1	0.45–0.52	Blue	For water body penetration, coastal mapping, soil and vegetation discrimination, forest type mapping
2	0.52–0.60	Green	For vegetation discrimination and vigour assessment
3	0.63–0.69	Red	Plant species studies cultural feature typology
4	0.76–0.90	Near IR	Vegetation studies, vigour assessment, delineation of water bodies
5	1.55–1.75	Mid IR	Vegetation and soil moisture studies. Snow/clouds discrimination
6	10.40–12.50	Thermal IR	Vegetation stress, analysis, soil moisture studies, thermal mapping
7	2.08–2.35	Mid IR	Discrimination of minerals and rock types. Studies of vegetation moisture content

Table 1.10 Heat Capacity Mapping Radiometer (HCMM) Specifications

Orbital altitude	620 km
Angular resolution	0.83 mrad
Instantaneous field of view	0.6 × 0.6 km at nadir (IR) 0.5 × 0.5 at nadir (visible)
Scan angle	60 degrees (full angle)
Scan rate	1.19 samples per resolution element at nadir
Sampling interval	9.2 μsec
Swath width	716 km
Information bandwidth	53 kHz per channel
Thermal channel	10.5 to 12.5 μm
Usable range	260 to 340 degrees K
Visible channel	0.55 to 1.1 μm
Dynamic range	0 to 100 percent albedo
Scan mirror	45 degrees elliptical, flat
Telescope diameter	20 cm
Calibration	
Infrared	View if space, seven-step staircase. Electronic calibration and blackbody calibration once each scan
Visible	Preflight calibration

remained in the range of receiving stations (NASA, 1980). During 28 months of satellite operation, over 37 600 standard images were obtained. Interpretation, analysis and experimental mapping were done in relation to geological structures, lithology and thermal inertia. Fundamental information about the HCMM mission and characteristics of results are presented in a publication by Short and Stuart (1982).

1.3.3 Seasat mission

Satellite Seasat, launched on 26 June 1978, was the first to carry the synthetic aperture radar (SAR), designed for civil applications. Seasat, which operated only 106 days, till 10 October 1978, was put on a near circular polar orbit with an inclination of 108 degrees and an altitude of 790 km. Orbiting period was 100 min giving 14.3 orbits per day. The complete cycle required 152 days.

The Seasat SAR used L-band, 23.5-cm wavelengths and HH polarization. Data obtained have an approximate resolution of 25 m.

Optically processed Seasat SAR data have been used to produce several uncontrolled mosaics of California, Florida, Jamaica, the

United Kingdom and Iceland. More than 300 digitally corrected scenes (100 × 100 km) and 400 optically corrected scenes (100 km by up to 4000 km) are available for users from the NOAA National Space Science Data Centre (NSSDC).

Seasat also carried four other sensors:

— radar altimeter to determine the sea surface conditions;
— radar scatterometer to measure wind speed and direction;
— microwave radiometer to measure sea surface temperature, rain rate and water vapour content;
— visible and IR radiometer to measure sea surface temperature and to image ocean and coastal features.

The radar altimeter provided data on topography of the ocean surface, with a relative height accuracy of up to 10 cm. The Seasat evaluation summary is presented in Table 1.11.

Additional information on Seasat can be obtained from the *Seasat Synthetic Aperture Radar Data User's Manual* and from publications by Ford *et al.* (1980) and Fu and Hold (1982).

1.3.4 The SPOT system

The SPOT system is a French contribution to the Earth observation satellite programme designed for the general remote sensing of Earth's resources and for cartographic applications. On 22 February 1986 France launched an Ariane rocket, the first of four satellites named Satellite Probatoire d'Observation de la Terre (SPOT), which obtained its operational level in May 1986. The SPOT programme is managed by the Centre National d'Etudes Spatiales (CNES), which is responsible for programme development and satellite operation. SPOT-2 was launched in 1990 as a back-up for SPOT-1.

The first generation of SPOT satellites operates from circular, near-polar, Sun-synchronous orbits at an altitude of 825 km and inclination of 98.7 degrees. Crossing of the equator is at 10:30 a.m. local solar time. The duration of SPOT cycle is 26 days. However, SPOT has increased revisiting capabilities due to the pointable optics of the system (Table 1.12).

The off-nadir viewing at the area of interest is possible because the image swath may be offset

Table 1.11 SEASAT Evaluation Summary

Sensor	Observables	Demonstrated accuracy	Demonstrated range of observables
Altimeter	Altitude	8 cm (precision)	< 5 m
	Wave height	10% or 0.5 m	0 to 10 m
	Wind speed	2 m/sec	0 to 10 m/sec
Scatterometer	Wind speed	1.3 m/sec	4 to 26 m/sec
	Wind direction	16 degrees	0 to 360 degrees
Scanning microwave radiometer	Sea surface temperature	1.0 degree C	10 to 30 degrees
	Wind speed	2 m/sec	0 to 25 m/sec
	Atmospheric water vapour	10% or 0.2 g/cm^2	0 to 6 g/cm^2
SAR	Wave length	12%	Wavelength < 100 m
	Wave direction	15 degrees	0–360 degrees

Table 1.12 SPOT Basic Characteristics

Orbit	Circular at 832 km
	Inclination: 98.7 degrees
	Descending mode at 10:30 a.p.
	Orbital cycle: 26 days
Sensor:	Two identical instruments
High resolution	Pointing capability: 27 degrees East or West of the orbital plane
Visible (HRV)	Ground swath: 60 km each at vertical incidence
	Pixel size:
	10 m in panchromatic mode
	20 m in multispectral mode
	Spectral channels (μm)
	Panchromatic: 0.51–0.73
	Multispectral: 0.50–0.59
	0.61–0.68
	0.79–0.89
Image transmission	Two onboard recorders with 24 min capacity, each direct broadcast at 8 Gz (50 Mbits/sec)
Weight	1759 kg
Size	2 × 2 × 3.5 m plus solar panel (9 m)

from the vertical by tilting a steerable mirror sideways (to the West or to the East), step by step from 1 to 27 degrees, allowing the scene centre to be targeted anywhere within a 950-km wide strip centred on the satellite track. This technique provides a quick revisit capability on specific sites. At the equator, the same area can be revisited several times during the 26 days of the orbital cycle, i.e. 98 times in one year, with an average revisit of 3.7 days. At latitude 45 degrees, the same area can be revisited 11 times in a cycle, i.e. 154 times in one year, with an average of 2.4 days, a maximum time lapse of 4 days and a minimum time lapse of 1 day.

Stereo imaging can be obtained by combining two images of the same zone but recorded during different orbits and at different viewing angles (Table 1.13).

The sensors of SPOT consist of two identical High Resolution Visible (HRV) Linear Array Sensing Systems, each one covering two 60-km wide strips of terrain overlapping by 3 km for the vertical position of sensors. Each HRV sensor can operate in either of two modes. In a panchromatic mode (P), HRV will provide 6000 pixels giving a 10 m ground resolution in the range of 0.51–0.73 μm and 6 bits quantization.

Table 1.13 Characteristics of SPOT Scenes

	XS mode	P mode
Scene dimensions (nadir viewing)	60 × 60 km	60 × 60 km
Pixel size	20 × 20 m	10 × 10 m
Number of spectral bands	3	1
Dimensions of preprocessed scenes:		
Number of pixels per line (raw scene to level 2)	3 × (3000–5200)	6000–10 400
Number of lines per scene (raw scene to level 2)	3 × (3000–4900)	6000–9800
Volume (8-bit bytes)	27–76.5 Mb	36–100 Mb

In the multispectral mode (XS), HRV will provide 3000 elements with the ground resolution of 20 m, recorded in the spectral ranges of 0.50–0.59, 0.61–0.78, 0.79–0.89 μm with 8 bits quantization.

SPOT data are transmitted to the receiving stations in Toulouse, France; Kiruna, Sweden; Prince Albert-Gatineau, Canada; Hyderabad, India and Maspalomas, Spain. Other ground receiving stations are foreseen or already operational in China, Bangladesh, Brazil, Argentina and Australia (Fig. 1.6).

For areas outside of receiving stations, SPOT data may be recorded on request for 20 min of worldwide imaging by the two types of recorders on board the satellite. These data are read out by the Toulouse receiving station. SPOT data are distributed to users by SPOT IMAGE as standard data on Computer Compatible Tapes (CCTs) recorded at 6250 or 1600 bpi and as films of the 241 × 241 mm format presenting the full SPOT scene at the scale 1:400 000 for Level 1. The basic format for Level 2 products is 350 × 350 mm film for scales 1:400 000 and 1:200 000 and format 700 × 700 mm for scales 1:200 000 and 1:100 000. Films can be ordered in black and white (for panchromatic P mode or for three spectral band presentation from XS mode), or as colour composites from the XS mode of operation.

The four basic levels of data correction and processing are used for the SPOT scene, with a nominal coverage of 60 × 60 km recorded in the panchromatic (P) and multispectral (XS) mode.

The central catalogue of all collected SPOT images for 1986 includes references to more than 2 000 000 scenes out of which 25 percent have less than 10 percent cloud cover and 30 percent have cloud cover of less than 25 percent. The *SPOT Newsletter* published by SPOT IMAGE provides updated information about SPOT missions.

SPOT data have been intensively tested and used for cartographic applications. The main advantage of SPOT data versus MSS or TM data refers to the following characteristics:

—broad range of viewing conditions and spectral mode of operation,
—stereoscopic imaging capability,
—higher spatial resolution needed for thematic mapping, particularly in areas with intensive subdivision of terrain and diversified land use/land cover conditions.

Results of cartographic applications of SPOT data have been reported in several articles: Welch (1985), Gugan (1987) and Rochon and Toutin (1986). Input from SPOT to GIS was described by Denègre (1987). A comprehensive review of results of SPOT after two years of operation was provided by Rivereau and Pousse (1988).

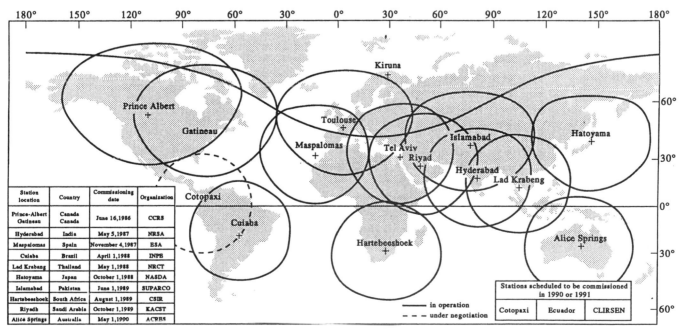

Fig. 1.6 SPOT receiving stations.

The SPOT system is scheduled at least until 2005, with improved specifications:

1. A better spectral resolution with a middle infrared sensor (MIR: 1.58–1.75 μm) with SPOT-4 (1996).
2. A vegetation monitoring instrument with a large field of view (2200 km) and a statistics-adapted resolution (1 km).
3. A better geometric ground resolution (5 m) with SPOT-5 (2001).
4. An along-track stereoscopy facility with SPOT-5.

1.3.5 MOS-1, Japan marine observation satellite

In February 1987 Japan launched its first MOS-1 satellite dedicated to the collection of oceanographic data. The satellite was put on the circular near-polar orbit with an inclination of 70 degrees and an altitude of 909 km (Matthews, 1988).

MOS-1 has the three following sensors:

—Four-channel imaging radiometer for the visible and near infrared spectrum. The coverage path is 78 km width with 45 m ground resolution.
—Four-channel radiometer with the one visible band 0.5–0.7 μm and resolution 870 m and the three bands in the range of 6.0–12.5 μm with the resolution of 2 600m.
—Two-channel microwave radiometer operating at the frequencies of 23 GHz and 31 GHz and used for monitoring sea surface temperature.

The new MOS-1 scanner MESSR operates in four bands, with a ground resolution of 50 m (Tsuchiya *et al.*, 1987) (Table 1.14).

1.3.6 Indian IRS satellite mission

On 17 March 1988 India launched from the Baikonour Cosmodrome in the USSR its satellite IRS-1. The launcher was the three stage Vostok rocket. The satellite is located on the Sun-synchronous orbit with 99.02 degrees inclination and 904 km altitude. It provides complete global coverage between 81 degrees North and 81 degrees South latitudes. The equatorial crossing is at 10:25 a.m. descending node. Cycle of satellite operation comprises 307 orbits in 22 days (Fig. 1.7).

The sensor payload consists of two push-broom cameras: Linear Imaging Self Scanning Sensors (LISS-II) of 36.25 m resolution and one camera (LISS-1) of 72.5 m ground resolution employing a linear charge coupled device (CCD) array of 2048 detectors. Each camera system images in four spectral bands in the visible and near infrared (0.45–0.86 μm). The ground swath for the images obtained by the LISS-I camera is 148.48 km, whereas for the LISS-II cameras image adjacent swath is of 74.24 km width each, with an overlap of 1.5 km across the satellite track. The energy sensed is quantized into 128 levels. Data products from LISS-I and II include black and white 70 mm and 240 mm film negatives or positives, 240 mm colour composites and nine-track 1600 bpi CCTs with digital data.

The facilities for the Data Product System (DPS) are located at three centres:

—Data Reception System (DRS) at Shadnagar, Hyderabad.
—Data Product System (DPS) at Balanagar, and at Ahmedabad (SAC).

The dissemination of IRS-I data started in May 1988 (IRS Project Team, 1988).

Table 1.14 MOS-1 Specifications

Item	Sensor		
	MESSR	VTIR	MSR
Measurement objectives	Sea surface colour	Sea surface temperature	Water content of atmosphere
Wavelength (μm)	0.51–0.59	0.5–0.7	231 K 311 K
	0.61–0.68	6.0–7.0	
	0.72–0.80	10.5–11.5	
	0.80–1.10	11.5–12.5	
Frequency (Gz)	—	—	23.8–31.4
Geometric resolution (IFOV in km)	0.5	0.9	32–33
Radiometric resolution	39 dB	55 dB 0.5 K	1 K 1 K
Swath width (km)	100	1500	320
Scanning method	Electronic	Mechanical	Mechanical

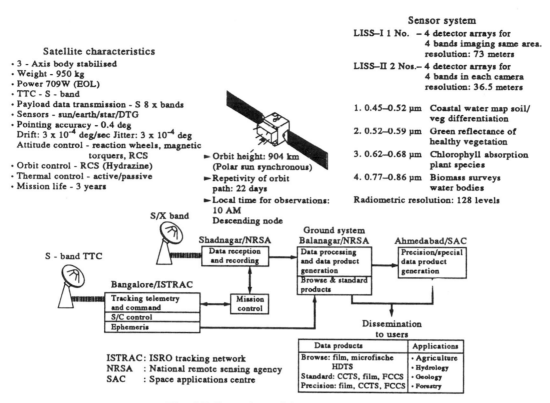

Satellite characteristics
- 3 - Axis body stabilised
- Weight - 950 kg
- Power 709W (EOL)
- TTC - S - band
- Payload data transmission - S 8 x bands
- Sensors - sun/earth/star/DTG
- Pointing accuracy - 0.4 deg
  Drift: 3 x 10^{-4} deg/sec Jitter: 3 x 10^{-4} deg
  Attitude control - reaction wheels, magnetic
              torquers, RCS
- Orbit control - RCS (Hydrazine)
- Thermal control - active/passive
- Mission life - 3 years

► Orbit height: 904 km
  (Polar sun synchronous)
► Repetivity of orbit
  path: 22 days
► Local time for observations:
  10 AM
  Descending node

Sensor system

LISS–I 1 No. – 4 detector arrays for
              4 bands imaging same area.
              resolution: 73 meters
LISS–II 2 Nos.– 4 detector arrays for
              4 bands in each camera
              resolution: 36.5 meters

1. 0.45–0.52 μm   Coastal water map soil/
                  veg differentiation
2. 0.52–0.59 μm   Green reflectance of
                  healthy vegetation
3. 0.62–0.68 μm   Chlorophyll absorption
                  plant species
4. 0.77–0.86 μm   Biomass surveys
                  water bodies

Radiometric resolution: 128 levels

S/X band

S - band TTC

Bangalore/ISTRAC

Tracking telemetry
and command
S/C control
Ephemeris

Mission
control

Shadnagar/NRSA
Data reception
and recording

Ground system
Balanagar/NRSA
Data processing
and data product
generation
Browse & standard
products

Ahmedabad/SAC
Precision/special
data product
generation

Dissemination
to users

Data products	Applications
Browse: film, microfische HDTS	• Agriculture
	• Hydrology
Standard: CCTS, film, FCCS	• Geology
Precision: film, CCTS, FCCS	• Forestry

ISTRAC: ISRO tracking network
NRSA : National remote sensing agency
SAC   : Space applications centre

Fig. 1.7 Overview of the IRS mission.

1.4 Russian environmental satellites and their sensors

Russian unmanned environmental satellites belong to the categories of Meteor-Priroda and Cosmos. Satellites Meteor-Priroda are equipped with optical-mechanical scanners, while Cosmos satellites provide acquisition of environmental information by using space photography. Space photography is also performed regularly from the manned missions of Soyuz-Salyut and from the Space Station Mir.

Investigations and monitoring of the Earth's resources by the Meteor-Priroda system started in 1974. Modified meteorological satellites from the Meteor series were renamed Meteor-Priroda. This system operated till 1980 from orbits inclined at 82 or 98 degrees and altitude of 900 km and 650 km respectively.

The first generation of satellites from the Meteor-Priroda series was equipped with two scanners of different resolution. Scanner MSU-M provided four-band data from the visible and near IR spectrum (0.5–1.0 μm). Resolution for the nadir line was 1700 m, for the nominal altitude of 650 km. Scannings were obtained for the path of 1930 km width.

The second scanner, coded MSU-S, was a two-band scanner operating in the range of 0.5–

0.7 and 0.7–1.0 μm. Ground resolution for the nadir line was 142 m, with a scanning path of 1380 km width. Data from Meteor-Priroda satellites was received and processed by the ground stations in Moskow, Novosibirsk and Chabarovsk. MSU-S data was also recorded by the unmanned receiving stations located in different parts of the USSR and by the orbiting Russian space stations.

In 1980, the USSR launched the first satellite of the second generation, Meteor 2-Priroda, which additionally had been equipped by the three following sensors:

—Optical/mechanical scanner MSU-SK designed as an experimental multispectral, four-band instrument operating in the visible and near IR portion of the spectrum (0.5–0.6, 0.6–0.7, 0.7–0.8 and 0.8–1.0 μm). Ground resolution was 243 m for the nadir line and 650 km altitude. The width of scanned path was 600 km.

—High resolution three-band scanner coded MSU-E which operated using a solid state array of sensors. Spectral ranges of bands were: 0.6–0.7, 0.7–0.8 and 0.8–1.0 μm. The IFOV of scanner was only 2.5 degrees, resulting in a narrow coverage path of 28 km. The ground resolution was 28 m.

—Eight-band experimental scanner named Fragment operated in the bands of: 0.5–0.7, 0.5–0.6, 0.6–0.7, 0.7–0.8, 0.8–1.1, 1.5–1.8 and 2.1–2. 4 μm. From an altitude of 650 km, Fragment provided a scanned path of 85 km in width, with a ground resolution of 80 m. Fragment was constructed with the technical assistance of Zeiss Jena in DDR.

Products from Fragment sensors were disseminated as CCTs and as black and white images at the scale 1 : 1 600 000.

Satellites from the Cosmos family designed for space photography are classified into two subcategories named RESOURCE 0 and RESOURCE F.

Satellites from subcategory RESOURCE 0 operate from the circular orbits inclined at 82 degrees. Altitudes of orbits range from 230 km to 240 km. Space photography is performed by two types of cameras: KATE-200 and KFA-1000.

Space camera KATE-200 is a multispectral camera comprising four lenses with the focal length $f = 200$ mm. Three cameras are used with black and white films with filtration providing recording in the range of 0.5–0.6, 0.6–0.7 and 0.7–0.8 μm. The fourth camera is used with spectrozonal colour film. The size of pictures is 180 × 180 mm. The approximate scale of negatives is 1 : 1 300 000, and the ground resolution 15–30 m. The quality of original negatives allows for enlargements.

Space camera KFA-1000 is a single-frame camera equipped with a lens of $f = 1000$ mm focal length. Photographs have a size of 300 × 300 mm. Space photography with KFA-1000 was originally performed using panchromatic film with the scale of *ca.* 1 : 240 000 and a ground resolution of 5–7 m (Fig. 1.8). High quality negatives allowed for enlargements even up to the scale 1 : 24 000. The KFA-1000 cameras have been used in a Trimetrogon configuration, with one camera oriented vertically down and two cameras in an oblique position. In this configuration it was possible to obtain a coverage path of 220 km in width.

With the improvement of the quality of spectrozonal colour films, the KFA-1000 camera is now used more intensively for spectrozonal space photography which provides information recorded on two layers of emulsion sensitized to ranges 0.57–0.68 and 0.68–0.81

μm. The ground resolution of the KFA-1000 with a spectrozonal film is approximately 10 m. KFA-1000 is also used in a twin camera configuration, with the cameras inclination of 19 degrees and partial overlapping. Twin camera configuration provides coverage for the path of 150 km width. Stereoscopic photography with an overlap up to 80 percent has been reported.

Satellites from the subcategory RESOURCE F are orbiting at the altitudes 180–450 km. Orbits are inclined at 82 degrees. On these satellites are the MK-4 space cameras. MK-4 is a multispectral, four-lens camera equipped with focal length $f = 300$ mm lenses. Photographs are in a 180 × 180 mm format and at the approximate scale 1 : 600 000. They are taken on black and white films with different filtration. The colour reproduction is achieved by the additive system, for example by the Rectimat-CM, an additive colour enlarger and precise rectifier for multispectral imagery. Simulated spectrozonal prints are used for interpretative purposes.

Many former USSR and international Interkosmos space mission photography was obtained from the MKF-6 multispectral camera built by the Zeiss Jena in DDR. The MKF-6 comprises six lenses with the focal length of 125 mm. Image size is 55 × 81 mm. The camera has forward image motion compensation. Experimental work related to the Soyuz 22 mission indicates that for the flight height of 250–260 km the scale of photographs was approximately 1 : 2.0 million and the ground resolution 13 m. Enlargements for mapping objectives up to the scale of 1 : 100 000 were satisfactory.

1.5 Missions related to the Space Shuttle launcher

Development of NASA's Space Shuttle as a reusable platform provided a convenient solution for experimental testing of new electronic and photographic sensors designed for future operational missions. From 1981 until the tragic explosion of Space Shuttle Challenger on 28 January 1986, experiments involved the use of the Shuttle Imaging Radar SIR-A and SIR-B, the Modular Optoelectronic Multispectral Scanner (MOMS), and the two space cameras: the Metric Camera and the Large Format Camera (LFC).

Fig. 1.8 Space photography by former USSR camera KFA-1000. Warsaw, Poland.

1.5.1 Space Shuttle Imaging Radar missions

The first Shuttle Imaging Radar experiment, SIR-A was conducted in November 1981 from the altitude of 260 km. SIR-A was using the L-band system (23.5 cm) with HH polarization. Acquisition of imagery provided the coverage swath of 50 km width with a ground resolution of 40 m × 40 m.

Approximately 10 million square kilometres were acquired and used mainly for geological mapping. SIR-A data is available for users from the NASA's National Space Science Data Centre (NSSDC). Information on SIR-A may be obtained from publications by Chimino and Elachi (1982), Holmes (1983) and Sabins (1983).

The SIR-B experiment was conducted in October 1984. The L-band and HH polarization were also used. The main difference between SIR-A and SIR-B refers to the antenna configuration, which in the case of SIR-B was movable and could be tilted to send radar signals to the Earth at varying look angles (range between 15 to 60 degrees). This provided an opportunity for the acquisition of stereo radar images. The resolution of SIR-B was 25 m (azimuth) and the range resolution was from 14 m at the look angle of 60 degrees to 46 m at the look angle of 15 degrees. Data from SIR-B is also available from the NSSDC.

1.5.2 Modular Optoelectronic Multispectral Scanner (MOMS)

The Modular Optoelectronic Multispectral Scanner (MOMS) was developed in the Federal Republic of Germany with the intention for usage either on aircraft or during spacecraft missions. MOMS was used in two Space Shuttle missions in June 1983 and February 1984.

MOMS comprises optical multilinear array sensors operating on the push-broom principle. MOMS has two channels: visible (0.575–0.625 μm) and near infrared (0.825–0.975 μm). Each scanning line is composed of 6912 pixels, providing a spatial resolution of 20 m. During experimental missions the recording time was limited to about 20 min. Recording resulted in 150 image frames located between latitudes 28 degrees North and 28 degrees South, covering approximately 1 800 000 square kilometres with a swath of 140 km.

Processing and archiving of MOMS data was carried out by the German Remote Sensing Data Centre at Oberpfaffenhofen. Available products include:

—quick look images,
—CCTs on standard CCT format 1600 bpi (raw and corrected data),
—black and white images (transparencies and paper prints).

Experimental applications of MOMS data for mapping objectives indicate that MOMS information is suitable for thematic mapping for the different geoscientific disciplines up to the scale of 1:50 000 (Bodechtel, 1986).

1.5.3 Spacelab Metric Camera

The Metric Camera represents a modification for the Spacelab System of the Zeiss RMK A 30/23 aerial survey camera. Specifications of this camera are given in Table 1.15.

The Metric Camera was used for the first time during a Spacelab mission launched on NASA's Space Shuttle flight in November 1983. The Shuttle carried the European built Spacelab suitable for work for five astronauts. Space photography represents one of the 37 major Spacelab experiments. During this flight the Metric Camera was used effectively for 3 hours

Table 1.15 Spacelab Metric Camera Specifications

Type	Modified Zeiss RMK A 30/23
Lens	Topar A 1 with 7 lens elements
Calib. focal length	305–128 mm
Max. distortion	6 μm (measured)
Resolution	391 p/mm AWAR on Aviphot Pan 30 film
Film flattening	By blower motor incorporated in the camera body
Shutter	Aerotop rotating disc shutter (between the lens shutter)
Shutter speed	1/250–1/1000 s in 31 steps
F/STOPS	5.6–11.0 in 31 steps
Exposure frequency	4–6 and 8–12 s
Image format	23 × 23 cm
Film width	24 cm
Film length	150 m = 550 images frames
Dimensions	46 × 40 × 52 cm
Dim. camera magazine	32 × 23 × 47 cm
Mass	54.0 kg
Mass camera magazine	24.5 kg (with film)

from an altitude of 250 km. Two types of films were used: 2443 Kodak False Colour Infrared and Kodak XX Aerographic black and white. The mission generated 550 colour and 480 black and white space photographs at the image scale of 1:820 000, covering approximately 11 million square kilometres.

Investigations of obtained space photographs focused on the analysis of photo applicability for topographic mapping, map revision and compilation of thematic maps. Photogrammetric evaluations of photographs have shown that planimetric position accuracy of less than 20 m can be achieved. The photographs provided contour lines every 100 m for steep terrain and 50 m for flat areas, which is suitable for mapping up to the scale of 1:100 000 (Schroeder, 1986; Konecny, 1986). Thematic mapping of photographs was hampered by the late season of photography (November–December).

1.5.4 Large Format Camera (LFC)

The Large Format Camera (LFC) is a special purpose mapping camera built for NASA for the Space Shuttle missions. LFC is a precision cartographic camera equipped with the 305 mm focal length lens and 230 × 460 mm image size with the long dimension oriented in the direction of flight. LFC has an image motion compensation system (Doyle, 1978).

The Large Format Camera was used for the first time during Space Shuttle mission No. 14 in October 1984. Operating from an altitude of 235 km and 375 km, the camera provided space photography with the ground coverage of 180 × 360 km and 285 × 570 km respectively. Stereoscopic coverage with an overlap of 20, 40, 60 and 80 percent was also obtained.

LFC photography is commercially distributed by the Chicago Aerial Survey Inc., 2140 Wolf Road, Des Plains, IL 60018, USA.

More specific information about contemporary space cameras and their performance is presented in proceedings from the two conferences:

—Symposium of Commission II ISPRS, "Photogrammetric and Remote Sensing Systems for Data Processing and Analysis", Baltimore, US, 26–30 May 1986
—Symposium of Commission I ISPRS, "Progress in Imaging Sensors", Stuttgart, FRD, 1–5 September 1986.

Comprehensive comparative data was also presented by Szangolies (1987) and Kromer (1987) (Table 1.16).

1.6 Synopsis

Current operational systems of satellites provide the following standard remote sensing data:

I. *Satellite Data and Images Useful for Thematic Cartography*
1. NOAA: twice daily, resolution 1 km.
2. Meteosat: every 30 min, resolution 1 km.
3. Landsat MSS: every 16/18 days since 1972, resolution 80 m.
4. Landsat TM: every 16 days since 1982, resolution 30 m.
5. IRS-1: every 22 days since 1988, resolution 36–72 m.
6. Space photography: MKF-6 camera since 1976, no periodicity resolution 15–30 m.

II. *Satellite Data and Images Useful for Thematic and Topographic Mapping*
1. SPOT digital data: every 26 days operational since 1986, programmed until 2005, resolution 10/20 m (and 5 m after 2000).
2. Space photography: experimental Metric Camera and Large Format Camera, KATE-200, KFA-1000.

In reference to the fundamental cartographic requirements for mapping at different scales, namely

—planimetric accuracy,
—elevation accuracy,
—detectability of objects,

the following comments adapted from Konecny's presentation (1989) seem to be appropriate:

"As planimetric accuracy does not represent a real problem, most concerns are regarding elevation accuracy and object detectability. The highest level of elevation accuracy is obtainable from SPOT data (approx. 5 m). Highest detectability is obtainable from space photography in which respect top resolution and detection of terrain observables is superior to digital sensors."

Table 1.16 Comparison of Technical Characteristics of Space Cameras (after Szangolies, 1987)

Cameras–Missions	(1)	(2)	(3)	(4)	(5)
Operation time	1976–82	1982–86	from 1987	1984	1983
Calibrated focal length (mm)	125	125	125	305	305
Image size (mm × mm)	55 × 81	55 × 81	55 × 81	230 × 460	230 × 230
Forward motion compensation	yes	yes	yes	yes	no
Spectral bands (μm)	0.48*	0.48*	0.48*	0.4–0.9	0.53–0.7
	0.54*	0.54*	0.54*		(for PLA)
	0.60*	0.60*	0.60*		0.53–0.9
	0.66*	0.66*	0.66*		(for MLA)
	0.72*	0.72*	0.72*		
	0.84†	0.84†	0.84†		
Resolving power (l/min)	150–220	150–220	150–220	80	30
Flight height (km)	250–260	350	240–300	220–370	250
Image scale (in Mio)	1/2.0	1/2.8	1/1.9	1/0.7	1/0.8
			1/2.4	1/1.2	
Area per photopair (in km)‡	70 × 160	100 × 225	70 × 150	220 × 165	125 × 185
			85 × 190	365 × 275	
Ground resolution (m)	13	9–13	9	10	20
Theoretical accuracy of height measurement (m)‡	42	58	40–50	7–12	16
Theoretical accuracy of coordinate measurement (m)	6	8	6–7	4–6	4
Enlargement for map scale to 1/100 000	20	28	19–24	7–12	8

(1) MKF-6 from Jena Soyuz 22-Mission, USSR. (2) Salyut 6-Mission, USSR. (3) MIR-Mission, USSR. (4) LFC–NASA Challenger Mission, USA. (5) Metric camera Option feintechnik, GmbH/FRG. Spacelab-mission, ESA.

* Amplitude: ± 0.02.
† Amplitude: ± 0.05.
‡ 66% overlap.

1.7 Future satellite missions and their cartographic significance

Future development in satellite remote sensing depends upon interaction of many technical, economical and political factors which change on global and regional scales at a pace difficult to predict. There is no doubt that improving the global political climate, reduction of military expenditures and an increased social awareness regarding environmental conditions of the biosphere provide an opportunity for intensification of the international and national satellite missions.

Developments will certainly refer to:

—sensor construction,
—platform design,
—data reception, processing and analysis,
—socio-economic framework: commercialization, international co-operation, joint venture missions.

Several future satellite missions have been forecast in the United States, Europe, Japan, Canada, Brazil, India, The Netherlands and China. Preliminary technical information and schedules are frequently revised.

The next decade will be characterized by the continuation of present missions with modification or addition of new sensors, by improvement of data acquisition, data processing and dissemination.

The future of satellite missions beyond 2000 is less certain; however, some concepts and ideas have been formulated.

In the following section we will briefly review future developments related to the three types of satellite systems:

—meteorological satellites,
—general environmental satellites,
—cartographic satellites.

1.7.1 Future meteorological satellites

Geostationary satellites are projected by the USA, ESA, Japan and the former USSR. In the United States NOAA is planning the next generation of GOES-NEXT satellites which are currently under development.

In Europe, ESA through its Meteosat Operational Programme (MOP) is planning two further Meteosat launches coded MOP-2 and MOP-3, with a full range of services to users until at least the end of 1995.

Japan indicates plans for launching GMS-5 for 1994. Satellites will provide information from the visible and infrared radiometers. The former USSR will continue its Meteor-2 series.

Polar meteorological satellites will also continue their operations. In the United States, NOAA will operate an advanced TIROS-N series (NOAA K, L, M, N, O, P). In addition, the Defense Meteorological Satellite Program (DMSP) will continue providing data.

1.7.2 *Future general environmental satellites*

This category represents the largest and most diversified group, which includes satellites with different sensors (active and passive), designed for general, multidisciplinary applications or for more specific objectives, e.g. monitoring of vegetation.

Proposed satellites will use either the near-polar, Sun-synchronous orbits, which will allow for repetitive coverage, and eventually joint international activities, like those proposed for the 1996 Earth Observing System (EOS), or low orbits suitable for manned missions with space platforms equipped with photographic and electronic sensors.

The United States plans to include continuation of the Landsat series, development of new sensors as payloads for the polar platforms representing elements of orbiting stations, and the future Space Shuttle missions.

The Landsat programme managed by EOSAT has approved schedules for Landsat 6 and 7, but details regarding specifications of platform and sensors are uncertain. One proposal for the Landsat 6 sensor refers to the Enhanced Thematic Mapper (ETM), similar to the TM sensor but with additional panchromatic band (0.5–0.85 μm) and ground resolution 15 m. Landsat 7 would eventually have ETM with multiband thermal infrared sensing capabilities. EOSAT is also investigating and assessing performance of the Multispectral Linear Array (MLS) for future applications. Several international organizations and EOSAT have reached a consensus that Landsat 7 should have stereomapping capabilities. Beyond Landsat 7, we know only that there is a general intention for continuing the Landsat-type data collection.

A number of significant sensors have been designed for the reactivated Space Shuttle missions. Completely new development includes:

MAPS: measurements of air pollution from space,
FILE: feature identification and location experiment,
SIR-C: Shuttle imaging radar with C-band,
SISEX: Shuttle imaging spectrometer,
LIDAR: light intensity detection and ranging.

An interesting Topex/Poseidon (USA–France) joint programme is focusing on the study of ocean surface topography and ocean circulation. The programme was proposed for 1992, with a three year duration. Advanced radar altimeter and tracking systems are anticipated.

The plan for a Space Station was initiated in the United States in 1984 with strong support from Europe, Canada and Japan. The European component of this programme is termed Columbus. Two polar platforms as elements of the Space Station will be built with the assistance of the Space Shuttle; one will be prepared by ESA and will operate from an orbit giving a morning equatorial crossing time at around 10:00 a.m. The second polar platform will be operated by NASA from a similar orbit but with an afternoon equatorial crossing time at 2:00 p.m. Polar platforms will operate from altitudes of 850 km. The long list of proposed instruments indicates the focus on meteorological, atmospheric, oceanic and land observations.

Future European (ESA) activities in this category of general environmental satellites culminate in the European Remote Sensing Satellite (ERS) programme. ERS missions will involve two satellites, ERS-1 launched in July 1991 and ERS-2 planned for 1994, with continuous operations till 1996. ERS missions are mainly oriented towards monitoring of polar ocean and ice. ERS-1 is on a low, Sun-synchronous orbit to an altitude of 800 km. The instrumentation of ERS-1 includes the following instruments:

—a radar altimeter,
—an active microwave instrument (AMI), containing a C-band wind scatterometer with a C-band synthetic aperture radar (SAR) (Table 1.17),
—an along-track scanning radiometer (ATSR).

Table 1.17 ERS-1 SAR Parameters

Frequency	L-band
Polarization	HH linear
Orbit	570 km, Sun-synchronous
Off nadir angle	35 degree
Swath width	75 km
Resolution	18 m × 18 m (swath centre)
Multilook number	3
Signal-to-noise ratio	7 dB
Signal-to-ambiguity ratio	14 dB
Output data rate	60 Mps

Transmitter and receiver

Frequency	1275 MHz
Band width	15 MHz
Pulse width	35 μs
PRF	1505–1605 Hz
RF peak power	1050 W (minimum)
Noise figure	4–6 dB (maximum
Gain control	70–92 dB

Antenna
Solar paddle deployment mechanism
Microstrip array for radiating elements

Japan, not affected so far by commercialization, is planning a substantial international contribution to the future monitoring of the global environment. For the second half of the 1990s, the Space and Technology Agency (STA) has recommended the following three projects:

—ADEOS (Advanced Earth Observation Satellite). ADEOS is proposed for 1995 with two core sensors: (1) AVNIR (Advanced Visible and Near IR Scanner), a four-band scanner with 8–16 m IFOV, with ± 40 degrees pointing capability and with two AO (Application of Opportunity) twelve-band sensors for the visible and thermal IR with 700 m IFOV and 1400 km swath width.

—NPOP-1. Two sensors of ITIR (Intermediate Infrared Radiometer) and AMSR (Advanced Microwave Scanning Radiometer). Both sensors are proposed for the NASA Polar Orbit Platform (NPOP-1) scheduled for 1996.

—TRMM (Tropical Rainfall Measuring Mission). Under study as a joint project with the United States, with approximate launch of satellite in 1996. Objectives: measurements of rainfall, water circulation and atmosphere–ocean interaction in the tropical zone.

Japan's J-ERS-1 project is a complementary

project to the MOS 1 and 2 missions with instrumentation including: L-band SAR combined with optical measurements by a short-wavelength infrared radiometer (SWIR) and a visible near infrared radiometer (VNIR).

In 1994 Canada is planning to launch Radarsat, the first Canadian Earth Observation Satellite. Radarsat will be launched on a Delta rocket into a circular, polar orbit with an inclination of 99 degrees. The altitude of the orbit will be 800 km, the cycle of operation 24 days on equatorial crossing, with 14.4 orbits per day. The payload will include a synthetic aperture radar (SAR) operating at 5.3 GHz (C-band), with the azimuth and range resolution of 30 m. Radarsat will provide flexibility in area coverage, the angle of observation and ability for zooming and collecting data with different levels of details. SLAR antenna will be able to point its beam within a swath of 500 km between 20 and 50 degrees off the side of the satellite. This mode of operation can provide complete coverage for Canada once every 72 hours and daily coverage for the Arctic.

The participation of the former USSR in the future global environmental satellite studies is hard to predict. From one side former USSR technical capabilities related to satellite launchers, platform and sensor design have been recently documented by the Energia rocket, the Mir space station, the Buran space shuttle and by a new radar satellite system. On the other side present economical and social adjustments may result in the revision of many space programmes, their priorities and time schedules. With continuation of the present political climate, we may expect more productive international co-operation, access to their remote sensing data and more joint space ventures.

Plans for future environmental satellites in Brazil, India and China are less visible. Changes of sensor specifications, rocket launchers, mission characteristics and objectives are frequent.

1.7.3 *Future cartographic satellite missions*

Future cartographic satellite operations can be described by two types of missions:

—future SPOT and other stereomapping satellite missions,
—future space photography from orbiting platforms and space stations (polar

platforms, Space Shuttle, Cosmos satellites, Buran, MIR stations).

Future SPOT missions: SPOT 3, 4 and 5 are scheduled for 1993, 1996 and 2001 respectively. Included are modifications of platforms and the change of sensor specifications. Major changes will refer to SPOT 4 and 5, while SPOT 3 will have a configuration similar to SPOT 2.

For the next generation of SPOT satellites (i.e. 4 and 5), it is proposed to add a 20-m resolution band in the mid-infrared portion of the spectrum (1.58–1.75 μm), which should improve monitoring of vegetation. Another proposal refers to the addition of a new, wide field of viewing instrument with a swath width of 2200 km and a ground resolution of 1 km for regional studies. The basic instruments for topographic applications will continue to be the HRV with 10 and 20 m resolution, but with along-track stereoscopic capabilities.

Future stereomapping satellite systems have been described by Light (1989) and Colvocoresses (1990). The optimal, desired specifications for future stereomapping satellites, according to Colvocoresses (1990), should be as follows:

Orbit. 918 km altitude preferred, lower altitude (to 518 km) acceptable. Circular, Sun-synchronous, daily contiguous-swath coverage, descending node crossing 9.00 to 9.45 a.m. local time.

Swath width. 180 km preferred, adjustable to smaller swath for high resolution stereoscopic and multispectral coverage.

Sensors. Three multispectral sets of linear arrays. Refractive optics looking forward, aft and vertical. Three or four spectral bands selected from the visible and near-infrared that do not require cryogenic cooling of detectors. Recording at 256 grey levels (8 bits). Nondestructive, onboard data compression to no more than 32 or 64 grey levels (5 or 6 bits).

Spatial resolution. 10 m pixels capable of onboard aggregation to lower (coarser) resolution. Use of stereo and the offset of spectral band footprints will improve effective resolution beyond the 10 m indicated by the pixel size.

Attitude control and determination. Control with 0.1 degree, basically to hold satellite vertically along ground track, but capable of off-track viewing when required. Spacecraft stability in the order of 10^{-6} degrees per second. Attitude determination to within 5 arcs seconds based on stellar sensing.

Spacecraft position. Provided by the Global Positioning System or equivalent to within 3–5 m in the three basic coordinates.

Data transmission. 50–100 Mb/s, at 32 or 64 grey levels (5 or 6 bits), X-band through a fixed omnidirectional antenna to the network of existing ground stations. Onboard storage on tape recorders or equivalent.

Described by Colvocoresses (1990), the stereomapping satellite system would provide data suitable for 1:50 000 scale mapping with 20 m contours. The realization of the project would cost approximately 10 years of effort and 1 billion US dollars.

Future space photography missions certainly represent a potential domain of activities. Existing space cameras, the Metric Camera, the Large Format Camera (LFC) and the former USSR's cameras KATE-200 and KFA-1000 may be used during future space platform missions, Space Shuttle operations, and the MIR Space Stations missions. Except for efforts of the former USSR, limited interest in space photography is expressed by other organizations involved in cartographic applications of satellites, despite the positive evaluation of experimental space photography from LFC and the Metric Camera.

1.8 Framework for future activities

Despite many efforts and international co-operation, particularly in the domain of meteorological studies, present activities in environmental and cartographic satellite remote sensing reflect to a great extent the old scenario of a politically and economically divided world.

Competing organizations, overlapping missions and many superficial bureaucratic activities hamper the real progress in global scientific co-operation. However, steps in the proper direction have been taken. Organizations like the World Meteorological Organization (WMO), the Food and Agriculture Organization (FAO), the UN Environmental Programme (UNEP), and others, such as the International Geosphere–Biosphere Programme (IGBP) will depend heavily upon satellite data and worldwide monitoring of the environment. Organizations like ESA, NASA and NOAA are planning more joint ventures. In the former USSR, new trends to commercialization has resulted in the selling by Priroda and Sojuzkarta of space photographs from multispectral MK-4 and KFA-1000 cameras. The global environmental concerns, which require coordination and international action, may contribute to better strategies and to joint co-operative missions.

Cartographic objectives of satellite missions retain their top priorities due to the recognition of the unsatisfactory global map coverage, the strong demand for efficient map revision and the need for compilation of the new thematic maps presenting the diversity of the geosphere. Cartographic goals of the satellite missions could be achieved faster if leading industralized countries, having technical and economic capabilities, would maintain their trust, solidify and coordinate better their contributions and support to satellite remote sensing and satellite cartography.

References

ADENIYI, P. O. and BULLOCK, R. A. (Eds.). 1988. *Seasonal Land Use and Land Cover in Northern Nigeria; An Atlas of the Central Sokoto-Rima Basin.* Department of Geography, University of Waterloo, Occasional Paper No. 8.

ALLAN, T. D. 1983. *Satellite Microwave Remote Sensing.* Chisterts. Ellis Horwood.

BARKER, J. L. (Ed.). 1984. *Landsat-4. Investigation Summary, including Deembo 1983 Workshop Results.* NASA-CP-2326, Vol. 1–2.

BARKER, J. L. (Ed.). 1985. *Landsat-4. Science, Characterization, Early Results.* NASA-CP-2355, Vol. 1–4.

BAUDOIN, A. 1992. Improvements of the SPOT system at the turn of the next century. *Proceedings 17th ISPRS Conference,* Washington.

BODECHTEL, H. 1986. Thematic mapping of natural resources with the modular optoelectronic multispectral scanner (MOMS). In K. H. Szekielda (Ed.), *Satellite Remote Sensing for Resource Management.* Graham & Trotman Ltd, London.

CAMPBELL, J. B. 1987. *Introduction to Remote Sensing.* Guilford Press, New York and London.

CARTER, D. J. 1986. *The Remote Sensing Sourcebook.* McCarta Ltd, London.

CERCO. 1988. The SPOT system and its cartographic applications. *Proceedings of the Conference,* Saint-Mandé, 6–15 June 1988.

CHEN, H. S. 1985. *Space Remote Sensing Systems.* Academic Press, New York.

CHEN-SHU-PENG. 1986. *Atlas of Geo-Science Analysis of Landsat Imagery in China.* Science Press, Beijing.

CIMINO, J. B. and ELACHI, C. (Eds.). 1982. *Shuttle Imaging Radar-A (SIR-A) Experiment.* Jet Propulsion Laboratory Publication 8277. Pasadena, California.

COLVOCORESSES, A. P. 1990. An operational Earth mapping and monitoring satellite system: a proposal for Landsat 7. *Photogrammetric Engineering and Remote Sensing,* Vol. **56**, No. 5, pp. 569–571.

DENÈGRE, J. 1987. Apport de SPOT aux systèmes d'information géographique. *Colloque SPOT-1. Image Utilization Assessment Results,* Paris, pp. 1459–1466.

DENÈGRE, J. (Ed.). 1988. *Thematic Mapping from Satellite Imagery: An International Report.* Elsevier Applied Science Publishers Ltd, Amsterdam.

Dierke Weltraumbild-Atlas. 1981. George Westermann Verlag, Braunchweig.

DOYLE, F. J. 1979. The large format camera for shuttle. *Photogrammetric Engineering and Remote Sensing,* Vol. 45, No. 1, pp. 200–203.

DRURY, S. A. 1990. *A Guide to Remote Sensing.* Oxford University Press.

EHLERS, M., EDWARDS, G. and BÉDARD, Y. 1989. Integration of remote sensing with geographical information system; a necessary evolution. *Photogrammetric Engineering and Remote Sensing,* Vol. 55, No. 11, pp. 1619–1627.

FISHER, P. F. and LINDENBERG, RICHARD E. 1989. On distinction among cartography, remote sensing and geographical information system. *Photogrammetric Engineering and Remote Sensing,* Vol. 55, No. 10, pp. 1431–1434

FORD, J. P. *et al.* 1980. *Seasat Views North America, the Caribbean, and Western Europe with Imaging Radar.* Jet Propulsion Laboratory Publication 80–87. Pasadena, California.

FU, L. and HOLT, B. 1982. *Seasat Views: Oceans and Sea Ice with Synthetic Aperture Radar.* Jet Propulsion Laboratory Publication 81-120 Pasadena, California.

GATLAND, KENNETH. 1989. *The Illustrated Encyclopedia of Space Technology,* 2nd ed. Salamander Books Ltd.

GUGAN, D. J. 1987. Practical aspects of topographic mapping from SPOT imagery. *Photogrammetric Record,* Vol. 12 (69), pp. 349–355.

GUYENNE, T. D and CALABRESI, G. 1989. *Monitoring Earth's Environment.* ESA-SP-1102.

HARRIS, R. 1987. *Satellite Remote Sensing.* Routledge & Kegan Paul, London and New York.

HOLMES, A. L. 1983. Shuttle imaging radar-A, information and data availability. *Photogrammetric Engineering and Remote Sensing*, Vol. 49, pp. 65–67.

HYATT, E. 1988. *Keyguide to Information Sources in Remote Sensing*. Mansell Publ. Ltd, London and New York.

IRS PROJECT TEAM. 1988. *The Indian Remote Sensing Satellite System. Remote Sensing Yearbook 1988/89*, pp. 59–72. Taylor & Francis Ltd, London and New York.

JENSEN, J. R. 1986. *Introductory Digital Image Processing*. Prentice-Hall, Englewood Cliffs, New Jersey.

KONECNY, G. 1986. First results of the European spacelab photogrammetric camera mission. In K. H. Szekielda (Ed.), *Satellite Remote Sensing for Resource Development*. Graham & Trotman Ltd, London.

KONECNY, G. 1989. Recent development in remote sensing. Invited paper, 14th World Conference of the International Cartographic Association, Budapest.

KROMER, J. 1987. Suitability of space photographs for map production and map revision. *Jena Journal for Photogrammetrists and Surveyors*, Vol. 1, pp. 12–16.

LIGHT, D. L. 1989. Remote sensing for mapping. American Society for Photogrammetry and Remote Sensing. *Proceedings ASPRS/ACSM Annual Convention*, Vol. 3, pp. 50–74.

LILLESAND, T. M. and KIEFER, R. W. 1987. *Remote Sensing and Image Interpretation*, 2nd ed. John Wiley & Sons, New York.

MATTHEWS, J. 1988. Images of remote sensing in Japan. In *Remote Sensing Yearbook 1988/89*, pp. 29–47. Taylor & Francis Ltd, London and New York.

MOIK, J. G. 1980. *Digital Processing of Remotely Sensed Images*. NASA-SP-431.

Multilingual Dictionary of Remote Sensing and Photogrammetry. 1984. American Society of Photogrammetry.

NASA. 1977. *Skylab Explores the Earth*. NASA-Sp-380.

NASA. 1982. *Meteorological Satellites: Past, Present and Future*. NASA-CP-2227.

NASA. 1986. *Earth System Science Overview. A Program for Global Change*. Washington, D.C. NASA.

NASA. 1988. *Earth System Science: A Closer View*. Report of the Earth System Science Committee, Washington, D.C. NASA.

PALUDAN, T. and CSATI, E. 1978. *Euludus Map; An International Land Resources Map Utilizing Satellite Imagery*. NASA-TP-1371.

RICHARDSON, J. A. 1986. *Remote Sensing Digital Image Analysis*. Springer Verlag, Heidelberg.

RIVEREAU, J. C. and POUSSE, M. 1988. SPOT after two years in operation. *CERCO. Proceedings of the Conference "The SPOT and its Cartographic Applications"*, 6–15 June 1988.

ROCHON, G. and TOUTIN, T. 1986. SPOT, a new cartographic tool. *International Archives of Photogrammetry and Remote Sensing*, Vol. 26 (4), pp. 192–205.

SABINS, F. F. 1983. Geologic interpretation of space shuttle radar images of Indonesia. *Amer. Associatia. of Petroleum Geology Bulletin*, Vol. 67, pp. 2076–2099.

SABINS, F. F. 1987. *Remote Sensing: Principles and Interpretation*, 2nd ed. W.H. Freeman and Co., New York.

SAGDAYEW, R. S., SALISHCHEW, K. A. and KOUTZLEBEN, ?. (Eds.). 1982. *Atlas of Interpretation of Multispectral Aerospace Photographs; Methods and Results*. Academia Verlag, Berlin; Publishing House "Nauka", Moscow.

SCHROEDER, M. 1986. Spacelab metric camera experiments. In K. H. Szekielda (Ed.), *Satellite Remote Sensing for Resource Development*. Graham & Trotman Ltd, London.

SHEFFIELD, C. 1981. *Earthwatch; A Survey of World from Space*. Macmillan, New York.

SHEFFIELD, C. 1983. *Man on Earth*. MacMillan, New York.

SHORT, N. M. 1982. *The Landsat Tutorial Workbook: Basics of Satellite Remote Sensing*. NASA-RP-1078.

SHORT, N. M. and BLAIR, ROBERT W. (Eds.). 1986. *Geomorphology from Space: A Global Overview of Regional Landforms*. NASA-SP-486.

SHORT, N. M. and STUART, L. A. 1982. *The Heat Capacity Mapping Mission (HCMM) Anthology*. NASA-SP-465.

SOUTHWORTH, C. S. 1985. Characteristics and availability of data from Earth imaging satellites. *US Geological Survey Bulletom*. 1631.

SZANGOLIES, K. 1987. Acquisition and use of space photographic data for mapping. *Jena Journal for Photogrammetrists and Surveyors*, Vol. 1, pp. 2–4.

SZEKIELDA, K. H. (Ed.). 1986. *Satellite Remote Sensing for Resource Development*. Graham & Trotman Ltd, London.

TSUCHIYA, K., ARAI, KOHEY and IGARASHI, TAMOTSU. 1987. Marine observation satellite. *Remote Sensing Reviews*, Vol. 3, pp. 59–101.

US GEOLOGICAL SURVEY. 1979. *Landsat Data User's Handbook*. USGS.

VAUGHAM, W. W. (Ed.). 1982. *The Conception, Growth, Accomplishments and Future of Meteorological Satellites*. NASA-CP-2257.

WELCH, R. 1985. Cartographic potentials of SPOT image data. *Photogrammetric Engineering and Remote Sensing*, Vol. 51, pp. 1085–1091.

Chapter 2

PRODUCTION PROCESSES FOR EXTRACTING INFORMATION FROM SATELLITE DATA

Compiled by
Donald T. Lauer

US Geological Survey, Sioux Falls, South Dakota 57198, USA

CONTENTS

2.1 Introduction

2.1.1 Purpose of section

Cartographers, as well as resource specialists, engineers, land managers, and planners, are making extensive use of satellite image data for thematic mapping. Applications include cartographic mapping, geologic and hydrologic investigations, land management and inventory, and environmental monitoring. The purpose of this section of the guide-book is to briefly review the different production processes for extracting thematic information from satellite data. These processes are presented in two parts. First, the principles, methods, and techniques of visual image interpretation are discussed, where the amount and quality of information extracted from the satellite data depends largely on the training and skills of the image analyst. Second, the principles, processes and procedures of computer-assisted image analysis are presented. In this case, the special powers of computerized data processing are employed by the image analyst to make thematic maps from satellite data. The materials presented in this section only briefly introduce these production processes, and the interested reader should pursue other references (such as the American Society of Photogrammetry and Remote Sensing's *Manual of Remote Sensing*, 2nd edition, Volumes I and II, 1983) for a more indepth discussion of the subject.

2.1.2 Image, line and thematic mapping

It is important to note that when satellite data are used for cartographic production, the methods employed can result in an image map of the landscape, or a simple line map of important Earth features, or a comprehensive thematic presentation of the land surface. For example, the production process may be limited to some form of computerized geometric correction and radiometric enhancement of the image that maintains the continuous tone of the Earth's surface and possibly combines the corrected and improved image with some cartographic information provided by other sources; but, the extraction of simple line or comprehensive thematic information from the "image map" is performed by the final user who must employ visual image interpretation techniques. On the other hand, satellite data can be transformed directly into a thematic map by means of a wide variety of production processes, which are discussed below — all of which involve the interaction between a skilled analyst and advanced, computerized data processing technology.

2.2 Visual image interpretation

2.2.1 Basic principles

Visual image interpretation is often a practical and cost-effective means for extracting useful thematic information from satellite imagery. It is important, however, to understand both the procedures used by a skilled image analyst when performing an image-interpretation task and the concepts on which those procedures are based.

The basic principles of satellite image interpretation can be stated as follows:

(1) a satellite image of the Earth is nothing more than a graphic form of data that provides a pictorial presentation of the pattern of the landscape;

(2) the pattern is composed of elements, or indicators, of features and events that reflect the physical, biological and cultural components of the landscape;

(3) similar conditions in similar environments reflect similar patterns, and unlike conditions reflect unlike patterns;

(4) the type and amount of information that can be extracted from a satellite image is proportionate to the knowledge, experience, skill and interest of the analyst, and an awareness of the limitations in the procedures used (Estes, 1980).

There is no magic involved in visual image interpretation. An understanding of the image-forming processes; image-pattern elements; Earth features, processes and phenomena; and an interest in learning are necessary to develop the art of interpreting satellite images for extracting thematic information. An image is only a tool and the interpreter has to extract the information through proper use of the tool. The skill of extracting information from satellite images is developed through experiences — in other words, one learns by doing.

2.2.2 Interpretation process

Visual image interpretation is defined as the process of detecting, delineating, and identifying features and/or conditions on images and judging the significance of those features and/or conditions (Colwell, 1987, 1965). The image elements that allow the interpreter to locate, delineate, identify, and evaluate subjects to be mapped are tone, colour, texture, pattern, shape, size, shadow, parallax and time-variant phcnomena (Avery, 1987). It is important to note that several factors relate directly to both the perception of these elements and the resulting interpretation of imagery. These factors are:

(1) sensitivity characteristics of the imaging system (camera and film, electro-optical device, or other type of detector);
(2) resolution characteristics of the imaging system;
(3) film exposure or data processing;
(4) season of year;
(5) time of day;
(6) atmospheric effects;
(7) image scale;
(8) image motion;
(9) stereoscopic parallax;
(10) visual and mental acuity of the interpreter;
(11) interpretation equipment and techniques;
(12) training aids.

Note that factors 1 through 9 mainly affect the quality of imagery, while 10, 11, and 12 reflect the human's ability to extract information from images. Certain combinations of these factors would better allow an interpreter to perform various interpretation tasks than other combinations. Consequently, a primary objective during the process of interpreting satellite images is to describe, to the best of one's ability, the optimum combination of factors needed to solve specific thematic mapping problems.

2.2.3   Image elements

2.2.3.1   Tone

Tone in a satellite image is the relative blackness or whiteness (i.e. brightness) and is the result of the amount of energy reflected and/or emitted by the feature being imaged. Tone is fundamental to interpretation of black and white satellite images and, when used with other recognition elements, is a primary element for feature identification and interpretation. The tones of satellite images are influenced by many factors, and the tones of familiar objects often fail to correspond to one's perception of those objects in nature. For example, a body of water may appear in tones ranging from white to black, depending on the position of the Sun and the number of wave surfaces reflecting energy to the sensor system. When the image interpreter understands the factors that govern tone, he or she regards the tones of objects of interest as major clues to their identity or composition. The soil scientist uses tonal variations to classify soils; the forester, to distinguish hardwood from coniferous forest types; the geologist, to map lithology and structure. In a satellite image, where the shapes of objects often cannot be resolved, or in stereoscopic pairs of satellite images (e.g. SPOT data), where the objects of interest might have little or no visible height, tone is particularly important. The image interpreter may capitalize on variations in energy reflectivity or emissivity by using images taken in spectral channels that best record the tone contrasts of the objects to be studied (Colwell, 1961).

2.2.3.2   Colour

A feature has colour when it reflects different amounts of energy in particular combinations of wavelengths. For example, vegetation appears green to the human eye because, in general, plants preferentially reflect a larger percent of green energy compared to blue and red. The human eye can distinguish about 1000 times as many presentations of colour as it can tones of grey, thus colour permits recognition and interpretation of a greater amount of information about the Earth's surface. In the interpretation of rocks, soils and plants where there may be an abundance of features whose natural colours are important, colour satellite images, as compared to black and white, can greatly facilitate image interpretation. Likewise, false-colour imagery, which combines spectral channels other than blue, green and red, has proved to be useful for special studies of plant conditions, vegetation distribution, soil-moisture conditions and other feature delineations.

2.2.3.3   Texture

Texture in satellite imagery is created by the frequency of tonal or colour change in groups of objects that are too small to be discerned as individual features. It follows that the size of the object required to produce texture varies with the resolution and scale of the imagery. For example, in high-resolution, large-scale aerial photographs, trees can be seen as individuals; their leaves or needles cannot be discerned separately, but contribute to the texture of the tree crowns as seen on the photographs. In aerial photographs of lower resolution and smaller scale and in most types of satellite imagery, the tree crowns contribute to the texture of the whole forest. Within a given range of scales, the texture of a group of objects (such as a forest) may be distinctive enough to serve as a reliable clue to the identity of the objects. Texture can be an important interpretative factor in using images acquired from orbital altitudes. For example, the relative erosional dissection of an area may be inferred only by its texture because most individual drainageways cannot be delineated on current types of satellite imagery. Drainage pattern is an important indicator of the type of surficial materials and bedrock. In other cases, features may be similar in tone and colour, but may exhibit considerable difference in texture (e.g. volcanic fields versus desert pavement).

2.2.3.4 Pattern

Earth scientists have emphatically stressed the importance of patterns, or spatial arrangements of objects, as clues to their origin or function. Geographers and anthropologists study settlement patterns and their distribution in order to understand the effects of diffusion and migration in cultural history. Outcrop patterns provide clues to geologic structure, lithology and soil texture. The varying relations between organisms and their environment produce characteristic patterns of plant association. Regional patterns that formerly could be studied only through laborious ground observation are instantly and clearly visible on aerial and satellite images. The images often capture many significant patterns that might be overlooked or misinterpreted by the ground observer. Innumerable variations in classic patterns can be seen and exploited by means of image interpretation. The trained observer responsible for making thematic maps appreciates the significance of satellite imagery chiefly through his or her understanding of patterns on the Earth's surface. Some patterns are primarily cultural, and others are primarily natural. There are, however, few parts of the world that have not been affected by mankind, and most of the

patterns visible in satellite imagery result from the interaction of natural and cultural factors. Patterns of historic settlement, mining and agricultural activity often are visible on images acquired from space, directly or through altered patterns of vegetation and erosion. Patterns formed by agricultural practices, fracture alignments, drainage networks and vegetation are among the most important factors in the interpretation of a satellite image. As mentioned above, intricate patterns may be reflected as a textural difference in space imagery. In many instances, regional patterns associated with other image elements allows for an image to be successfully interpreted.

2.2.3.5 Shape

The shapes of objects or features seen in vertical view are sometimes surprisingly difficult to interpret. The plan or top view of an object is so different from the familiar profile or oblique view, that inexperienced interpreters often fail to recognize the image of the building in which they are working, as seen on a vertical aerial photograph. The ability to understand and make use of the plan view has to be acquired like another language, but it then becomes a powerful tool. To the satellite image interpreter who is experienced in industrial studies, the vertical view of an urban area tells as much about its character as does a walk or drive through the area. Likewise, the vertical view of a forest from orbital altitudes may reveal information about its condition, or the vertical view of a landform may show spectacular effects of tectonic processes.

2.2.3.6 Size

The size of an object is one of the most useful clues as to the object's identity. By measuring the size of an unknown object on a satellite image, the interpreter can eliminate from consideration whole groups of possible identifications. It is always advisable when faced with an unknown object to measure it, and when working with satellite images of variable scale, the interpreter should make frequent measurements of the objects of interest. Thus, the interpreter working with space imagery must always be conscious of image scale when assessing the significance of size, and, of course, larger features may occupy only a small portion of the image area because of the synoptic view provided from orbital altitudes.

2.2.3.7 Shadow

Shadow is a familiar phenomenon, and in ordinary life the size and shape of objects are often judged by observing the shadows they cast. The shadows present in conventional aerial photographs sometimes help the interpreter by providing him or her with profile representations of objects of interest. Shadows are particularly helpful if the objects are very small or lack tonal contrast with their surroundings (US Department of Agriculture, 1966). Under these conditions, the sharp tonal gradients of the shadows may enable the interpreter to identify objects which themselves are just at the threshold of recognition. If the interpreter is interested not in a particular class of objects, but in the landscape as a whole, as so often is the case when using satellite imagery, he or she must forego some of the advantages of shadow identification in order to see as much as possible of the ground surface. To work with the vertical view, the interpreter must revise his or her ideas of the external world and acquire new habits of observation. Moreover, objects are imaged in most satellite scenes at very small scales. Because of the vertical view and the small scale, some elements of appearance have greater importance in satellite images than they have in the ground view (e.g. shadows which enhance topographic features).

2.2.3.8 Parallax

In stereoscopic pairs of conventional aerial photographs, the observer sees objects in three dimensions and distinguishes close from distant objects; that is, he or she perceives angles of parallax. In observations of objects in nature, the angle of parallax is determined by the distance between the pupils of the observer's eyes, or eye base. In aerial photography, the distance between successive exposures (air base) corresponds to the human eye base. The air base, as imaged on the photographs, is the photo base and is much greater than the human eye base. The impression of depth is much exaggerated in stereoscopic pairs of aerial photographs, but the skilled interpreter learns to allow for this exaggeration of depth. The vertical exaggeration in overlapping satellite images often is not apparent because of the relatively low angles of parallax. Satellite images that do not overlap cannot be viewed in three dimensions, and the shape of features are

apparent in plan view only. However, shape is still a primary interpretative factor and very useful in identification of major landform features (for example, lava fields, volcanic cones, alluvial fans, dune fields, beaches, bays, lakes, mountains, etc.). Nevertheless, with improved sensor system performance (such as resolution, overlap, etc.) available in current and future Earth-orbiting systems, stereoscopic observations, and measurements of parallax are of increased importance to the interpreter of satellite imagery. To the skilled image interpreter, the value of three-dimensional shape is that it delimits the class of features and objects to which an unknown must belong, it frequently allows a conclusive identification, and it aids the understanding of significance and function.

2.2.3.9 Time-variant phenomena

Very often the temporal aspects of Earth phenomena are overlooked by the image interpreter. For certain applications of thematic mapping, considerable additional information is provided by comparing images taken over a period of time. Agricultural analyses can be more complete by using images that have been acquired during various stages of the growing cycle. Winter wheat has a different growing cycle than other agricultural crops. Grassland has different stages of growth than planted crops. Other time-variant phenomena that can be observed on satellite imagery include assessments of coastal erosion, sedimentation in water bodies, surface-water distribution, snow cover, etc.

2.2.4 Interpretation techniques and output products

Certain image-interpretation techniques, when properly applied, can improve the quality and quantity of useful thematic information extracted from satellite imagery (Colwell, 1987). These techniques include using:

(1) methodical procedures;
(2) efficient searches;
(3) knowledge of factors governing image formation;
(4) the background and training of the interpreter;
(5) the concept of "convergence of evidence";

(6) the "conference system";
(7) information available in analogous areas;
(8) reference materials;
(9) simple and sophisticated equipment;
(10) field data.

The image interpreter must have a full understanding of how the image was formed, what the image elements represent and the basic Earth processes and phenomena that are present on the image. A systematic approach to the problem under study is perhaps the most important consideration. In general, a systematic analysis of satellite imagery involves the convergence of empirical evidence drawn from the imagery through regional analyses (geographic aspects, physiographic aspects, geology, climate); local analyses (landforms, drainage patterns, erosion, image patterns, vegetation, spectral features); compiling collateral information (reports, maps, field data); summarizing interpretation results; and field checking key areas. Additional benefits and information are also obtained by using a multidisciplinary/interdisciplinary team approach.

The results of this type of systematic analysis can then be presented in a thematic map format either by making an interpretive overlay, or a series of overlays, directly on the satellite images or by transferring the interpreted information to an existing base map using appropriate optical or electronic data transfer equipment.

2.3   Computer-assisted image analysis

2.3.1   Basic principles

Since computers can be used to rapidly and accurately process, measure, and compare large quantities of numerical data, they are playing an increasingly important role in the analysis of multispectral satellite imagery for thematic mapping. The rationale for using computer-assisted image analysis techniques relates to a few basic principles:

(1) the satellite data often are already in digital format;
(2) errors introduced by sensors or processing systems usually can be corrected;
(3) adjustments for differing amounts of illumination can be made;

(4) each individual picture element, or pixel, can be analysed;
(5) sophisticated statistical and mathematical processes can be applied to the data;
(6) the analysis is objective and repeatable;
(7) numeric sensitivity is greater than that of visual image analysis;
(8) large quantities of data can be analysed;
(9) other types of data (maps, statistics, other sensor, etc.) are easily integrated or merged with digital satellite data;
(10) output can be generated using various formats (maps, tabular, images, etc.).

In general, computer-assisted image analysis is more applicable than visual interpretation when analysing large areas that require detailed information classes. Nevertheless, several disadvantages to using these techniques may exist and often include

—a large capital investment;
—requirements for specially trained programmers and analysts;
—higher data costs;
—less reliability in complex environments;
—the results may not be fully understood by the user.

A decision on whether or not to use computer-assisted image analysis techniques for making thematic maps from satellite imagery will depend largely on the size of the study area, the detail desired in the final results, the availability of computer hardware and software and the special skills of the analyst.

2.3.2   Image preprocessing

2.3.2.1   Radiometric corrections

Bad data lines, line segments, or pixel dropouts are often referred to as intermittent striping problems. When intermittent striping problems occur in satellite digital data, classification errors will occur since the bad data will not correspond to the proper training set statistics. One technique used to insert new data is to replace the bad data line with the preceding line. A second technique commonly used is to replace the bad data by interpolating between the brightness values from the pixels in the line before and after the bad data line (Jensen, 1986). Although the differences between the two techniques seems slight, the latter tends to provide a more accurate representation of the correct brightness values.

Radiometric striping often appears as systematic light or dark lines across an image and is caused by each detector in the sensor device having a slightly different response to the incident radiation falling upon it, and hence, for the same intensity of radiation, a slightly different output voltage. Specifically, the detectors have different gains and offsets (Richards, 1985). Radiometric striping introduces anomalous variations in the spectral signatures of features on the Earth's surface and will influence classification performance either due to high variances calculated for training data, or lack of adequate training data to characterize signatures of pixels occurring in lines with striping. Also, if an objective of an analysis session is to enhance image contrast, any slight striping will also be enhanced. It is possible that when severely striped images are enhanced, the spatial distribution of landscape features will be destroyed due to the enhanced striping effect.

Several techniques are available to normalize the data and minimize effects of striping. One technique, referred to as histogram normalization, calculates the mean brightness value for each detector in each band. A normalization factor is calculated by ratioing either the maximum mean or minimum mean to the mean of each detector. The normalization factor for a given detector is then multiplied by the brightness value of each pixel recorded by the detector, and the pixel is assigned a new brightness value.

Correction for solar illumination is applied to adjacent scenes of satellite data that are to be digitally mosaicked and that were acquired under different illumination conditions. This type of correction is also necessary for comparison of spectral properties of features for scenes acquired under different conditions of solar illumination. The Sun elevation angle (Sun angle) correction involves multiplication of all brightness values in the scene by a constant that is a function of the Sun angle. It should be noted that this correction does not remove the effects of topography, nor does it correct for different azimuths of solar illumination.

Lastly, the atmosphere affects data acquired from Earth-orbital altitudes in two ways: scattering and absorption. Atmospheric scattering is a function of molecules (Rayleigh)

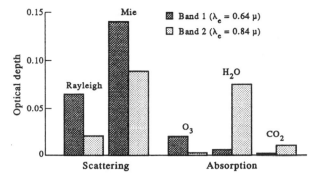

Fig. 2.1 The effects of atmospheric scattering and absorption on bands 1 and 2 of Advanced Very High Resolution Radiometer data (courtesy of ST Systems Corporation).

and aerosols (Mie) in the atmosphere (McCartney, 1976). Rayleigh effects are easy to model as they are largely invariant in time for a particular location. Mie scattering involves more complex models requiring aerosol measurements, which are difficult to obtain. Atmospheric absorption is largely a function of several gases in the atmosphere, such as water vapour, carbon dioxide and ozone (Asrar, 1989). Absorption models are generally straightforward to apply, but rely on measurements, which can be difficult to obtain. Atmospheric scattering and absorption greatly influence the measurement of spectral characteristics of Earth features and conditions (see Fig. 2.1). If comparisons are to be made between ground-based spectral measurements and satellite measurements, then corrections for atmospheric effects must be applied.

2.3.2.2 *Geometric corrections*
(This aspect is considered in more detail in Chapter 3)

Systematic and predictable geometric errors in satellite imagery include skew caused by rotation of the Earth under the satellite and variable line length if the sensor system employs a mirror scanning mechanism. Variable and measurable errors include distortions caused by variations in spacecraft velocity, altitude and attitude (Williams, 1979).

Skew distortions are introduced as a result of the Earth's rotation and satellite movement in orbit as an image is acquired. The amount of skew present in a scene is largely a function of latitude and spacecraft heading. Each image is de-skewed by an algorithm that translates scan lines to the right by a calculated number of pixels, depending on the estimated Earth latitude for the scan line, so that terrain features are in proper position from top to bottom of the scene.

Systems that use an oscillating mirror surface create variations in line lengths due to the varying velocity of the mirror. A correct line length is calculated as a function of the mirror scan rate, detector sampling rate and swath width.

When geometric errors in satellite imagery are introduced that are variable in nature, the effects are not predictable; thus to correct the errors, their effect must be measured. This correction can be accomplished by measuring the apparent displacements of ground control points, detectable and recognizable features in an image whose geographic positions are known. Once the displacement of control points is known, mathematical functions of mapping transformations are used to calculate the coordinates of a pixel in the corrected image. These coordinates will be found at some location in the distorted image. This location seldom coincides with a pixel in the distorted image, thus the distorted image must be resampled to determine the brightness value that should be assigned to the pixel in the corrected image.

Three commonly used resampling techniques include nearest neighbour, bilinear interpolation and cubic convolution (Simon, 1975). In nearest-neighbour resampling, the brightness value of a pixel in the corrected image is determined from the brightness value of the pixel nearest the location of the pixel in the input image. Bilinear interpolation is another resampling technique that determines the brightness value of a pixel in the correct image by interpolating between the brightness values from four nearest pixels around a given location. Cubic convolution resampling is a higher order resampling technique that is often used. The brightness value of a pixel in a corrected image is interpolated from the brightness values from a larger population of the nearest pixels around the location of the corrected pixel.

When an image is being acquired, degradations can occur that reduce apparent resolution. For example, an image can be blurred as a result of poor optics or sensor motion. These phenomena are modelled as a sequence of "filterings" by the imaging device, with each "filter" representing the effect of a single degrading influence, such as optical blurring. As in the case of spatial enhancement, these models can be applied either in the frequency domain, where an image device's "transfer function" allows low frequencies (large area "trends") to be passed preferentially to high frequencies (fine detail), or in the spatial domain as a moving window process, where the scene is "convolved" with the imaging device's "point spread function".

Image restoration is the process of removing these degradations to achieve the maximum possible image sharpness. Restoration techniques generally involve modelling sensor response, computing a correction filter, then applying the filter to the degraded image. Correction methods such as "inverse filtering" or "Wiener filtering" are used in image restoration (Andrews and Hunt, 1977). These correction filters may be applied either in the frequency or spatial domains, or can be included in image resamplers as part of the geometric rectification process.

2.3.3 *Image enhancement*

The purpose of image enhancement is to allow for improved visual image interpretation by amplification of important spectral or spatial characteristics while suppressing nonessential characteristics of redundancy. With good quality satellite imagery, very little "general enhancement" is possible. Enhancement procedures should be applied to improve imagery for specific interpretation problems. Unfortunately, however, improvement of an image for interpreting one feature often makes interpretation of other features more difficult. Characteristics of the human eye/brain receiver should be considered in enhancement processing. For example, most people can detect many more shades of green than shades of blue; also, the human optical system performs an edge enhancement operation that is difficult (and usually unnecessary) to duplicate using digital techniques.

2.3.3.1 *Contrast enhancement*

Contrast enhancement (or stretch) is performed by amplifying the brightness range or selected subrange(s) of an image. Some general enhancement is possible by determining the minimum and maximum image brightness values and then adjusting this range to correspond to the "straight line" portion of the film density range on which the image will be recorded. More dramatic enhancement is possible by truncating data from the minimum and maximum tails of the distribution, but once performed, the truncated information will be recorded at the same density as the retained minimum and maximum data values. If a different function is used for each spectral band, the tonal balance of the color composite may change. This may sometimes be used to advantage, but would not be useful, for example, if images were to be mosaicked.

In linear contrast enhancement, each brightness value is multiplied by a constant and then adjusted with an additive constant or bias.

In order to enhance one portion of the density range more than another, non-linear enhancement may be used. Non-linear enhancements may be used to approximate an equalized distribution (or a ramp cumulative distribution) so that approximately equal numbers of pixels are recorded at each film density. Non-linear enhancements are quite often effective in increasing contrast for specific features (usually the densities that predominate the scene), but may be detrimental in interpretation of other (usually spatially small or infrequently occurring) features (Pinson and Lankford, 1981). Many non-linear enhancement functions use systematically derived probability density functions, such as logarithmic, Gaussian or sinusoidal enhancement. Other enhancement functions are data-dependent and require histograms to be created before the function can be generated. Examples of data dependent enhancements include the ramp cumulative distribution function, probability distribution function stretch, stretch proportional to frequency distribution and histogram equalization.

2.3.3.2 Spatial enhancement

Spatial enhancements adjust the brightness value of each pixel by comparison to the values of neighbouring pixels. For example, edge enhancement may be performed by exaggerating the difference between a pixel and its neighbours. On the other hand, image smoothing may be performed by reducing the difference between a pixel and its neighbours. Edge enhancement is a high-frequency passing, or high-pass filter. Smoothing is a low-frequency passing, or low-pass filter (Moik, 1980).

A more sophisticated spatial enhancement technique actually transforms the data into the frequency domain (such as Fourier transform). Once mapped into the frequency domain, the image may be filtered to allow only certain frequencies to be passed. The filtered data are then transformed back to the spatial or image domain. Much the same effect can be obtained in the spatial domain by using a "moving window", or "moving average" technique. Since this technique is simpler to implement and operate, it is more commonly used for spatial enhancements.

High-frequency enhancement is typically employed on satellite imagery to exaggerate the edges between contrasting Earth features, or to bring out linear trends of geologic significance (Mayers *et al.*, 1988). For this reason, it is often used for mapping lineaments and drainage patterns. Figure 2.2 shows an example of the application of an adaptive boxcar filter to band 4 of Landsat Thematic Mapper data.

Low-pass filters can be used to smooth, generalize or "defocus" an image. Since generalization is of limited use in enhancement, one common technique ignores differences from the neighbourhood unless it is large enough to indicate erroneous data. A threshold is established so that smoothing is not performed unless a pixel varies greatly from its neighbours, indicating noise (for example, a loss in data transmission). When detected as noise, the pixel may be reassigned to the average value of its neighbours. This technique is called noise suppression or noise filtering.

A low-pass, post-classification filter may be used to generalize and refine a classification result image by assigning each pixel to the dominant class within the neighbourhood (Fosnight, 1988). This method is useful in generating thematic maps with a minimum mapping unit larger than a pixel. Use of a larger neighbourhood will yield a more generalized image with a larger minimum mapping unit (see Fig. 2.3).

2.3.3.3 Arithmetic combination (ratio and difference)

Simple arithmetic combinations of satellite images provide a method for enhancement of temporal change, reduction of topographic and overall brightness effects and for combining data to a smaller number of spectral channels (Schowengerdt, 1983).

A ratio image is calculated by dividing the

Original band 4

Adaptive boxcar filter

Fig 2.2 The application of a high-pass adaptive boxcar filter to band 4 of Landsat Thematic Mapper data. The unfiltered image is at the top and the filtered image is at the bottom (Mayers *et al.*, 1988).

brightness value of each pixel in a band by the corresponding brightness value from another band or group of bands. In this process, any multiplicative constants that exist in both bands will be cancelled. For example, the effects of topographic slope and aspect on incoming solar flux creates an overall brightness effect that can, in large part, be cancelled using ratio methods.

Another form of ratioing is the making of an "index image", whereby, on a pixel-by-pixel basis, one spectral band is subtracted from another, and this value is divided by the sum of the two bands. When this technique is applied to the near-infrared and red bands of multi-spectral satellite data, it provides an excellent indication of the presence, or lack of presence, of green vegetation — and is often referred to as a Normalized Difference Vegetation Index (NDVI) image. Differencing is accomplished by subtracting one image from another, leaving only differences between the images. This technique is useful for identifying change information from temporal data of correspond-ing wavelength bands. Since only the change information is preserved, this method has limited use, but is quite effective in graphically

displaying either land use or phenological change (Gallo and Daughtry, 1987).

2.3.3.4   *Spectral transformation (rotation)*

Spectral transformation provides a method for systematically reducing highly correlated data to a smaller number of nonredundant channels that retain much of the variance information contained in the original data (see Fig. 2.4). As an enhancement technique, images from several dates and several bands may be combined and the information concentrated into a single colour image. An added benefit from spectral transformation is the computational advantage of a reduced number of channels. Spectral transformation is a powerful tool for reducing the information from many images to a few nonredundant, uncorrelated images. A draw-back to the use of this technique is that colour tones are largely unpredictable and are difficult to interpret as one would interpret a natural or false colour image. One type of transformation is the Principal Components Transform. This transformation technique rotates the axes to align with the calculated eigenvectors, then scales and translates the data to maximize

Original classified image

40 unit distance threshold
8 grid-cell area threshold

6 unit distance threshold
50 grid-cell area threshold

2.5 unit distance threshold
250 grid-cell area threshold

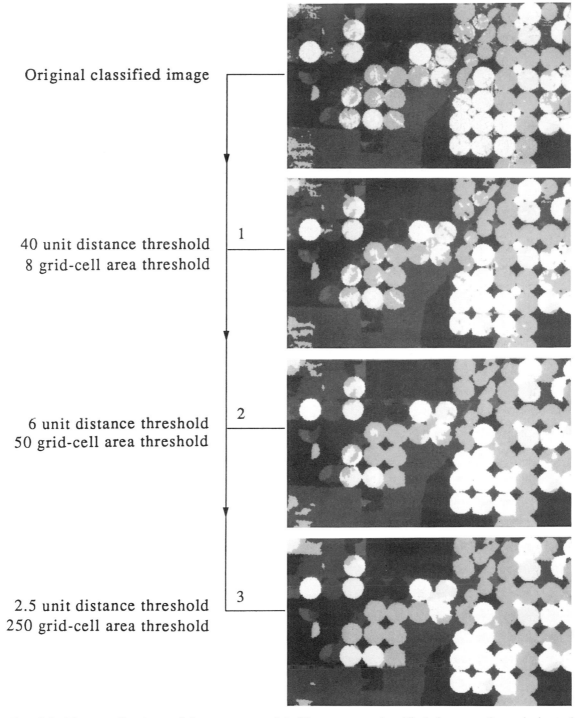

Fig. 2.3 The application of low-pass spatial filters to a classified image of an irrigated agricultural area (courtesy of the US Geological Survey, EROS Data Center).

resultant contrast, and orders the components such that the first component contains the largest proportion of the total scene variance (Fontanel *et al.*, 1975). The Canonical Transform differs from the principal components method in that statistics are extracted from each "class" delineated by the analyst. Rotation axes are calculated such that spectral distances are maximized among the classes and are minimized within the classes. The attempt is to increase the spectral differences between the classes to be interpreted, while minimizing intraclass variability that might cause confusion.

2.3.4 Image classification

Computer-assisted classification of digital satellite imagery is the grouping of a large

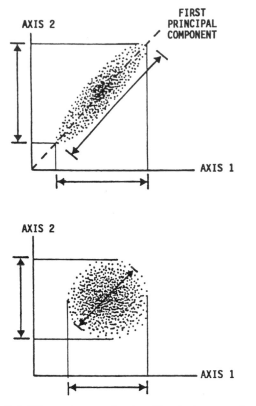

Fig. 2.4 The top chart illustrates principal component analysis of highly correlated data, where the projection on the first principal component is longer than the projection on either original axis. The bottom chart illustrates principal component analysis of uncorrelated data, where the projection on any new axis will be equal to the projection on the original axis (Jenson and Waltz, 1979).

number of individual pixels into a small, more manageable number of thematic classes or categories. The thematic classification assignment of an unknown pixel is based on the similarity of that pixel's spectral characteristics with those characteristics of pixels of a known category. A major assumption is that each thematic category has unique spectral characteristics that can be derived from multispectral satellite imagery. Thematic classification schemes, for example, may be descriptive of vegetation or land cover, land use, surficial geologic groups, or surficial hydrologic groups, as examples. Because of the other contributing factors (slope, aspect, relative frequency and density of the feature, etc.), the spectral response for a given thematic category is not a unique spectral pattern, but a varying spectral pattern. When the spectral variations are greater between, than within, thematic categories, successful classifications are possible (Lillesand and Kiefer, 1979).

2.3.4.1 Classification decision

Spectral pattern recognition involves two conceptual steps. First is the identification of spectral patterns for each thematic category. Second is the classification decision procedure that assigns a pixel to a thematic category on the basis of that pixel's spectral characteristics compared to known spectral patterns developed in the first step (Schowengerdt, 1983). Various classification decision rules (see Fig. 2.5) are available for determining classification assignments of unknown pixels based on their spectral characteristics:

Minimum distance to the means — The unknown pixel is assigned to the class whose mean spectral pattern, determined from a sample of known pixels, is closest to that of the unknown pixel.

Minimum distance to the nearest member of a class — The unknown pixel is classified into the same class as that of the nearest known pixel.

Parallelepiped — A parallelepiped is a "multi-dimensional rectangle", defined by the upper and lower spectral limits of known pixels. The unknown pixel is assigned to the class whose parallelepiped contains the unknown pixel.

Maximum likelihood — With this classification decision rule, known pixels are used to

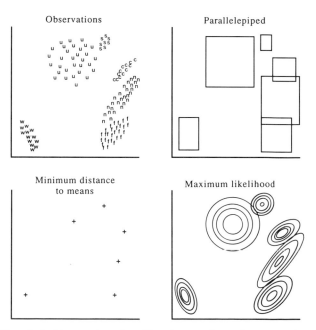

Fig. 2.5 Example classification rules often used to assign a pixel value to a thematic map category (courtesy of the US Geological Survey, EROS Data Center).

estimate statistical properties of each thematic category. Each thematic category is described statistically by its multivariate probability density function. The density function represents the probability that the thematic category's spectral pattern falls in a given spectral region. The maximum likelihood algorithm classifies an unknown pixel to the class with the greatest probability density at the spectral region of the unknown pixel. This algorithm is the most commonly used classification decision rule in most interactive digital image processing systems.

2.3.4.2 Training statistics

Training sets are statistical descriptions of the multispectral characteristics of thematic categories used to "programme" the classifier with the digital characteristics of categories. Training areas are sampled portions of the scene, randomly or purposely selected, used to derive training statistics, and must encompass the spectral variability in the multispectral scene and be inclusive of all thematic categories in a classification scheme. Several approaches to deriving training statistics can be used and include supervised, unsupervised, supervised clustering and controlled clustering.

Supervised development of training statistics involves the purposive selection of training areas that are known and homogeneous thematic categories. An unsupervised approach involves the random selection of training areas without regard to thematic composition (Jensen, 1986). The portion of the scene sampled, when using either the supervised or the unsupervised approach, depends on the number of resource categories and spectral variation in the scene, but is typically 5–10 percent.

The calculation of unsupervised training statistics employs a clustering algorithm that divides the pixels into a number of homogeneous spectral groups, again without regard to resource composition (Bryant, 1978). The maximum number and maximum variability of spectral groups along with other parameters must be specified to the cluster algorithm program. Statistics are determined for each cluster of pixels. Training statistics derived by the clustering algorithm are used in the maximum likelihood algorithm to classify the entire scene. At this point, each cluster is identified as representing a resource category

with the use of supporting ground or photographic data. Ideally, each resource category will be described by at least one cluster, and each cluster will represent only one resource category. This approach is useful when it is difficult to define homogeneous spectral and resource training areas, which is very frequently the case in mapping inaccessible areas and only broad resource ground truth is available.

The supervised clustering approach involves the selection of a number of known and homogeneous training areas for each thematic class, and a clustering algorithm is used to derive the training statistics by clustering all training areas for each thematic category collectively. Finally, the controlled cluster approach uses purposive selection of training areas generally encompassing several thematic categories in each training area, and the calculation of training statistics employs the same clustering algorithm, clustering training areas separately or collectively. Each cluster is identified as representing a thematic category with the use of supporting data, and training statistics for clusters representing a mix of thematic types can be eliminated (Bryant, 1978). In any case, training statistics are evaluated by using the maximum likelihood algorithm to classify several training and test areas and comparing the results to supporting aerial photography or ground data. If preliminary classification results are not acceptable, new training statistics need to be derived or the classification objectives need to be examined.

2.3.4.3 Classification

With the maximum likelihood classification decision rule, each pixel is assigned to the spectral class that has the greatest probability density function for the multispectral values of the pixel. Each and every pixel can be assigned to only one spectral class. Categorization is achieved by grouping results so each class is a single, unique thematic category. Image stratification often is used to improve classification results with the use of ancillary data and may be done prior to or after original classification is accomplished. Stratification may be based on one or several types of information including topographic, environmental data, or administrative boundaries. Also, multispectral data from different images could comprise different strata. Finally, spatial smoothing is often used to improve the appearance of a final map product and to reduce classification errors (Fosnight, 1988).

2.3.4.4 Classification accuracy

An unbiased assessment of the accuracy of classification results is often obtained by sampling to determine the proportion of pixels correctly classified. An estimate of overall classification accuracy of a thematic map made from satellite data can be obtained as well as accuracies for individual map categories.

A simple random sample of points may be useful for obtaining estimates of overall accuracy. Stratified random sampling, where each resource category may be considered a strata, is useful for obtaining individual category as well as overall classification accuracy estimates. Pixels to be sampled are selected in a random fashion, but often need to be limited to areas where supporting data, such as aerial photography, are available, or areas accessible on the ground. Selected sample points are located on aerial photographs by transferring sample points from the satellite image to the aerial photo, or by applying an established geometric link between the satellite image and a map. Sample points are interpreted on the photo or map, the interpretation is compared to the classification results, and correctly classified pixels are tabulated. The proportion of correctly assigned pixels is estimated by comparing the number of sampled pixels correctly classified to the total number of sample points. The accuracy of the proportional estimate is determined by calculating the confidence interval about that proportion. An assessment of the classification results can also be summarized in a contingency table that compares, by thematic map categories, the digital image classification result to the actual correct classification. In this manner, the nature and patterns of misclassification can be identified and evaluated.

2.3.4.5 Classification output products

Satellite data are geometrically distorted, and it becomes necessary to correct for these distortions to provide classification results in a useful format. Geometric control is established with the use of control points that can be identified on both the satellite image and a map base. Transformation equations are developed from control points to establish a link between the image, map and geographic coordinates. The equations are applied in an appropriate manner to provide geometrically corrected map overlays and film products of classification results, geographic coordinates of selected sample points, and corresponding image location of map data. Generation of thematic map overlays involves first, determining the portion of the classification results corresponding to the map, and secondly, correcting the position of pixels so they correspond to the proper geographic position on the map. Map overlays depicting classification results can be generated with a number of devices including line printers, flat-bed plotters and film recorders.

Tabular area summations can be made by simply tabulating the number of pixels assigned to each thematic map category and multiplying the result by a factor to yield surface area (acres, hectares, square miles, square kilometres, etc.). Often it is desirable to obtain this result within some stratification, for example, ownership or administrative strata. A geometric transformation equation can be used to determine the corresponding image location for strata boundaries, and these boundaries are plotted on a map sheet. Similar acreage summations can then be made within strata. A sampling grid can be established to provide the framework for allocating sample plots to derive estimates of resource attributes. For example, weighted and non-weighted summaries of pixels assigned to each category in each cell can be obtained. This information is particularly useful if sample cells are to be selected with variable probabilities. Geographical coordinates of selected cells can be determined with the use of the proper geometric transformation equation.

2.4 Summary and conclusions

A variety of different production processes for extracting thematic information about the Earth's surface from satellite data are described in this part of the guide-book. Two general methods, namely visual image interpretation and computer-assisted image analysis, that can support cartographic production are outlined. The reader should be advised, however, that these methods can be applied in varying combinations and at different levels. For instance, a first level could be the generation of a geometrically corrected and radiometrically enhanced image that receives little additional interpretation and serves a wide number of users primarily as an "image map"; a second level is where the image, either in raw or enhanced form, is visually interpreted to support specific applications; a third level might involve the creation of an index image based on a ratio or other mathematical combinations of spectral channels (e.g. the Normalized Difference Vegetation Index) with little or no additional visual interpretation; a fourth level could be a thematic map derived from an unsupervised classification that only shows spectral differences in landscape features, but where the features are not identified; a fifth level might involve a controlled clustering approach to show spectral differences in landscape features and a full interpretation and grouping of the features into identified thematic map categories together with an estimate of the thematic map's accuracy. Regardless of the level or combination of methods employed, there is almost always a requirement for effective man–machine interaction. Furthermore, the quantity and quality of thematic information that can be extracted from satellite data increases when combinations of various types of satellite data are used and when the satellite data are combined with other types of graphic and tabular data (Nyquist, 1987; Lauer, 1986)—which is the subject of the next part of this guide-book.

References

ANDREWS, H. C. and HUNT, B. R. 1977. *Digital Image Restoration.* Prentice-Hall, Englewood Cliffs, New Jersey.

ASRAR, G. 1989. *Theory and Applications of Optical Remote Sensing,* chapter 9, The atmospheric effect on remote sensing and its corrections. Kaufman, Y. J. (Ed.). John Wiley & Sons, Inc., New York.

AVERY, T. E. 1987. *Interpretation of Aerial Photographs,* Burgess Publishing Company, Minnesota.

BRYANT, J. 1978. *Applications of Clustering in Multi-image Data Analysis.* College Station, Texas, Department of Mathematics, Texas A & M University, Report No. 18.

COLWELL, R. N. 1961. Some practical applications of multiband spectral reconnaissance. *American Scientist,* Vol. 49, No. 1, March 1961, pp. 9–36.

COLWELL, R. N. 1965. The extraction of data from aerial photographs by human and mechanical means. *Photogrammetria,* Vol. 20, pp. 211–228.

COLWELL, R. N. 1987. Remote sensing—Past, present and future. *Proceedings, Study Week on Remote Sensing and Its Impact on Developing Countries,* Vatican City, Italy, 16–21 June 1986, Pontifical Academy of Sciences, pp. 3–141.

ESTES, J. E. 1980. Attributes of a well-trained remote sensing technologist. *Proceedings, Conference of Remote Sensing Educators* (CORSE-78), Stanford University, California, 26–30 June 1978; NASA Scientific and Technical Information Office, Conference Publication 2102, 1980, pp. 103–118.

FONTANEL, A., BLANCHET, C. and LALLEMAND C. 1975. Enhancement of Landsat imagery by combination of multispectral classification and principal component analysis. *NASA Earth Resources Surv. Symp.* July 1975, Houston, Texas. NASA-TMX-58168, pp. 991–1012.

FOSNIGHT, E. A. 1988. Applications of spatial post-classification models. *International Symposium on Remote Sensing of Environment, 21st,* Ann Arbor, Michigan, October 1987. Ann Arbor, Environmental Research Institute of Michigan, pp. 469–485.

GALLO, K. P. and DAUGHTRY, C. S. T. 1987. Differences in vegetation indicies for simulated Landsat-5 MSS and TM, NOAA-9 AVHRR, and SPOT-1 sensor systems. *Remote Sensing of Environment,* Vol. 23, pp. 439–452. Elsevier Science Publishing Company, Inc., New York.

JENSEN, J. R. 1986. *Introductory Digital Image Processing: A Remote Sensing Perspective.* Prentice Hall, New Jersey.

JENSON, S. K. and WALTZ, F. A. 1979. Principal components analysis and conconical analysis in remote sensing. *Proceedings American Society of Photogrammetry/American Congress of Surveying and Mapping Annual Meeting,* Washington, D.C., 18–23 March 1979.

LAUER, D. T. 1986. Applications of Landsat data and the data base approach. *Photogrammetric Engineering and Remote Sensing,* Vol. 52, No 8, pp. 1193–1199.

LILLESAND, T. M. and KIEFER, R. W. 1979. *Remote Sensing and Image Interpretation.* John Wiley & Sons, Inc., New York.

MAYERS, M., WOOD, L. and HOOD, J. 1988. Adaptive spatial filtering. *Proceedings American Congress on Surveying and Mapping (ASCM), American Society for Photogrammetry and Remote Sensing (ASPRS),* Fall Convention, Virginia Beach, Virginia, September 1988. Falls Church, Virginia, ASPRS, pp. 99–105.

MCCARTNEY, E. J. 1976. *Optics of the Atmosphere: Scattering by Molecules and Particles.* John Wiley & Sons, Inc., New York.

MOIK, J. G. 1980. *Digital Processing of Remotely Sensed Images.* NASA Scientific and Technical

Information Branch. NASA-SP-431. US Government Printing Office, Washington, D.C.

NYQUIST, M. O. 1987. The Integration of remotely sensed data into a geographic information system—rediscovered!?? *Proceedings 21st International Symposium on Remote Sensing of the Environment*, Ann Arbor, Michigan, 26–30 October 1987, pp. 487–493.

PINSON, L. J. and LANKFORD, J. P. 1981. *Research on Image Enhancement Algorithms.* Tullahoma, Tennessee, Technical Report RG-CR-81–3. University of Tennessee Space Institute.

RICHARDS, M. E. 1985. An evaluation of a new statistical approach to traditional linear de-striping. *Proceedings American Society of Photogrammetry Annual Meeting, 51st*, Washington, D.C. March 1985. Falls Church, Virginia, American Society of Photogrammetry, vol. 2, pp. 557–575.

SCHOWENGERDT, R. A. 1983. *Techniques for Image Processing and Classification in Remote Sensing.* Academic Press, New York.

SIMON, K. W. 1975. Digital image reconstruction and resampling for geometric manipulation. *Proceedings International Symposium of Machine Processing of Remotely Sensed Data, 1st.* West Lafayette, Indiana, 1975. West Lafayette, Indiana, Purdue University, pp. 3A1–3A11.

US Department of Agriculture. 1966. *Foresters Guide to Aerial Photo Interpretation.* US Forest Service, Agricultural Handbook 308, Washington, D.C.

WILLIAMS, J. M. 1979. *Geometric Correction of Satellite Imagery.* Farnborough, Hants, United Kingdom. Technical Report 79121, Royal Aircraft Establishment.

Chapter 3

METHODS FOR COMBINING SATELLITE-DERIVED INFORMATION WITH THAT OBTAINED FROM CONVENTIONAL SOURCES

Compiled by
Sten Folving* and Jean Denègre†

*Institute for Remote Sensing Applications, Joint Research Centre, Ispra, Italy
†Conseil National de l'Information Géographique, Paris, France

Translated by
Allan Moore

Institut Géographique National, France

CONTENTS

3.1 Introduction

The availability since 1972 of satellite observations of the Earth has led to considerable innovations in the content, expression and perception of geographical information in general. At the same time it has created equally new problems for cartographers, because it has appeared necessary to combine satellite-derived data with data from conventional sources which often appear to be complementary. In fact, satellite-derived data rarely provide all the information sought and, in any case, in order to be usable, they need to be combined with a minimum of topographical or thematic reference elements to make it possible for a reader to find his bearings and analyse the contents of the "space-map" so obtained.

If one attempts to classify, in a very general way, the different cases where that combination arises, one must, in the first place, consider the forms in which satellite data are supplied. Ranging from the most elementary to the most elaborate, one can distinguish three levels, already described above (Chapter 2):

(i) "Image" level, where there is no data-interpretation

(ii) "Classified image" level, where the interpreted satellite-derived information is supplied in the form of pixels assigned to a limited number of classes.

(iii) "Line map" level, where the interpreted satellite information is supplied in the form of vectors or content zones (thematic polygons).

In the last case, the cartographic problems of the combination with data obtained from other sources (in the field, aerial photographs, etc.) do not differ in any particular way from the normal problems of compiling documents, well known to cartographers.

Consequently, the cartographic problem only arises in a new way in cases (i) and (ii) where the satellite information conserves its "image" character in the form of pixels, whether the latter are interpreted (case (ii)) or not (case (i)). This problem is new in so far as the satellite image is so detailed that any addition of cartographic overprinting takes place to the detriment of the image-provided details; this problem already exists with photo-maps, but it is amplified here by the normal smallness of the scale, which accentuates the difference between the real details and the artificial symbols, between the positions and conventional displacements, between image and map. One must add, for greatly enlarged satellite imagery, the raster structure in large-size pixels which is not easily combined with topographic elements in the vector mode.

If one now considers the two main families of maps where satellite imagery plays a role, i.e. general (or topographic) maps and thematic maps, one is finally led to consider three types of documents where that imagery is combined with outside data (see Fig. 3.1):

A. *General maps*, constituted by an uninterpreted satellite-imagery background with topographic overprinting (objects, place-names, etc.).

B. *Thematic maps*, constituted by a classification (or a photointerpretation) of satellite imagery, with reduced topographic overprinting (objects, place-names, etc.) and minimum references.

C. *Thematic maps*, constituted by a classification (or a photointerpretation) of satellite imagery, with an uninterpreted satellite image itself as map background, giving a general reference, generally monochrome, and completed, where necessary, by a few topographic elements.

To each of these three types of documents correspond different problems and specific methods for combining the available information. In order to describe these methods, depending on the problems that arise, the following aspects will be discussed:

1. Geometric adjustment of the satellite information and the conventional information.
2. Assembly and cutting of the image along the map edges.
3. Processing of the image to form the map background.
4. Adding the conventional information to the satellite data (interpreted or not).

1 and 2 concern types A, B and C.
3 concerns types A and C.
4 concerns types A and B.

3.2 Geometric corrections

Satellite images, compared to maps, have geometric deformation of various origins:

A. General image maps
(uninterpreted image background)

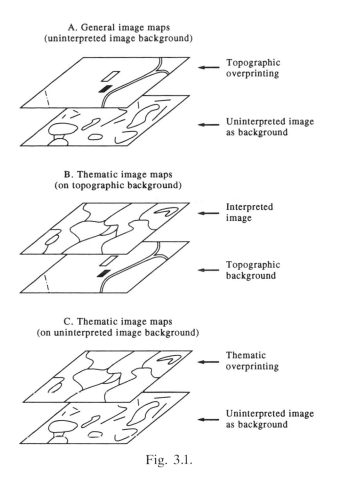

← Topographic overprinting

← Uninterpreted image as background

B. Thematic image maps
(on topographic background)

← Interpreted image

← Topographic background

C. Thematic image maps
(on uninterpreted image background)

← Thematic overprinting

← Uninterpreted image as background

Fig. 3.1.

systematic effects due to the recording device and to the Earth's rotation, satellite attitude defects, etc. (see Chapter 2).

With certain exceptions, it is the conventional information which provides the geometric reference in the form of existing maps or control points obtained from other sources.

The deformations are generally modelled by a polynomial function (image line and column coordinates in accordance with the cartographic coordinates, e.g. UTM) whose coefficients are determined by introducing into the calculations the coordinates of points identified in both the image and reference systems, which may be a map or an other image.

3.2.1 X–Y coordinates

The coordinates of the image control points are obtained by pointing at the selected points at the image-screen, they will most often be given in image coordinates, the cartographic co-ordinates are found by means of a digitizing table—or, evenly by simple map reading of the cartographic coordinates in the system chosen. The success of this kind of modelling evidently

depends on a good identification of the control points, the pointing accuracy and the quality of the map used for these types of correction.

3.2.1.1 Image-to-map registration

The tie-points, cartographic coordinates from the map and line and column number from the image(s), are used for modelling by means of a deformation grid. A case of four tie-points could be modelled by a polynomial of the form:

$$x = a_0 + a_1 x' + a_2 y' + a_3 x'y' + a_4 (x')^2 + a_5 (y')^2$$
$$y = b_0 + b_1 x' + b_2 y' + b_3 x'y' + b_4 (x')^2 + b_5 (y')^2$$

where x and y are the coordinates of the original—distorted—image points and x' and y' are the coordinates of the final corrected image. If the power 2 terms are set to zero, we will have the possibility shown below (see Fig. 3.2).

A piecewise geometric approximation can be obtained by means of a continuous net of quadri-lateral tie-points if different polynomial approx-imations are defined for each set (see Fig. 3.3).

If the squared terms have to be kept, a least square fit has to be performed, which will result in residuals. A simple affine transformation will account for most of the cases shown above, but it is recommended firstly to run the fitting a couples of times as an editing of the tie-points very often will be necessary before the whole data set is processed.

Remark: If a standard image-processing

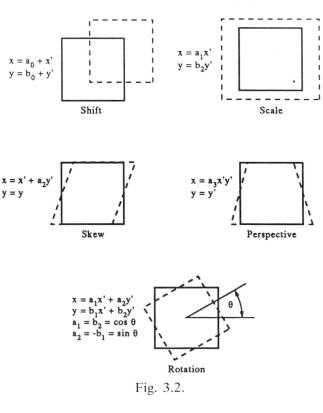

$x = a_0 + x'$
$y = b_0 + y'$

Shift

$x = a_1 x'$
$y = b_2 y'$

Scale

$x = x' + a_2 y'$
$y = y$

Skew

$x = a_3 x'y'$
$y = y'$

Perspective

$x = a_1 x' + a_2 y'$
$y = b_1 x' + b_2 y'$
$a_1 = b_2 = \cos \theta$
$a_2 = -b_1 = \sin \theta$

Rotation

Fig. 3.2.

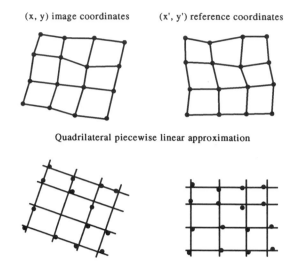

(x, y) image coordinates   (x', y') reference coordinates

Quadrilateral piecewise linear approximation

Least squares global linear polynominal approximation

Fig. 3.3.

software package is available, the UTM map coordinates can be translated into image coordinates as:

$$\text{line} = (N_0 - N)/\text{pixels} - 1$$
$$\text{column} = (E - E_0)/\text{pixels} - 1$$

where N_0 and E_0 are the UTM coordinates of the upper left corner of the corrected "image map". The size of the pixels has to be measured in the same units at the map coordinates used.

3.2.1.2 *Image-to-image registration*

Precise image-to-image registration is very often as necessary as image-to-map registration. Normally the operator simply has to point to the approximate location of the tie-points on the slave imagery, the points on the master image being defined. A correlation procedure can then be used for optimization of finding the correct corresponding tie-points; the pixels with the reported highest spatial correlation are then

used for the polynomial transformation as in the image-to-map registration.

3.2.2 *Interpolation of radiometric information*

For each column and line (i,j) in the processed image a corresponding coordinate (c,l) in the original, uncorrected image must be found (see Fig. 3.4).

The radiometry of the (i,j) pixel is obtained by searching for the homologous position (c,l) in the original image via the deformation function and by calculating a radiometric value (grey level) based on the pixels occupying that position. Because the polynomium seldom points to exactly one original pixel some sort of choice and approximation must be made in order to define the radiometric value of the new pixel. Often the nearest neighbour is used, but, first order interpolation may also be used (see Fig. 3.5).

More complicated methods, as for instance cubic convolution, can also be used. This solution would in many cases be well suited for cartographic purposes as it tends to produce a slightly smoothed result, whereas it cannot be recommended for digital image processing aiming at information extraction and compilation (see Fig. 3.6).

3.3 Mosaicking (assembling and cutting along the map edges)

If the image does not completely cover the image domain it is necessary to use several images; in the case of contiguous images obtained from the same satellite ground track on the same day, the raw images are rigorously identical over their common zone and one can edge-match them before any other processing.

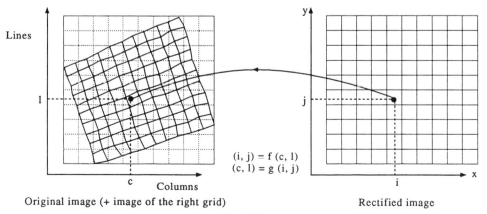

Lines

l

$(i, j) = f (c, l)$
$(c, l) = g (i, j)$

c   Columns

Original image (+ image of the right grid)

y

j

i   x

Rectified image

Fig. 3.4.

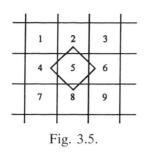

Fig. 3.5.

In all other cases, two contiguous scenes never have exactly the same radiometry in their overlapping zone. For each channel one has to define an image-radiometry transformation-law for equalizing the radiometry of that image with those of the image selected as the reference one over their common zone. That law is then applied to the entire image (see Section 3.4).

Despite everything common zone radiometries are slightly different and a brutal, straightforward, joining will be visible.

3.3.1 Simple case

If an interactive image display system is available, the join-line in the common zone of the two images can be defined by interactive plotting on the screen. One should in this case be careful to use "natural" break points, e.g. edges or breaks in the landscape. Once the join-line has been defined, mosaicking becomes a simple—either automatic, or programmable—I/O procedure. This method, however, requires a considerable overlap between the adjacent images (see Fig. 3.7).

3.3.2 Complex mosaicking

When the side-lap between images is small or when the orbits are not parallel and far from in accordance with map North–South direction, which is often the case when preparing (very) small-scale maps, a special method can be used. The single images are separately geometrically registrated as if covering the entire region. A corresponding dicotome mask is produced. A "0" indicates no image data for the corresponding map area and a "1" means that there exist image data. The corrected images are added and the masks are likewise added. The final map is then produced by dividing, pixel by

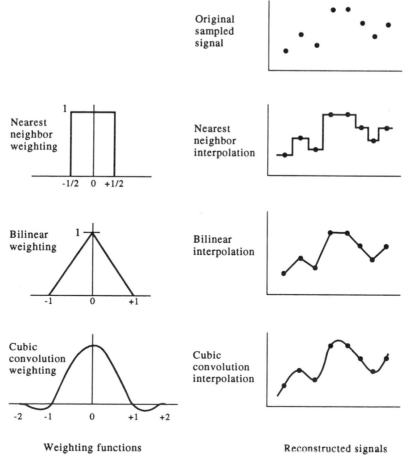

Fig. 3.6. (From Niblack, Digital Image Processing.)

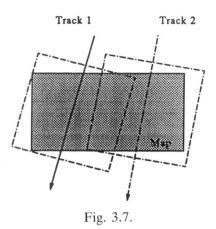

Fig. 3.7.

pixel, the sum of the images by the sum of the masks (see Fig. 3.8).

Naturally radiometric corrections are required before this procedure is of any use, and even then very small differences in radiometry will be readily distinguishable; therefore, a smoothing procedure should be considered before the final mosaicking. Actual cut-points may be blurred by procedures which calculates some sort of running mean along the edge of cutting. These values are then used instead of the original values from the border zone of the two images being mosaicked.

Once the map has been completely covered by an image assembly, the latter can be cut in accordance with the geographic boundaries of the map, i.e. in general according to the meridians and parallels. This can also be the occasion for producing the corresponding geometric marginal information, i.e. framework, graticule intersection and graticule ticks, e.g. every 10′, and grid ticks, e.g. every 10 km.

3.4 Radiometric processing

Many factors are influencing the data quality of remote sensing imagery. Recording conditions will always vary from one data set to another. An overview of those different factors has been presented in Chapter 2. Irradiance will depend on pixel–Sun relation: Sun elevation and azimuth. The topographical location of the pixel is important in defining the geometrical relations between irradiance onto the pixel and the reflection of energy towards the sensor. The composition of the atmosphere is crucial for the modulation of the ground information contained in the signal.

(Striping of the imagery—caused by differences in the sensor sensitivities and/or miscalibration—should be removed either by histogram equalization or by averaging of adjacent lines, unless the single sensors can be calibrated separately, as described in the following paragraphs.)

3.4.1 Primary processing

As the main purpose is to use remote sensing data as image background for other map information it must be of high importance to report the landscape or land-surface as accurately as possible. This again means that the spectral proportions of the imagery should be as close as possible to reality. It is, therefore, often convenient to recalculate the digital counts normally supplied into values with a physical meaning—for instance, watts per areal unit and

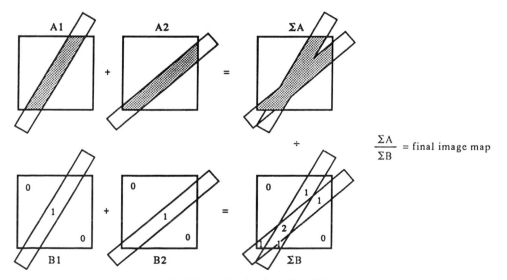

Fig. 3.8. (From Carle, Satellite Mapping.)

reflectance percentage. The radiance in mW/cm^2/sr can be calculated according to:

$$P_r = \frac{D_n}{D_{max}} (L_{max} - L_{min}) + L_{min},$$

where P_r is the radiance of the pixel in question, D_n is the digital count (value) of the pixel, D_{max} is the maximum value of the dataset used (256 for 8-bit data); L_{max} and L_{min} are scanner parameters giving the radiation with which the sensor is saturated and the minimum detectable radiance. By multiplying P_r with π ($= 3.1416$) and dividing by incoming radiance, E_0—as top of atmospheric radiance for the spectral band in question—multiplied by sine to the Sun elevation:

$$R_g = \frac{\pi P_r}{E_0 \sin a},$$

the ground reflectance, R_g, is obtained.

3.4.2 Atmospheric correction

It is highly recommended always to avoid images which have not been recorded during near to perfect atmospheric conditions. The atmosphere should be as clear as possible. If this qualification is met, the data can be used as the before mentioned primary processing should be sufficient. Of course clouds should be avoided (one has to remember also that the shadow from the clouds prevents us from obtaining information from the ground).

If optimal atmospheric conditions cannot be found for any of the available scenes, some atmospheric corrections should be made. Atmospheric corrections without contemporary measurements are rather complicated. A commonly used method could be proposed.

Consider the method for preprocessing given in Section 3.4.1. It is easily seen that we are dealing with a linear transformation. If we use offset and slope of a given calibration function for the sensor we can obtain the spectral radiance in channel i, $P_r(i)$, from $D_n(i)$ by:

$$P_r(i) = C_a(i) + C_b(i) \times D_n(i)$$

where C_a and C_b are calibration function offset and slope for channel i.

Surface reflection is :

$$R_g = \frac{1}{A_0(i)} [\pi P_r(i)/E_0(i) - A_a(i)].$$

A_0 and A_a are atmospheric functions which

relate top of atmosphere reflectance, R_a, to ground reflectance, R_g:

$$R_a = A_0 + A_a R_g.$$

The atmospheric function will most often have to be estimated. One way of doing this could be to estimate A_0 from shadow areas or by using clear water reflectance as obtained from the single channels. The atmospheric transmission, A_a, can be related to general visibility, which can be obtained from meteorological stations.

3.4.3 Filtering

The purpose of the map and the sort of map information that will be "put on top of" the satellite imagery will determine how the satellite data will be presented. It may be necessary to enhance features or it may sometimes be necessary to reduce the dynamic variation of the satellite data. By filtering the data, it is possible to achieve the result which will suit the purpose of the map.

Many sorts of filters can be used, some very computing-time demanding; the two main types—smoothing and edge enhancing—will be mentioned.

3.4.3.1 Smoothing

A smoothing, or a low-pass, filter can be considered as a means with which to perform a sort of two-dimensional running average of the data. The result will often be a blurred image, as strongly contrasting features are suppressed. The number of cells in the filter and the weights will determine the "blurredness" of the result.

Examples of low-pass filters:

$$\frac{1}{9} \times \begin{matrix} 1 & 1 & 1 \\ 1 & 1 & 1 \\ 1 & 1 & 1 \end{matrix} = \begin{matrix} \frac{1}{9} & \frac{1}{9} & \frac{1}{9} \\ \frac{1}{9} & \frac{1}{9} & \frac{1}{9} \\ \frac{1}{9} & \frac{1}{9} & \frac{1}{9} \end{matrix}; \quad \frac{1}{25} \times \begin{matrix} 1 & 1 & 1 & 1 & 1 \\ 1 & 2 & 2 & 2 & 1 \\ 1 & 2 & 5 & 2 & 1 \\ 1 & 2 & 2 & 2 & 1 \\ 1 & 1 & 1 & 1 & 1 \end{matrix}.$$

[The filter is placed in the top left corner, the cell weights are multiplied by the corresponding image pixel values, added and placed in the new, filtered, image at the position of the central filter cell. The filter is then moved one or more steps (pixels)—the computation is repeated for the new position, and so on.]

3.4.3.2 Edge enhancement.

High-pass filtering or edge enhancement can be used for both general emphasis of linear features in the landscape and for highlighting lines with specific directions.

Either the filtered result is added to the original image or a high-boost filter is used. In high-boost filtering the original image is multiplied by a constant, K, in the filtering process.

High-boost filtering:

$K \times$ original + low-pass filter =
$$(K + 1) \times \text{original} - \text{high-pass filter}.$$

Examples of high-pass filters:

$$\frac{1}{9} \times \begin{array}{ccc} -1 & -1 & -1 \\ -1 & 8 & -1 \\ -1 & -1 & -1 \end{array} = \begin{array}{ccc} \frac{-1}{9} & \frac{-1}{9} & \frac{-1}{9} \\ \frac{-1}{9} & \frac{8}{9} & \frac{-1}{9} \\ \frac{-1}{9} & \frac{-1}{9} & \frac{-1}{9} \end{array} ; \quad \frac{1}{25} \times \begin{array}{ccccc} -1 & -1 & -1 & -1 & -1 \\ -1 & -1 & -1 & -1 & -1 \\ -1 & -1 & 24 & -1 & -1 \\ -1 & -1 & -1 & -1 & -1 \\ -1 & -1 & -1 & -1 & -1 \end{array} .$$

Examples of directional filters:

$$\begin{array}{ccc} -1 & 0 & 1 \\ -1 & 0 & 1 \\ -1 & 0 & 1 \end{array} ; \quad \begin{array}{ccc} -1 & -2 & -1 \\ 0 & 0 & 0 \\ 0 & 2 & 1 \end{array} .$$

3.4.3.3 Topographical inversion

The difference in illumination caused by topographical variation can often cause misinterpretation by the map reader/user who is accustomed to the cartographical shadowing of the slopes facing the reader/user. As most satellite imagery from the Northern hemisphere considered candidate for map background is day-recorded (i.e. with a real illumination from the South instead of a conventional one from the North-West) this problem will be present in any area with a stronger relief.

By means of digital terrain models, DTM, the satellite images can be corrected for the variation in illumination. Eventually it is possible to invert the natural illumination difference, thus obtaining a cartographic correct topography enhancing map background by shadowing/shading.

If no digital terrain model is available, and one wants to get rid of the shadows and illumination variations of the terrain, rationing may be used. By dividing the channels, one by one (chan.1/chan.2; chan.2/chan.3), the illumination is pseudo-normalized, which occasionally is better than to use the "wrong" relief shading. However, as ratio-images very often tend to be, noisy a filtering procedure is necessary to get a useful result.

3.5 Data presentation for map-background purposes

In the case where the satellite image (uninterpreted) serves as a map background (cases A and C above) it is best to process the entire radiometry in order to produce a readable and expressive representation of the added cartographic informations. In particular, in case C (thematic maps with image background), the map must associate as harmoniously as possible the "natural" pixel zones, the symbol colour areas, the outlines or, more generally, the superposed lines. The selected colours must make it possible to distinguish perfectly between the different overprintings between themselves and to distinguish them from the background.

3.5.1 General case

If the background is monochrome, the main problem arises with selecting the most expressive (or most contrasted) channel compatible with good legibility of the overprinting. Printing in black or bistre is most frequently used.

If the background is coloured, two main options are generally considered:

(i) Representation in infrared colour ("false colours").

(ii) Representation in "pseudo-natural" colours.

(i) Infrared (IR) colour (like that used for aerial photographs) is well known. Three multispectral image channels are represented:

(a) Either in subtractive synthesis (paper printing):
  —Green channel, e.g. SPOT 0.5–0.59 μm printed in magenta
  —Red channel, e.g. SPOT 0.61–0.68 μm printed in magenta
  —IR channel, e.g. SPOT 0.79–0.89 μm printed in cyan-blue with a printing density inversely proportional to the pixel radiometry in each channel: water zones have nil radiometry in the infrared and the corresponding pixels will be printed in cyan-blue symbol colours (100%).

(b) Or in additive synthesis (display on colour screen):
  —Green channel in violet-blue (complementary of yellow)

—Red channel in green (complementary of magenta)

—IR channel in red (complementary of cyan-blue) with an intensity proportional to the pixel radiometry in each channel.

(ii) Representation in "pseudo-natural" colours consist of trying to find conventional map colours, i.e. vegetation in green, water in blue, etc.; in order to achieve this, one generally adopts the following rules (in additive synthesis).

Green colour obtained from the green channel.

Red colour obtained from the red channel.

Blue colour obtained artificially (for Landsat MSS and SPOT which do not have blue channels) by linear combination of the three red, green and infrared channels—see paragraph below.

3.5.2. *Particular case of the blue colour (in pseudo-natural colours)*

The linear combination mentioned above has to be calculated differently according to the themes to be represented in *blue* (water, bare ground, vegetation), which leads to making a summary classification of the image in the radiometry space (X = red-green and Y = infrared-red).

For each quadrant the blue is defined by the linear combination:

$$\text{Blue}\,(X, Y) = A_i X + B_i Y + C_i$$
($i = 1$ to 8 depending on the quadrant)

and the coefficients $A_i B_i C_i$ are determined interactively, whilst at the same time ensuring

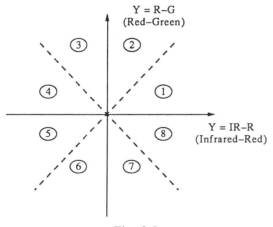

Fig. 3.9.

blue continuity at the frontiers between quadrants.

However, this processing gives results which are often unsatisfactory for water. Consequently, it is best to isolate that theme and apply special processing to it.

3.5.3 *Special case for water areas*

As the linear combination, mentioned above, rarely gives satisfactory results for water, and as the radiometry from water bodies is dependent on factors, as for instance, depth, turbidity and algae concentrations, a classification could be considered. This classification would require training areas representing any variation of the water area. The classification result would build the background for forming a mask with which to adjust the blue colour. By means of Principal Component Analysis, a mask can very often be made which is extremely powerful for delineating the water areas.

The results from either classifications or Principal Component Analysis provide masks for a dynamic stretching, specific for each channel, which is determining for the application of blue in the presentation of the hydrographic background.

3.5.4 *Colour transformations—special enhancements*

As mentioned above, Principal Component Analysis can be used to transform various colour enhancements. It is a method—or perhaps rather a collection of methods—very powerful for analysing multispectral and multitemporal data sets.

Principal Component Analysis used for providing image background for various types of satellite maps can be applied either in order to reduce the dimensionality i.e. to extract a maximum of variation from multichannel images into a few image planes. Often more than 70–80 percent of the total variation of a multispectral data set can be represented in one data layer, the first principal component; the second and third principal components will typically contain 15–20 and 3–10 percent of the variation respectively. By using principal components directly for map production, one secures that the maximum amount of the variation is presented in the map. On the other hand, the colour presentation will bear no resemblance to natural colours, which can make life more difficult for the map user and which of course demands much effort from the map maker to provide a well-prepared legend for the map.

Many multispectral data sets often are highly correlated. Principal Component Analysis can be used to perform a decorrelated stretch of the data. First, the original, preprocessed data set is transformed, or rotated; then, the various components are stretched independently; finally, the stretched principal components are again returned back to the original dataspace (see Fig. 3.10).

1. Channels *y* and *x* are highly correlated. A Principal Component Analysis transforms the data into a new coordinate set where the first coordinate carries much of the total variation, the second principal coordinate—perpendicular to the first—carries less of the total variation.
2. The data after the principal component transformation.
3. The data is stretched independently along the principal component axis in order to use most of the available range.
4. After the retransformation the data set will be decorrelated and contains a higher dynamic data range.

The method of decorrelation stretching can be used for any number of channels; when only three channels are to be used, for instance the first three bands of Landsat TM (blue, green and red), a HSI transformation can be performed. By transforming the three RGB colour components into Hue, Saturation and Intensity, it becomes possible to stretch (and decorrelate) the saturation and intensity image planes. The stretched HSI data sets can again be transformed back into RGB colour space.

Another possibility is to substitute the intensity data plane by data from other images

than the one used in the first transformation and thereby to include accessory data in the RGB presentation/printing of the images. This method will often be used when one wants to include black and white, one-channel satellite—or digitized air photo data—into lower resolution multispectral data sets.

One could use the three first bands from Landsat TM for the RGB data sets and perform a HSI transformation. As the TM data represents a geometrical resolution of 30 m, it could be fruitful to include SPOT 10 m resolution one-channel panchromatic data. By proper adjustment of the grey levels of the SPOT data to the intensity data of the HSI TM transformation, the SPOT data can be substituted for the TM intensity data when transforming back into RGB colour space.

This procedure requires extremely careful geometrical co-registration of the data sets.

3.5.5 *Graphic output*

After the geometric and radiometric processing described above, the image background is ready to be plotted in a graphic form in such a way as to be able to serve as a support for the topographic completion and the addition of outside information.

In the case where offset printing is planned, the calculation of a subtractive synthesis (for determining the printing percentages of a cyan-blue, magenta and yellow three-colour procedure) must be based on the digital image obtained generally on the screen by additive synthesis.

It is evidently advantageous, in that case, to output directly on to monochrome film (using a photoplotter) the corresponding master printing plates, screened and ready for reproduction for the printing.

The other solution consists of producing a

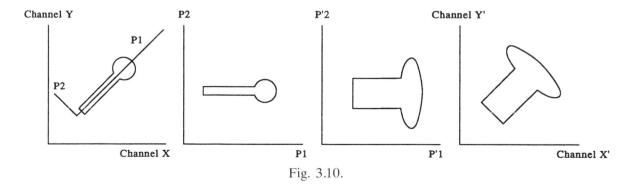

Fig. 3.10.

colour image on film (colour photoplotter) or a colour plot on paper (electrostatic or ink-jet plotter), and to apply to that document the normal techniques of colour separation and printing.

3.6 Adding conventional information to satellite images

To supplement or complement remotely sensed imagery with "normal" cartographic information has become an extremely important part of the work done in the use of remote sensing. It is and will still be difficult to interpret the multispectral satellite-derived information, most of all because the users more and more tend to expect the satellite data to explain or at least provide "all" information. Frustrations arising from the fact that remote sensing data does indeed contain more information that readily interpreted has also furthered the development of the use of combined remote sensing and geographical information systems approach to the scientific use of satellite images, but also to the applied use of such data.

The addition of normal cartographic information aims at aiding users to read satellite imagery, by giving them reference points or the references basically needed for interpretation. Cartographic-type information can be obtained from different sources:

—either existing maps,
—or a partial visual interpretation of satellite imagery, e.g. roads, rivers, built-up zones,
—or existing geographic information systems (GIS).

Regardless of their origin, the information must be represented in a form which is compatible with the image characteristics. Several criteria arise in that compatibility:

   (i) Richness of the cartographic information added to the image (density, readability).
   (ii) Geometric reliability of the added information.
   (iii) Thematic reliability of the added information.
   (iv) Up-to-dateness of the added information.
   (v) Aesthetics of the cartographic representation.

Those different criteria each play a direct part in the quality of the finished product. As regards the production methods, the choice of sources and the selection of the information depend directly on the objectives (or the themes) of the map to be produced, and on the availability of existing sources. Consequently, it is impossible to make general rules for these procedures. As concerns production techniques, two aspects are discussed below:

   (i) Compilation (graphic semiology): discussed in Chapter 4.
   (ii) Plotting and production of the master printing-plates for final printing purposes: this topic is discussed below.

3.6.1 Analogue and digital methods

In order to produce the plots of cartographic overprinting and the master plates for final printing, two methods can be used: the first uses normal photomechanical techniques (analogue methods), the second is based on the digital nature of the satellite image and uses the normal information in digital form also in order to combine them in accordance with an entirely computerized procedure.

The first method uses normal map-editing procedures and consequently is well known. There is no need here for special comments, except as concerns master printing-plate production. As a multispectral satellite image is often printed in three or four colours (with the usual colours), it will be advantageous to adopt the same colour separation for the cartographic overprinting: this makes it possible to combine them directly with films of the image, and so to conserve the same number of master plates for printing purposes.

It should be noted that the analogue method, entirely independent of image-processing (except for the final phase, just mentioned), makes it possible to use the complete range of normal map-editing resources, e.g. semiology, letter-types, map-extract reproductions, etc. That situation again contrasts with the digital method, where map-editing is limited by the raster mode (resolution of the elementary screen). However, that difference tends to disappear with continuous improvements in the performances of computer graphics in the raster mode and with the digital method in all its aspects.

The digital method leads, on the other hand, to the generalized use of geographic information systems (GIS), and it evidently represents the most likely convergence for the future of the three technological sectors involved: image-processing, computer graphics and databases. It is discussed again in the following chapter.

3.6.2 Computerized methods (with GIS)

It has to be stressed in this context that RS/GIS is not primarily geared to mapping and traditional cartography: the digital map in the RS/GIS connection is primarily a screen-oriented means of new and combined data analytical and data synthetical tool. Rarely are the products meant for cartographic production. (This is almost the same in the technical application of GIS which is used primarily for supply-line management.) If copied and distributed, it will mainly be in a digital form. The screen content can of course be transferred to a hard-copy device such as a colour ink jet plotter; however one can hardly talk about map production in this connection. (In this paragraph we will not be dealing with computerized printing in the traditional cartographical sense— these are considered analogue methods as the superimposition is only carried out in the printing process.)

3.7 Role of geographic information systems (GIS)

3.7.1 Definitions and objectives

The concept of GIS, resulting from the extension of database management systems (DBMS) to all types of geographic data, began early in the 1980s. Satellite remote sensing offers something completely original as far as geographic data is concerned, which emphasizes the specific nature of GIS compared to information systems generally.

An American definition has been issued by the "Federal Interagency Coordinating Committee on Digital Cartography" (FICCDC, 1988):

> "System of computer hardware, software, and procedures designed to support the capture, management, manipulation, analysis, modeling, and display of spatially referenced data for solving complex planning and management problems."

In France, the SFPT (the French Society of Photogrammetry and Remote Sensing) put forward (in October 1989) the following definition:

> "Data processing system which, using various sources, collects and organizes, manages, analyzes and combines, elaborates and presents geographically located information, contributing in particular to the management of spatially referenced data."

Thus, the following characteristics may be retained:

—management of multiple located data (including RS),
—capability of producing cartographic documents,
—aptitude for complex processing in view of land management, and therefore in order to aid decision-making.

Here, as application examples, are some common operations in the management of spatially referenced data:

—issuing planning permissions (houses, factories, etc.),
—land taxes,
—choice of highway, freeway and rail routes,
—definition of areas which may and may not be built upon (land classification regulations),
—agricultural planning (drainage, irrigation, etc.),
—urban, industrial or tourist planning, etc.,
—regional planning,
—monitoring environmental changes (desertification, deforestation, climatic changes, etc.),
—prevention of major natural or man-made hazards (flooding, landslides, pollution, etc.),
—intervention in the event of disasters,
—help to computer-assisted car-driving, etc.

A geographical information system does not in any way imply or depend on use of remotely sensed data. If a GIS is defined as a tool for planning and development through legal, administrative and economic decisions, it consists of partly a database with georegistered information related to infrastructural and economic aspects, and to environment and landscape-related information; it is used for systematical data collection, updating and data handling and for various presentations of data and derived results.

Mutual relationships between an GIS and remote sensing systems are obvious: remote sensing forms a potentially vast source of geographic data, of which only a very small part has been exploited and integrated until now; likewise, GIS, set up using existing maps or ground data, can considerably aid the interpretation of remote sensing images, using computer-assisted classifications.

3.7.2 *General architecture and components*

The overall organization represented below is the result of the above-mentioned definitions (from FICCDC, 1988) (see Fig. 3.11).

The database management system (DBMS) is a set of software which stores and uses all system data. A GIS DBMS must, in particular, be capable of being used for operations linked to data topology (location, adjacency, inclusion, etc.) and its transcription (cartographic or otherwise).

The user interface is formed from all the tools which can be used by users to communicate with the system, i.e. with all the databases to which it has access, and with all available application modules. These tools comprise a range of software products, such as function "menus", screen messages, graphic displays, etc.

Given the preceding definitions and characteristics of geographic information, a certain number of general functions can be attributed to GIS, specifying their main fields of use.

A GIS must a priori be capable of meeting the following general functions from the user's viewpoint (not exhaustive and not in order of importance):

—**providing** overall geographic information concerning a given area (inclusive of all objects and themes) (e.g. inventory of all objects or themes present on one commune);

—**retrieving** all objects or themes answering a given question (for example what are all the 4 m wide roads in a particular region?);

—**cross-checking** available information to look for correlations between different themes or objects according to predefined algorithms;

—**combining** specific data provided by the user with those of the GIS;

—**supplying** answers in appropriate forms to problems dealt with: digital files, texts, maps (on screen or paper), diagrams, images, or even in voice form;

—**updating** at any time, in addition to introducing new data or new topics.

Within a GIS all information is or must be made accessible in a homogeneous form suitable for handling and processing and which provides the final cartographic expression of the results.

3.7.3 *Integration of remote sensing and GIS*

"Integrated" systems, managing both traditional geographic data (most often in vector mode) and remote sensing images, are rare, or are still in the development stage, as certified by numerous reports (Boursier, 1986; Huguet, 1987, etc.). Now, this integration and this integration alone can be used for simultaneous processing within the same system (see Fig. 3.12).

Consequently, the general coordination of such a system can be presented as follows (see Fig. 3.13).

This type of model, where the images are stored in rectified form once they have been processed, thus allowing direct registration with

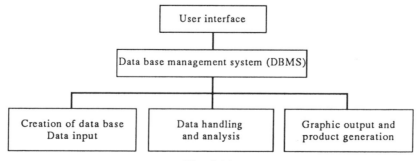

Fig. 3.11.

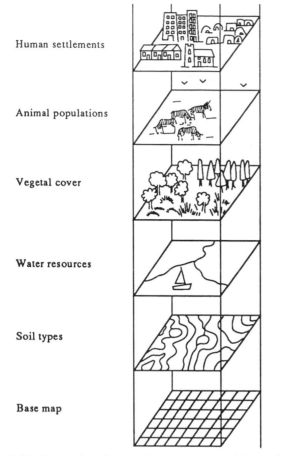

Human settlements

Animal populations

Vegetal cover

Water resources

Soil types

Base map

Fig. 3.12 Example of data integration within a GIS (from GRID, United Nations, Geneva).

other geographic data, corresponds to a so-called geo-coded system (case of SEP systems in France or LDIAS of the Canadian Center for Remote Sensing, or the IBIS system of the Jet Propulsion Laboratory of the Californian Institute of Technology). Another approach can consist in not storing rectified images a priori but in performing registration processes (geometric rectifications) only when requested by the user, and in a way which is, moreover, transparent for him (for example, the climatological information system of the Scientific Centre of IBM-France, with NOAA and Meteosat images managed with the SQL/DS DBMS).

It is easily seen that RS data can be integrated in two ways: either as preprocessed data (as an image) or as classified (primary) results (as a map). Whether the former or the latter form is selected depends on the intended use (and GIS software used).

3.7.4 *Technological aspects of GIS*

The data in most administrative Geographical Information Systems (often called Land Information Systems when purely administrative) will most commonly be vector-oriented,

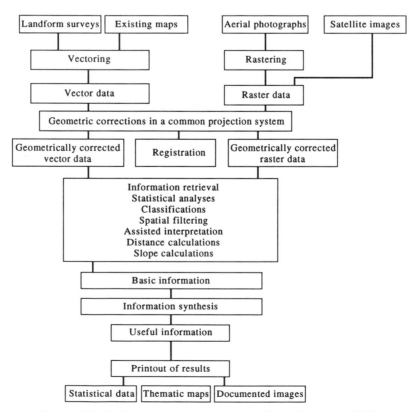

Fig. 3.13 Hybrid system vector–raster (from Huguet, 1987).

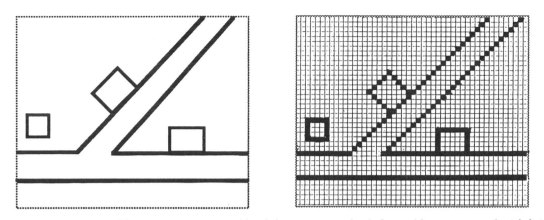

Fig. 3.14 Example of features represented both in vector mode (left) and in raster mode (right).

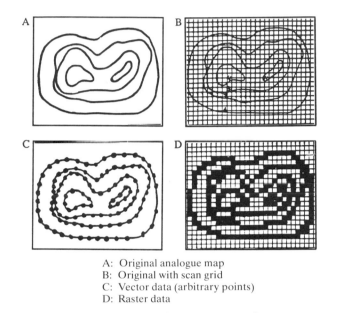

A: Original analogue map
B: Original with scan grid
C: Vector data (arbitrary points)
D: Raster data

Fig. 3.15 Raster/vector conversion.

whereas systems for research and development will be both vector- and raster-oriented.

A vector-oriented GIS will treat data as points in a cartesian or geographical coordinate system. The dataset is normally stored as a long string of coordinates together with information pertinent to the points and/or the surrounding areas.

A raster-oriented GIS will in principle have data stored as large continuous-area covering-data carrying points. Thus, but again in principle, a raster database requires much more storage space than a vector-oriented one (see Fig. 3.14).

The raster-oriented system will have its origin in the upper left corner, the Y-axis will be opposite "normal" coordinate systems whereas the X-axis will be like the vector data system which starts in the bottom-left corner. The raster database (on the figure) will require at least 30×30 storage data—unless packed in a special way; in the vector database we will need (1) two pairs of x,y for each straight line and (2) depending on wanted resolution, enough pairs of x,y to describe the curved lines. Naturally the vectorial lines and curves can loose in resolution (= cartographic neatness) when presented in raster mode.

Capturing the data for the raster-oriented GIS is of course no problem as for the satellite data. The data already is on raster format. If the database is vector-oriented, the main problem of data capturing will be connected to the transformation of satellite data into vectorized form, which is impossible for a whole satellite scene. A full raster-grey level image cannot be vectorized; either classified or otherwise processed derivatives have to be presupposed (see Fig. 3.15).

The transformation of vector data into raster format will require a transformation into a new coordinate system and a calculation to find which points will have to be projected onto the raster data plane. The position will have to be set to some value.

When raster-formatted data is transferred into vectorial data things become more complicated. The necessary number of nodes (X,Y-coordinates) have to be found and transferred into the new coordinate system; this often requires a thinning of the linear features of the raster data in order to find the optimal vector nodes. The original raster image is restored in vector image by simply joining the nodes by edges.

Once formatted in the structure relevant for the processing system in question, both the satellite data and the conventional map data can be superimposed, analysed and synthesized as appropriate for the given task.

Many routines used for analysing data in a GIS are the same as used for pure image analyses; these are analysis that mostly operate on single points, as for instance arithmetic, Boolean, Fourier and classification operations.

Polygon- and neighbourhood-related operations are of specific greater interest in GIS analysis. These operations concern:

—Label, areal and perimeter analysis—used for reclassification and relabelling regions, calculation of area and perimeter, often also shape and form classifications—eventually combined area, shape hierarchical classifications.
—Analysing polygon correlations, defining new polygons, merging and combining, masking.
—Distance and neighbourhood analysis.
—Contouring and interpolations.
—Surface/topography analysis.
—Optimizing of flow directions, path selection and pointing unto specific locations; extraction of potential areas, points and corridors.
—Synthesis.

3.7.5 Graphic production of digital image maps

Once the geographic data have been combined with the image using the preceding procedures and displayed on a screen, it remains to produce the graphic outputs if one wishes to conserve the result on a permanent support.

Different cases can arise: for a simple conservation on paper, an output on a colour hard copy (photographic) or on an electrostatic or coloured ink-jet plotter is sufficient. The procedure is suitable for an average cartographic rendering and a limited number of copies (from 1 to 20 or 30). On the other hand, if one requires very good cartographic quality and especially a fairly large number of copies it is preferable to output a document on to reproducible film and if possible on colour-separated plates.

In Section 3.5.5 mention was made of the operations required for the graphic output of the image alone: the operations are exactly the same for a combined image so long as that for the cartographic overprinting there has been planned the same types of colour separations (in subtractive synthesis and printing percentages).

References

BOURSIER, P. 1986. The integration of cartographic and image data into geographic information systems. *International Electronic Image Week*, IGN, France.

CARLE, C. 1982. *Satellite Mapping, Geometric Correction of Remote Sensing Images.* Meddelelser No. 12. Institut for Landmaaling of Fotogrammetri, Danmarks Tekniske Hoejskole.

CASTLEMAN, K. R. 1979. *Digital Image Processing.* Prentice-Hall, New Jersey.

FICCDC. 1988. *A Process for Evaluating Geographic Information Systems.* Federal Interagency Coordinating Committee on Digital Cartography, USGS, Reston, USA.

HUGUET P. 1987. *Traitement et analyse thématique d'images; de cartes et de données associées.* Actes du Forum Fi3G-Lyon Conseil National de l'Information Géographique, Paris, France.

NIBLACK, W. 1985. *An Introduction to Digital Image Processing.* Strandberg, Denmark.

SCHOWENGERDT, R. A. 1983. *Techniques for Image Processing and Classification in Remote Sensing.* Academic Press, New York.

STUCKI, P. 1979. *Advances in Digital Image Processing.* Plenum Press, New York.

Chapter 4

DESIGN AND SEMIOLOGY FOR CARTOGRAPHIC REPRESENTATION

Compiled by
Janos Lerner* and Jean Denègre†

*R. Eötvös Lorand University, Department of Cartography, Kun Bela ter 2, Budapest, Hungary
†Conseil National de l'Information Géographique (CNIG), 136 bis rue de Grenelle, 75700 Paris, France

Translated by
Allan Moore

Institut Géographique National, France

CONTENTS

4.1 Introduction

Presentation of remotely sensed and processed data generally appear in a maplike form. How should these output data look? This is the question that had to be answered in the short three decades of history of satellite cartography. The Earth's surface has been the input, with all its relief, land cover, colours, shades and the atmosphere. How to distinguish, how to recognize? These questions were raised shortly after the first Gemini and Apollo photographs were published during the 1960s. When the problem was solved, the Landsats came, then SPOT and the others—new possibilities with always the same questions and different answers.

The first photographs had shown a strange, unusual view of the Earth's surface and all the details were explained by a traditional, well-known feature—a map. The strange pictures became popular, appeared on the front covers of books, atlases and in calendars, as new and unique maplike representations. After the appearance of false colour images, in a short time they became not only popular but familiar as well. And this was the time when you didn't have to explain every detail; separate maps could be omitted, the image was covered by symbols and letters. Landsat assured geometric conditions for producing "traditional" photo-maps, SPOT's stereoscopic capability offered the possibility to supply these maps with height data.

In remote sensing two output phases exist. One preceding the classification procedure and one presenting the results of interpretation. Satellite cartography concentrates on this latter one. If the result is a sketch or a map, we have to face a strange contradiction: a modern "new concept" process results in one of the oldest forms of representation.

Computer techniques try to overcome this problem. Printer/plotter maps convert original pixel (spectral) data to map category (colour) information. Maps express information by symbols. Satellite images give data through pixels and density distribution. How to make a compromise? This is the task for satellite cartography.

4.2 Selection of information for satellite cartography

4.2.1 Image data

In Chapter 2 a detailed description of image information extraction procedures has been given. Analogue and digital interpretation techniques produce a variety of data useful for cartographic presentation of results. Even "raw" images (i.e. true colour, false colour and B/W

photographs, single-band scanner images, radar images, etc.) are used for cartographic purposes. Preprocessed, enhanced images, multiband composites, are sometimes mentioned as "maps", though not giving any classified information. For cartographic purposes the enhancement of radiometric data (see image elements in Chapter 2, Section 2.3) is most important.

4.2.2 Thematic data

These are the essential pieces of information to be represented. Thematic data include all the scientific data already known before image analysis ("background" informations), the final categories of classification process and location data (i.e. geographical names and coordinates).

Results of visual-analogue interpretation (see Chapter 3) appear in graphic form (i.e. symbols, lines and surface colours). Digital image interpretation gives a pixel-based map, where all the pixels are coded by colours. A great variety of sciences are using satellite images for thematic mapping, existing applications covering almost one hundred disciplines. Land use/cover and Earth sciences are of primary applications. The *International Report on Thematic Mapping from Satellite Imagery* (edited by J. Denègre, Elsevier, 1988) gives an overview of applications.

Representation methods of thematic data are traditional. Design of satellite cartography products is mainly influenced by the application of the right representation method. They all belong to cartographic data—the main element of selecting right information for satellite cartography.

4.2.3 Elements of cartographic representations

Thematic maps have a well-defined and widespread system of cartographic representation. To express reality, maps are using methods of classification, simplification and symbolization.

Classification means to create categories, artificial groups representing characteristic elements. Simplification or generalization is a process of selection, reduction, smoothing and combining data important for the purpose of mapping. Using symbols means to express reality in an abstract way through a process of transformation. Geographic information becomes a point, a line or a surface characterized by a certain size, shape, pattern and colour.

From a cartographic point of view, representation methods of thematic maps can be distinguished by the graphic appearance (i.e. points, lines, areas), what they express (quality only or quantity as well), whether they show any direction, distribution, frequency, what is the surface like they relate to (point, line, continuous or discontinuous surface, etc.). Quality categories are generally expressed by the shape or colour of symbols; quantity is usually expressed by size, pattern or density. Those rules have been developed conceptually and practically by Bertin (1983).

4.3 Types of space maps products

In order to be clearer, one can distinguish two major product-families of thematic cartography derived from satellite imagery:

— *Line maps*, where all the information is plotted and represented with the usual graphic variables, and where all the evidence of the initial imagery has disappeared (except, in certain cases, the shape of the pixels which remain visible).
— *Image maps* (maps with image-backgrounds) where the map background uses all or part of the initial satellite imagery, and where the thematic information is represented in various ways.

4.3.1 Line maps

The typology of thematic line maps, when they are derived from satellite imagery, is not fundamentally different from that of normal thematic maps, except perhaps as concerns the raster-type maps. Most frequent types of maps are:

— Simple sketch maps with line symbols (e.g. tectonic features).
— Sketch map with reference map (i.e. some topographic features like roads, drainage, settlements are indicated—generally in single colour—and thematic features appear in colour spots and lines (see, for example Chapter 5, Applications 8, 15, 20, etc.).

— Computer printer maps are usually in B/W colours with printer characters relating to surface categories.
— Computer plotter maps are pixel-based colour sketch maps, where colours represent the same classification groups as characters in printer maps (see, for example, Applications 5, 10, 13, etc.).
— Polygonal (choropleth) maps with thematic zoning in colour, which in no way distinguishes them from the same type of normal map (see, for example, Applications 6, 7, 9, etc.).

4.3.2 Image maps

Creating image maps we meet the three types of cartographic data-combination mentioned already in Chapter 3, where the technical procedures of these combinations were summarized.

Colour plotter graphics can be supplied by single colour (in some cases multi-coloured) reference maps (showing boundaries, roads, geographic grids, place names) (see, for example, Chapter 5, Applications 8, 12, 17, etc.). Reference maps can be compiled by the analogue method, by the digital method or they are got directly from an original topomap. In this latter case, a geometric rectification is necessary to adjust image and map information, i.e. image coordinates are converted into the map's projection (see Section 3.1).

Cartography has a well-defined, traditional system of graphic representation giving symbols, dots, lines (single ones or in an areal distribution), shades and colours and their certain combination. Images offer a special pixel-based representation built up from black–grey–white or coloured pixels. For the image user, pixels as elementary image details do not have primary importance. The user is interested primarily in the "photographic" appearance of the image, i.e. in a certain "aesthetic" distribution of pixels. He does not want to see a mosaic-like, image but an image resembling the reality, i.e. a photo-like representation where certain features are easily recognizable or explained. Image maps need a certain distribution of traditional cartographic data and image data. This combination can be approached from the aspect of content (see Chapter 3) or from the aspect of cartographic design.

4.4 Image map design

An image map is a certain combination of graphic (map) and image elements. What does an image map have to look like ? There is no general rule for "standard" image maps. However, there exist standard formats associating image data in raster mode and cartographic data in vector mode (e.g. GIS-GEOSPOT or SDTS in the USA, EDIGEO in France). But there are a series of possibilities, i.e. different representation methods of combining graphic elements with image elements (tones, colours), depending on how much of the image content needs to be preserved for a given scientific purpose (e.g. geology, land use). Considering the "background-content" ratio, there are three groups of image maps:

—photo-enhanced image maps,
—graphics-enhanced image maps,
—subject-enhanced image maps.

4.4.1 Photo-enhanced image maps

In this type of map photographic background dominates over graphic elements (see, for example, Chapter 5, Applications 2, 3, etc.). B/W single-band images, false colour, true colour images or mosaics are enhanced for giving an ideal, good quality. In most cases graphic content appears in a single colour in a form of symbols, dots, lines or line systems. When the categories are of linear features (e.g. tectonic lines, drainage types), they can appear in different types of line symbols and/or in different colours. If multicoloured symbols are used, the background is generally a single coloured image.

Advantages
—a representation close to reality,
—possibility for further interpretation,
—possibility to control previous interpretation,
—aesthetic appearance.
Disadvantages
—difficult readability for non specialists because of the lack of unified legend,
—if surface categories do not correspond with land cover categories, there is a disturbing "noise" effect because of the inhomogeneities of categories.

4.4.2 Graphics-enhanced image maps

If the graphic content is dominating over the image background we speak about graphics-enhanced image maps. The image is always monochrome, giving a real "background" or "base" for the thematic content. The monochrome image generally appears in grey or light brown, in a neutral colour. In some cases the background image is a processed image enhancing a characteristic parameter (e.g. linear elements for geological content or vegetation/forest cover for land-use content). Thematic content appears in colour symbols and surface colours. For relief representation, contour lines can be applied. Roads and settlements are also indicated by separate symbols. Water surfaces are enhanced by blue areal colour. In some cases a technique of "insel-map" is applied where thematic content is enhanced only for certain parts of the monochrome (in some cases multicoloured) image map (see, for example, Applications 10, 14, etc.).

Advantages
—possibility for detailed map information,
—no "blank" areas in the map (like in "traditional" maps),
—sometimes really nice, aesthetic appearance.
Disadvantages
—image is not interpretable anymore, as thematic information depresses image information and there is a lack of contrast,
—in some cases image information totally disappears because of full colors, applied mainly to hydrography and settlements.

4.4.3 Subject-enhanced image maps

The general problem of image maps is how to find a balance between image background and thematic (graphic) content. In Section 4.4.1 image content and in Section 4.4.2 graphic content dominated the other one. How can we make an image map to overcome the disadvantages indicated above? A procedure for creating an image map like this is illustrated in Fig. 4.2. According to the classified surface categories and colours chosen for final appearance a mask series is contructed. Through a traditional "pre-press" masking process, screened colour separations are produced where an original screened single-band image is used as raster screen. A digital

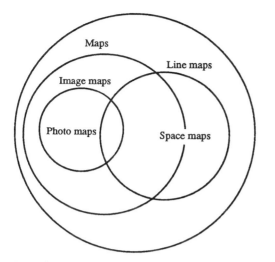

Fig. 4.1 Line maps and image maps (CNIG, 1989).

Advantages

—the original pixel (spot) and spectral ratios (contrast) are not lost, the image can be interpreted,

—there are well-defined surface categories to be represented well in a legend,

—surface categories are easily separable, even if they are not corresponding with land cover.

Disadvantages

—creating subject-enhanced image maps is a longer and more expensive procedure compared to the previous ones,

—best results are got in land cover/land use maps. Sometimes contrast inside one category can be disturbing because of diversified land cover types,

—sometimes colour surface categories dissect the harmony of the map, making the interpretation difficult. To obtain a better interpretable image the classified image can be combined (merged) with a "neutral" single-coloured image of the same band or

method to create subject-enhanced image maps is to order colour hues to the classified pixel data according to the pixel values of one of the original spectral bands. Graphic content appears in a single colour and can be combined well with image colours.

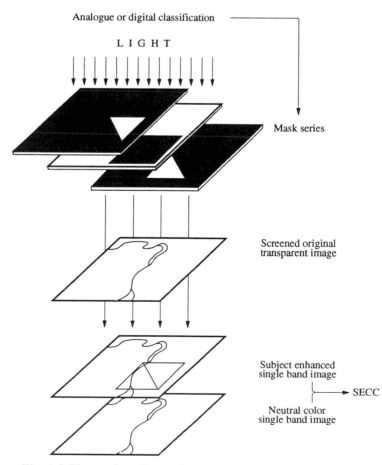

Fig. 4.2 Thematic-enhanced image map production process.

another one, depending on the subject of classification.

4.4.4 One particular case: toponymy

Correct toponymy is a general problem in cartography: How to put geographical names in the right place without disturbing map content and, at the same time to make the map readable? How to use different types of letters to give the place names a meaning? In traditional cartography the problem is more or less solved, as you can find (or create) blank areas for the names, you can use different letter types and different colours to distinguish, for example, names of towns from those of rivers. But how to overcome the problems for image maps?

"Blank" areas—i.e. spots with single colour or tone—are rare (e.g. large water surfaces, deserts, forests, snow covered areas). To create blank spots for lettering is not the best solution, as three times more information of the image content is lost in this way than in the simple lettering process. How to enhance readability of place names in image maps? You have to give up some cartographic conventions and rules! You can (must!) use different shades or colours for the names of the same category (e.g. white or yellow letters for dark areas and black or brown letters for light ones), but you have to keep the convention of using only one letter type for the same category. Figure 4.3 giving some typical examples of methods used in image map toponymy.

4.5 Image map typology

In a general way, and taking into account all preceding chapters, the following five criteria seem to be able to characterize an image map:

Metric criteria
1. Value of the geometric processing of the image used as background.
2. Coherence of the cartographic representation system, sheet-lines, format, marginal information, legend.

Cartographic criteria
3. Richness of the cartographic information added to the image.
4. Mounting and reliability of the cartographic information added to the image.

5. Quality of the cartographic aesthetics of the edited/published final product.

For each criterion, product quality can be characterized, as for conventional maps, by a level defined as follows:

Regular (or basic) image map

Preliminary image map
(Intermediate class, arising quite often: the product does not have all the characteristics of a map or else, if it has them all, they are not in conformity with the quality required for a basic map.)

Image document
Those definitions are made more explicit in the following section, which describes, for each level of the above five criteria, the corresponding typology (see Fig. 4.4).

4.5.1 Basic image map typology

In contemporary cartography, basic maps imply rigorous correspondence between the represented positions and the real positions in space.

A basic image map must therefore have the following general characteristics.

4.5.1.1 Criterion no. 1. Geometric processing of the image (used as background)

The image is corrected geometrically in such a way as to satisfy a planimetric accuracy with:

Standard deviation of 0.3 mm on the printed document (i.e. statistically: *90 % of the points better than 0.5 mm* from their exact positions, a standard often used outside France; or again: *99 % of the points better than 0.8 mm* from their exact positions, a standard in general use in France, called "tolerance").

When the data are defined uniquely in the digital form, those values must evidently be stated in terrain units. One then deduces from them the maximum representation scale to be respected in order to obtain a basic map.

Depending on the sensor or vector used, the data to be processed and the procedure to be used can vary quite considerably. In certain cases (defined in accordance with the preceding parameters) one can obtain a correct geometry by photographic rectification or geometric correction on control points (case of rectified photographs or preprocessed SPOT imagery at Level 2). In other cases, particularly for zones

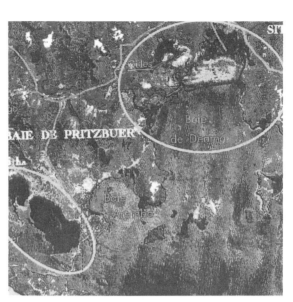

Photomechanical composition : simple black overprinting on the image background.

Computer-assisted composition in raster mode : colour negative overprinting (i.e. text in yellow on dark parts of the image, in white on red parts of the image, etc.) Chap. V, application 15.

Photomechanical composition in black with a slight negative delineation around the letters (for a better lecture)

Computer-assisted composition in raster mode : negative "pixelized" letters in arabic, with a slight shading (light from the N-W).

Fig. 4.3 Samples of place name compilation methods for satellite image maps.

with a pronounced relief, correction with a digital terrain model is necessary (case of orthophotographs or SPOT imagery pre-processed at Level 3).

4.5.1.2 Criterion no. 2. Representation system

—For maps in a series, the sheet-line system must be preferably standardized or at least regular, and based on rectangular or geographic axes. As a rule, the format must be standardized for basic maps.

—For products which do not form part of a series, sheet lines and format can be adapted to the general configuration of the mapped zone. The entire group must be coherent.

Plane representation system. For basic maps, the representation of the elements of the coordinate system must be complete: peripheral grid-ticks and sheet corners numbered, grid intersections. It is also useful to indicate, at medium and small scales, the geographic grid.

When the complete representation of the

	EVALUATION CRITERIA	VALUE	CHARACTERISTICS	IMAGES			PRELIMINARY MAPS		BASIC MAPS	
				Raw image	Uncontrolled mosaic	Controlled mosaic	Topographic preliminary map	Thematic preliminary map	Topographic basic map	Thematic basic map
METRIC PROPERTIES	Nr.1 IMAGE GEOMETRIC PROCESSING	3 2 1	Corrections of all distortions (sensor, relief) / Orthogonal projection rectification / Partial corrections / No orthogonal projection / No correction / Mean scale adjustment	1,5 to 1	2 to 1	3 to 2	3 to 2	3 to 2	3	3
	Nr.2 REPRESENTATION SYSTEM-DRESSING	3 2 1	MAP INDEX — PROJECTION — LEGEND / International system — Represented — Complete / Local system — Reduced representation — Reduced / No system — Not represented — No legend	1	2 to 1	3 to 1	3 to 2	3 to 2	3	3
CARTOGRAPHIC PROPERTIES	Nr.3 COMPLETENESS OF ADDED INFORMATION	3 2 1	Overprint with reliable and complete data / Overprint with approximate and generalized data / No overprint	1	1		2	2	3 and 2,5	3 and 2,5
	Nr.4 LOCATION OF ADDED INFORMATION	3 2 1	Standardized cartographic data and processes / Cartographic data and processes adapted to local conditions / Facultative cartographic data and processes	1	1	1	3 to 2	3 to 2	3	3
	Nr.5 CARTOGRAPHIC COSMETICS	3 2 1	Image: correct edges between images and sheets. Line mapping: in accordance with normal cartographic standards / Image: edges not improved / Line mapping and place names in simpler way / Raw image	1	2	2,5 to 2	3 to 2	3 to 2	3 and 2,5	3 and 2,5
	TOTAL			5 to 5,5	6 to 8	2 to 11,5	10 to 14	10 to 14	14 to 15	14 to 15
	DOCUMENTS ON IMAGE BACKGROUNDS (IMAGE MAPS)			Raw image	Uncontrolled mosaic	Controlled mosaic	Topographic preliminary map	Thematic preliminary map	Topographic basic map	Thematic basic map

Fig. 4.4 Provisional typology of image map products with quality classification according to five criteria.

coordinate system is not necessary for the user, e.g. tourist maps, it can be reduced. The information supplied, nevertheless, must make it possible to define unambiguously the plane coordinates of any represented point.

Marginal information-legend. A certain number of basic indications must appear in the margins, whether they be general (title, scale, authors, printing date, etc.) or specific:

—Acquisition date, origin of images, acquisition-scale when photographs are involved and processing applied (if necessary).
—Legend making it possible to interpret the image: insets or image-extracts showing typical examples of the vegetation, hydrography, etc.
—Complete legend of the conventional signs added to the image.

4.5.1.3 Criterion no. 3. Richness of the cartographic information (added to the image)

The elements represented, either by their image or by an additional sign, are those listed in general specifications for national mapping or in special documents, e.g. Calls for Tenders (list of special technical clauses).

Topographic information (planimetry)
—Landscape elements with explicit images need not be overprinted by a sign.
—Major details required to be shown at the adopted scale, which are not visible on the image, or are only visible in a very discontinuous manner, or difficult to identify, are represented by a sign combined with the image.
—Lettering (place names, indications) respect as much as possible the criteria of density, position, classification, of normal maps at the same scale.
—In general, the density and graphic expression/mode of the elements combined with an image must be harmoniously balanced in order that the entire group remains clear and readable and that the information supplied by the image is entirely conserved.

Topographic information (altimetry). The altimetry is represented by contours and/or spot heights depending on the landscape. The contour-interval, point-density are those of general reference standards or calls for tenders.

Thematic information. The thematic information added to the image cames from either an interpretation of the image itself (photo-interpretation, computer-assisted classification, etc.), or other information sources (existing maps, field observations, etc.).

—The richness of that information (number of themes) must be compatible with a good readability of the image map.
—The theme represented has priority over the additional planimetric representation, but the latter must remain sufficiently visible for positioning the phenomena.

4.5.1.4 Criterion no. 4. Positioning and reliability of the cartographic information added to the image

Reliability of the geometric position. The geometric quality of the image having been evaluated in Section 1.1 and the value of the added information, in the qualitative sense, having been estimated in Section 1.3, one examines the cartographic value, in the geometric sense, of that information. One must therefore evaluate here the entire group of data and the technical procedures used for mapping those elements, and the accuracy of their positioning in the light of international standards or special specifications (planimetric and altimetric accuracy).

That positioning must offer the same geometric quality as the image (see Section 4.5.1.1).

Information up-to-dateness. The date of the data added to the image must be in keeping with the date of the latter. In particular, if those data come from an existing map, the latter should be updated beforehand in accordance with the usual standards. In any case, the origin, printing date and/or that of the updating of the cartographic data added to the image must be clearly mentioned.

Reliability of the thematic information. One must indicate the origin of the thematic information and its production procedure (photo-interpretation, computer-assisted classification, external data, etc.), as well as the nature and the value of result-evaluation criteria, e.g. in the form of a confidence interval or a confusion matrix.

4.5.1.5   Criterion no. 5. Cartographic aesthetics

Image used as background. Its recording date must be selected within a favourable period (atmospheric conditions, vegetation, etc.); it must have been processed so that its general quality (definition, average density, contrast, etc.) is conserved, or even improved.

- —If a sheet is composed of several images with the same date, their join-lines must be practically invisible. The density or the radiometry on either side of the join-lines must be balanced.
- —Between adjacent sheets, the image-background must not show any great differences.
- —If images with different dates have to be used (or if certain aspects of the landscape are considerably different), differences are tolerated between them (density, colours, etc.).

Editing and printing. The complete group of these phases must conserve the image-quality and offer a level of quality sufficient for the information added to the image and the marginal indications and information.

The lettering must have the usual cartographic quality regardless of its origins: typography, photocomposition, incremental photo-plotter. Those items which will be combined with the image must be made readable by any process (colour, thickening, kept white, etc.).

That overall quality must be homogeneous throughout the production.

4.5.2   Preliminary image map typology

Some of the characteristics required for basic maps are no longer respected.

4.5.2.1   Criterion no. 1. Geometric image-processing

The quality level of the entire data-processing is not sufficient for considering the resulting image as an orthogonal projection of the terrain. For example, Level 2—correction of SPOT imagery of zones with pronounced relief.

4.5.2.2   Criterion no. 2. Representation system

No official cartographic system specifications are applicable. Sheet-lines and format are adapted to the shape of the zone to be mapped. The projection system can be represented in a concise way: sheet corners, grid ticks and grid intersections in the sole rectangular grid. The legend is reduced, even incomplete.

4.5.2.3   Criterion no. 3. Richness of the added cartographic information

The overprinting of information is reduced. However, one must not confuse reduced information with low-density information arising from the semi-desert character of a mapped region.

4.5.2.4   Criterion no. 4. Positioning and reliability of added cartographic information

Its quality is not guaranteed in the cartographic sense, i.e. it is not based on precise information (field completion, basic-map compilation, etc.). Its up-to-dateness is no longer guaranteed, e.g. compilation from an old map without any updating. Likewise, the reliability of the thematic information represented is not characterized by the usual evaluation criteria.

4.5.2.5   Criterion no. 5. Cartographic aesthetics

The overall quality is good, but can vary slightly between adjacent sheets (different procedures, etc.). Within a given sheet, image-joins may be visible but discontinuously.

4.5.3   "Image" document typology

None of the specifications defined above are used. Only the origin of the basic data and the procedures are indicated.

References

BERTIN, J. 1983. *Semiology of Graphics.* University of Wisconsin Press, USA.

ICA-ACI DENEGRE, J. *et al.* 1988. *International Report on Thematic Mapping from Satellite Imagery.* Elsevier, U.K.

CNIG. 1989. National Council for Geographic Information. *Typology and Terminology of Image Maps.* Paris.

GALTIER, B. 1992. Le rôle de la spatiocarte en tant que produit cartographique. *Proceedings 17th Conference ISPRS.*

Chapter 5/Chapitre 5
EXAMPLES OF APPLICATIONS/EXEMPLES D'APPLICATIONS

CONTENTS/TABLE DES MATIERES

GENERAL SATELLITE CARTOGRAPHY

SPATIOCARTOGRAPHIE GENERALE

Application Title
Demonstration of differences in spatial resolution on the same scene

Titre de l'application
Démonstration des différences de résolution spatiale sur la même scène

Localization/**Localisation**
Canada (Toronto)/**Canada (Toronto)**

RESPONSIBLE ORGANIZATIONS   ORGANISMES RESPONSABLES

Canada Center for Remote Sensing
and Ontario Center for Remote Sensing

Tel: 1/416 733 5071

(Contact person: David B. White)

1. Demonstration of differences in spatial resolution on the same scene

—Landsat MSS (20 September 1985) with resolution 80 m.

—Landsat Thematic Mapper (20 September 1985) with resolution 30 m.

—SPOT Multispectral (MLA) (25 June 1987) with resolution 20 m.

—SPOT Panchromatic (PLA) (25 June 1987) with resolution 10 m.

Section of the National Topographic Map 1 : 50 000.

Data supply: Canada Centre for Remote Sensing.

Compilation of data and preparation for printing: Ontario Centre for Remote Sensing.

1. Démonstration des différences de résolution spatiale sur la même scène

—Landsat MSS (20 septembre 1985), résolution de 80 m.

—Landsat Thematic Mapper (20 septembre 1985), résolution de 30 m.

—SPOT Multispectral (XS) (25 juin 1987), résolution de 20 m.

—SPOT Panchromatique (P) (25 juin 1987), résolution de 10 m.

Extrait de la carte topographique nationale à 1:50 000.

Origine des données: Centre Canadien de Télédétection.

Compilation des données et préparation pour l'impression: Centre de télédédétection de l'Ontario.

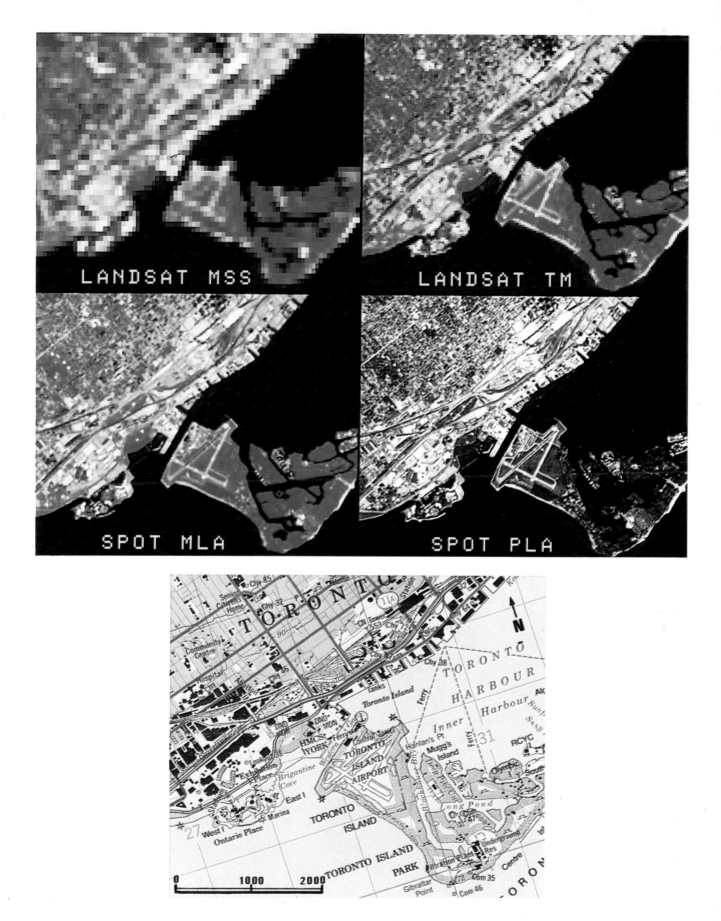

Canada (Toronto) / Canada (Toronto)

Application Title
Image mapping with merged Landsat/SPOT data, USA

Titre de l'application
Iconocarte par combinaison des données Landsat et SPOT

Localization/Localisation
USA (Sun City, Arizona)/EUA (Sun City, Arizona)
Argentina (Viedma)/Argentine (Viedma)

RESPONSIBLE ORGANIZATIONS   ORGANISMES RESPONSABLES

U.S. Geological Survey,
National Mapping Division,
EROS Data Center,
Sioux Falls, South Dakota 57198, USA

Tel: (1) 605/594-6114   Telex 910-668-0301

(Contact person: Bruce K. Quirk)

2. Image mapping with merged Landsat/SPOT data, USA

1. Objectives

The US Geological Survey has been producing image maps from digital data for several years. In the past, image maps have been made primarily from the Landsat multispectral scanner (MSS), return beam vidicon (RBV) and thematic mapper (TM). In addition, some experimental large-scale image maps have been made using colour infrared aerial photography that had been converted into a digital format. In both cases, the data sets are geometrically corrected, mosaicked and digitally enhanced before being printed at standard map scales ranging from 1:1 000 000 to 1:12 000. Traditionally, these image maps have been made using data from a single sensor such as the Landsat TM.

The introduction of the French SPOT satellite and its 10 m resolution panchromatic and 20 m resolution multispectral data have stimulated research into data-merging techniques or procedures for combining data sets of different spatial and/or spectral resolutions into a single data set. The new data set exhibits the best characteristics of both the original data sets. The use of the intensity, hue and saturation (IHS) transformation has been one of the techniques used in combining or merging these data sets.

2. Methods used

In order to merge the images from two different sensors, the data sets must first be spatially registered. The data set with the desired spectral resolution, in this instance Landsat TM, is then contrast enhanced and transformed into IHS colour space from the traditional red, green and blue (RGB) space. The higher resolution spatial band, SPOT panchromatic, can then be substituted for the I band of the IHS data set

and the new data returned to the RGB domain. The high-resolution band may be contrast enhanced before the substitution, so the histogram of the new I band is similar to the histogram of the I band it is replacing (see Fig. 5.2.1).

3. Results

This technique was used to merge Landsat TM and SPOT panchromatic data over Sun City, Arizona. The SPOT data set was acquired on 3 April 1986, and the Landsat TM on 5 April 1986. The data sets were then registered to each other and both contrast enhanced. Landsat TM bands 5, 4 and 1 were then transformed into IHS space. The SPOT panchromatic data was then remapped using the cumulative histogram percentages of the I band. The remapped SPOT data was then used as the new I band and the data set transformed back into RGB space. The figure provides an example of the Landsat TM (*top*) and SPOT panchromatic (*bottom*) scenes before merging and the final merged data set (*middle*). The other figure is an experimental Landsat/SPOT image map over Viedma, Argentina, made using the IHS data merging techniques.

4. References

CHAVEZ, P. 1986. Digital merging of Landsat TM and digitized NHAP data for 1:24 000 scale image mapping. *Photogrammetric Engineering and Remote Sensing*, vol. 52, No. 10, pp. 1637–1646.
HAYDN, R. *et al.* 1982. Application of the IHS color transform to the processing of multisensor data and image enhancement. *Proceedings of the International Symposium on Remote Sensing of*

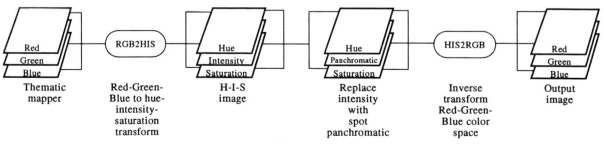

Fig. 5.2.1.

2. Iconocarte par combinaison des données Landsat et SPOT

1. Objectifs

Le Service Géologique des Etats-Unis produit des iconocartes issues de données numériques depuis plusieurs années. Dans le passé, les iconocartes ont été réalisées d'abord à partir du scanneur multibande (MSS) de Landsat, du "Return Beam Vidicon" (RBV), et du "Thematic Mapper" (TM). En outre des iconocartes expérimentales à grande échelle ont été produites à partir de photographies aériennes en couleur infrarouge, préalablement numérisées. Dans les deux cas, les données ont été corrigées géométriquement, mosaïquées et accentuées numériquement, avant d'être restituées à des échelles standards allant du 1 : 1 000 000 au 1 : 12 000. Traditionnellement ces iconocartes ont été produites avec des données d'un seul capteur comme Landsat TM.

L'introduction du satellite français SPOT et de ses données panchromatiques à résolution 10 m, et multibandes à résolution 20 m, ont stimulé la recherche visant à combiner les données ou à produire un seul fichier à partir de données à différentes résolutions spatiales et/ou spectrales. Le nouveau fichier présente les caractéristiques optimales des deux fichiers d'origine. L'utilisation de la transformation intensité, fréquence, et saturation (IHS) est l'une des techniques pratiquées pour combiner ou mélanger ces fichiers.

2. Méthodes utilisées

Afin de combiner les images de deux capteurs différents, les données doivent être d'abord calées géométriquement. Les données possédant la résolution spectrale souhaitée, en l'occurrence Landsat TM, sont ensuite accentuées en contraste et transformées dans l'espace couleur IHS à partir de l'espace habituel rouge, vert, bleu (RVB). La bande à meilleure résolution spatiale, SPOT panchromatique, peut être ensuite substituée à la bande I du fichier IHS et les nouvelles valeurs sont renvoyées dans le domaine RVB. La bande à haute résolution subit une accentuation de contraste avant substitution, de sorte que l'histogramme de la nouvelle bande I est semblable à celui de la bande I qu'il remplace (voir fig. 5.2.1).

3. Résultats

Cette technique a été utilisée pour combiner les données Landsat TM et SPOT panchromatique sur Sun City, Arizona. Les données SPOT furent acquises le 3 avril 1986, et Landsat TM le 5 avril 1986. Les images ont été ensuite calées l'une sur l'autre, et accentuées chacune en contraste. Les bandes Landsat TM 5, 4 et 1 ont été transformées dans l'espace IHS. L'image SPOT panchromatique a été ensuite recalculée en utilisant les pourcentages de l'histogramme cumulatif de la bande 1. Cette nouvelle image SPOT a été utilisée comme bande I, et le fichier complet retransformé dans l'espace RVB. L'illustration couleur montre un exemple de scènes Landsat TM (*en haut*) et SPOT panchromatique (*en bas*) avant combinaison (*au milieu*). L'autre illustration couleur est une iconocarte expérimentale sur Viedma (Argentine), réalisée avec les techniques de mélange IHS.

4. Références

Chavez, P. 1986. Digital merging of Landsat TM and digitized NHAP data for 1 : 24 000 scale image mapping. *Photogrammetric Engineering and Remote Sensing*, vol. 52, No. 10, pp. 1637–1646.

Haydn, R. *et al.* 1982. Application of the IHS color

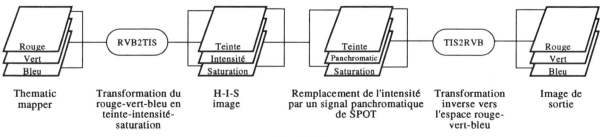

Fig. 5.2.1.

the Environment, Cairo, Egypt, January 1982, pp. 599–616.

FEUQUAY, J. 1987. Data integration using color space transforms. *Proceedings Pecora XI Symposium on Satellite Land Remote Sensing— Current Programs and a Look to the Future,* Sioux Falls, South Dakota, May 1986, p. 326.

WELCH, R. and EHLERS, M. 1987. Merging multiresolution SPOT HRV and Landsat TM data. *Photogrammetric Engineering and Remote Sensing,* vol. 53, No. 3, pp. 301–303.

transform to the processing of multisensor data and image enhancement. *Proceedings of the International Symposium on Remote Sensing of the Environment*, Cairo, Egypt, January 1982, pp. 599–616.

FEUQUAY, J. 1987. Data integration using color space transforms. *Proceedings Pecora XI Symposium on Satellite Land Remote Sensing— Current Programs and a Look to the Future*, Sioux Falls, South Dakota, May 1986, p. 326.

WELCH, R. and EHLERS, M. 1987. Merging multiresolution SPOT HRV and Landsat TM data. *Photogrammetric Engineering and Remote Sensing*, vol. 53, No. 3, pp. 301–303.

SPOT Image (c) 1991 CNES

Figure 2a. A mosaic of Sun City, Arizona, that shows both the original data sets and the result of merging the Landsat TM bands 5, 4, and 1 (top), with the SPOT panchromatic image (bottom), to form the merged image (middle). Image scale : 1:100,000.

Figure 2a. Une mosaïque de Sun City , Arizona , qui montre simultanément les données originales et le résultat de la combinaison des bandes 5, 4 et 1 d'une scène Landsat TM (en haut) et de la scène SPOT Panchromatique (en bas), pour former l'image combinée (au milieu). Echelle de l'image : 1:100 000

SPOT Image (c) 1991 CNES

Figure 3. Color reproduction of a digitally produced image map of the Viedma, Argentina 1:50,000-scale quadrangle. SPOT panchromatic data was merged with the 3-band Landsat TM data set of bands 2, 3, and 4.

Figure 3 b. Reproduction colorée d'une iconocarte numérisée de Viedma, Argentine, à l'échelle du 1:50.000. Une scène SPOT panchromatique a été combinée avec les trois bandes d'une scène Landsat TM, à savoir les 2, 3, et 4.

Application Title
The production of high quality satellite image maps by merging and mosaicking of multisensor data

Titre de l'application
Production d'iconocartes satellitaires de haute qualité par fusion et mosaïquage de données multicapteurs

Localization/Localisation
Germany (Berlin)/Allemagne (Berlin)

RESPONSIBLE ORGANIZATIONS   ORGANISMES RESPONSABLES

Technical University of Berlin
Department of Photogrammetry and Cartography
135, Strasse des 17. Juni
Sekr. EB9
1000 Berlin 12
Germany

Tel: 49/30 314 33 31

(Contact person: Prof. Dr Ing. J. Albertz)

3. The production of high quality satellite image maps by merging and mosaicking of multisensor data

The combination of multispectral data with additional data of high geometrical resolution has achieved increasing importance since the launch of the French satellite SPOT in 1986. The HRV system onboard of this satellite offers a resolution of 10 m in its panchromatic mode which has proved to be very useful for maps at a scale of 1:50 000. But the optimal interpretation potential can only be achieved by using additional data with multispectral informations. The Thematic Mapper system with its seven spectral bands offers high flexibility for the generation of image maps. Furthermore, the Thematic Mapper is the only sensor system allowing a natural colour rendition by using the first three spectral bands.

However, in order to preserve the high resolution of SPOT data as well as the multispectral information of the Thematic Mapper data, it is necessary to apply special processing techniques. These processing steps include not only merging of multisensor data but also mosaicking of the initial data if more than one scene is concerned. In most cases the production of satellite image maps needs mosaicking of several scenes in order to match the sheet-line system of topographical maps. For this purpose, a software package must be available which provides algorithms allowing mosaicking and merging of multisensor data sets without loss of information and quality.

Such a software system has been set up at the Technical University of Berlin in recent years. This system made it possible to produce the satellite image map "Berlin 1:50 000 and its vicinity", which required mosaicking of six SPOT scenes and two TM scenes and subsequent combination of both data sets.

Data processing of this satellite image map can be subdivided into the following procedures which are typical for the production of high-quality satellite image maps:

Preprocessing. Preprocessing of the raw satellite data includes mainly contrast stretching, the elimination of effects by calibration and transmitting errors as well as edge enhancement. Geometric preprocessing is not necessary because the distortions caused by several system parameters are already corrected by the ground receiving stations. In order to obtain a satellite image map with high quality in partly cloud-covered regions, a method of partial substitution has to be applied.

Geometric rectification. Merging and mosaicking of multisensor data requires high precision in geometrical rectification not only for one scene but also in the whole mosaic. The best results for this purpose can be derived by a simultaneous determination of all transformation parameters in a least squares adjustment, making use of the multiple information in overlapping regions of adjacent scenes. If mountainous regions are concerned, differences in height must be considered.

Radiometrical mosaicking. The differences of radiometric information between adjacent scenes can be eliminated by transformation of all images in a homogeneous common grey value system. For this purpose again, the multiple information in overlapping regions is used.

Combination of multisensor data. Merging of both data sets is carried out by IHS transformation, which has proved to be very useful. After the transformation of the TM data into the colour system IHS, the intensity component containing the relevant bright and dark contrast has to be modified. In the most simple case the intensity can be substituted by the panchromatic SPOT image data. However, in order to achieve an appropriate colour rendition, it is necessary to carry out a radiometrical adjustment of the SPOT data onto the intensity component which was derived from the TM data. Afterwards the re-transformation into the RGB system can be carried out.

Postprocessing. Due to different spectral properties of both sensor systems, not all information can be preserved after merging. Another reason for this effect consists in the processing of many scenes for large mosaics which are acquired at very different dates. Therefore specific postprocessing techniques have to be applied for particular object classes in order to avoid areas showing unnatural colours.

Cartographical processing. The result of the geometrical and radiometrical processing is a rectified, mosaicked and enhanced set of image data for a map sheet. However, in addition to

3. Production d'iconocartes satellitaires de haute qualité par fusion et mosaïquage de données multicapteurs

La combinaison de données multispectrales avec des données additionnelles à haute résolution géométrique a acquis une importance croissante depuis le lancement du satellite français SPOT en 1986. Le système HRV embarqué sur ce satellite offre une résolution de 10 m en mode panchromatique, ce qui s'est avéré très utile pour les cartes à l'échelle du 1:50 000. Cependant, on ne peut obtenir un potentiel optimal d'interprétation qu'en recourant à des données additionnelles contenant des informations multispectrales. Le système du Thematic Mapper notamment, avec ses sept bandes spectrales, offre une considérable flexibilité pour l'élaboration de documents cartographiques. Par ailleurs, le Thematic Mapper est le seul système de capteur autorisant une reproduction naturelle des couleurs par l'utilisation des trois premières bandes spectrales.

Cependant, pour conserver la haute résolution des données obtenues par SPOT ainsi que l'information multispectrale de celles fournies par Thematic Mapper, il est nécessaire de recourir à des techniques spéciales de traitement. Ces méthodes de traitement comprennent non seulement la fusion des données acquises par capteurs multiples, mais également l'assemblage des données initiales si plus d'une scène est concernée. Dans la plupart des cas, la production de documents cartographiques satellitaires impose l'assemblage de plusieurs scènes pour correspondre au découpage des cartes topographiques. A cette fin, il faut disposer d'un logiciel qui fournisse des algorithmes permettant l'assemblage et la fusion de données multicapteurs sans perte d'information ni de qualité.

Un tel logiciel a été mis au point à l'Université Technique de Berlin ces dernières années. Il a permis de produire l'iconocarte satellitaire "Berlin 1:50 000 et ses environs", pour lequel il a fallu assembler six scènes SPOT et two scènes TM et combiner ensuite les deux types de données.

Le traitement des données de cette iconocarte satellitaire est constitué des procédures suivantes, qui sont typiques pour la production d'iconocartes satellitaires de haute qualité:

Pré-traitement. Le pré-traitement des données satellitaires brutes consiste surtout à étaler le contraste, à éliminer les effets de la calibration et des erreurs de transmission, à accentuer les contours. Un pré-traitement géométrique n'est pas nécessaire, les distorsions causées par plusieurs paramètres du système étant déjà corrigées par les stations de réception au sol. Pour obtenir une iconocarte satellitaire de haute qualité dans les régions partiellement couvertes de nuages, il est nécessaire d'appliquer une méthode de substitution partielle.

Correction géométrique. La fusion et l'assemblage de données par multicapteurs imposent une extrême précision au niveau de la correction géométrique non seulement pour une scène, mais également pour l'ensemble du document. Les meilleurs résultats à cette fin peuvent être obtenus par une détermination simultanée de tous les paramètres de transformation dans un ajustement selon la méthode des moindres carrés en recourant aux informations multiples dans les zones superposées des scènes adjacentes. Dans le cas de régions montagneuses, il convient de tenir compte des différences d'altitude.

Assemblage radiométrique. Les différences d'informations radiométriques entre scènes adjacentes peuvent être éliminées par transformation de toutes les images en un système homogène de valeurs de gris. Là encore, on recourt aux informations multiples des zones superposées.

Combinaison de données multicapteurs. La fusion des deux types de données est effectuée par la transformation IHS, procédé qui a fait ses preuves. Une fois les données TM converties dans le système coloré IHS, la composante intensité contenant le contraste clair/obscur doit être modifiée. Dans le cas le plus simple, l'intensité peut être remplacée par les images panchromatiques de SPOT. Cependant, pour obtenir une reproduction appropriée des couleurs, il est nécessaire de procéder à un ajustement radiométrique des données SPOT sur la composante intensité, dérivée des données TM. On peut ensuite effectuer la reconversion dans le système RVB.

Post-traitement. En raison des différences des propriétés spectrales des deux systèmes de

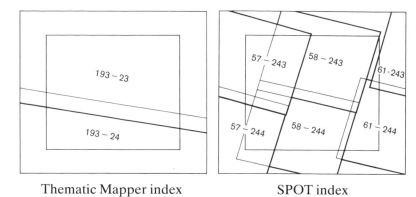

Thematic Mapper index   SPOT index

Fig. 5.3.1 Mosaicked and merged scenes of the satellite image map "Berlin 1:50 000 and its vicinity".

the image itself a map requires graphical elements such as lines, symbols, letters, masks for specific object classes etc. representing topographical features. Those elements are derived with the aid of digital image processing and stored in a separate file.

Digital screening. After the generation of such graphical data files the entire set of data is ready for the generation of printing originals. For this task a large format high resolution raster plotter system is used. By means of such systems the grey value of the image data in each spectral band is converted to a printing screen by digital computation. Therefore, the full information content of the data is maintained during this last step of data processing.

Merging of multisensor data is an effective tool to improve the quality of satellite image maps by combining the high resolution of panchromatic SPOT data and colour information from TM data. The methods applied yielded a very encouraging result despite the variations of the data due to different dates of acquisition.

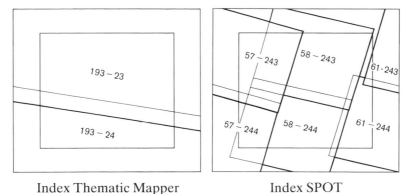

Index Thematic Mapper   Index SPOT

Fig. 5.3.1 Scènes assemblées et fusionnées du document cartographique satellitaire "Berlin 1 : 50 000 et ses environs".

capteurs, il est impossible de conserver toutes les informations après la fusion. Une autre raison est l'acquisition, à des dates très différentes, de nombreuses scènes pour de grands documents. Il est donc nécessaire d'appliquer des techniques spécifiques de post-traitement pour des classes particulières d'objets afin d'éviter d'avoir des zones présentant des couleurs non naturelles.

Traitement cartographique. Le résultat de ce traitement géométrique et radiométrique est un ensemble d'images rectifiées, assemblées et contrastées en vue de l'impression carto-graphique. Mais en plus de l'image elle-même, une carte doit comporter des éléments graphiques tels que lignes, symboles, lettres, masques pour classes spécifiques d'objets, etc., qui représentent des caractéristiques topo-graphiques. Ces éléments sont dérivés à l'aide du traitement numérique de l'image et mémorisés dans un fichier distinct.

Photogravure numérique. Après la création de ces fichiers graphiques, l'ensemble des données se prête à l'élaboration de planches d'impression. On utilise à cette fin un traceur à trame grand format et à haute résolution. Grâce à ces systèmes, la valeur de gris des images dans chaque bande spectrale est convertie en un film d'impression par calcul numérique. L'intégralité du contenu de l'information est ainsi conservée pendant cette dernière phase du traitement des données.

La fusion des données multicapteurs est un instrument efficace pour améliorer la qualité des iconocartes satellitaires en combinant la haute résolution des données SPOT panchromatiques et l'information couleur des données TM. Les méthodes appliquées ont donné un résultat très encourageant malgré les variations des données dues aux différentes dates d'acquisition.

Germany (Berlin) / Allemagne (Berlin)

Application Title
The Rheinhessen 1 : 50 000 thematic satellite image map

Titre de l'application
Iconocarte thématique satellitaire: Rheinhessen au 1 : 50 000

Localization/**Localisation**
Germany/**Allemagne**

RESPONSIBLE ORGANIZATIONS   ORGANISMES RESPONSABLES

Institute für Angewandte Geodäsie (IfAG)
Division of Photogrammetric Research
Richard-Strauss-Allee 11
D-6000 Frankfurt am Main 70

Tel: 49/69 63 331

(Contact person: B. Pfeiffer)

4. The Rheinhessen 1:50 000 thematic satellite image map

The annexed map shows a central section of the 1:50 000 Rheinhessen Thematic Satellite Image Map which was produced in 1989 by the Photogrammetric Research Division of the Institut für Angewandte Geodäsie in Frankfurt am Main and which has been published in the Series I of the *Nachrichten aus dem Karten- und Vermessungswesen.* This map demonstrated that it is possible to produce in a minimum of time new thematic maps by means of an adequate processing of the nowadays available satellite image data. The method consisted of combining the image data of two different sensor types each with different geometric resolution (30 m for the LANDSAT Thematic Mapper (TM) and 10 m for the SPOT High Resolution Visible (HRV) Sensor); thus a new digital image product was created showing the advantages of both systems (high three-dimensional resolution for SPOT and high spectral resolution for TM) in an optimal form. The result is a colour image product which can give the colours from the TM data and the high geometric resolution of the panchromatic SPOT data, i.e. with a pixel size of 10×10 m. So it is possible to produce image maps in the scale of 1:50 000.

Multisensor images from satellite image data are digitally produced using methods and procedures of digital image processing. The central part of the different procedures used is the colour space transformation according to intensity, colour frequency and colour saturation, the so-called IHS transformation. Each processing step is presented below for the production of the multisensor image map using the DIPIX-ARIES III image processing system for the L6114 Mainz Topographic Map sheets.

The image content is based on some important sections of two satellite photographs. The SPOT 1 HRV data in a panchromatic mode $(0.51-0.73\ \mu m)$ were obtained on 25 September 1986 (scene 050–249). A false colour composite has been chosen from the channels 3 $(0.63-0.69\ \mu m)$, 4 $(0.76-0.90\ \mu m)$, and 5 $(1.55 -1.75\ \mu m)$ using the multispectral LANDSAT 5 TM data of 17 August 1987 (scene 195–25). An RGB colour reproduction with channel 4 as red component, channel 5 as green component, and channel 3 as blue component leads to the colour impression as presented at the multisensor image map. The chosen colouring is

particularly well suited to indicate slight differences in land use in the case of regions covered with vegetation.

The relative rectification of the TM data onto the SPOT geometry was made by means of 32 control points determined in an interactive way. The transformation parameters of a polynomial of third degree have been used to transform the TM data with a cubic convolution resampling. The mean errors at the control points amounted to 1.44 pixels at right angles to the flight direction, and to 1.05 pixels in the flight direction.

Using the respective histogram, the grey value distribution of each TM channel was changed individually before the beginning of the IHS transformation in a way that we get a colour picture with a maximum of contrast. The line structures of a SPOT image whose grey value has also been extended have been extracted using a high-pass filter. These input data were taken to produce the multisensor image according to the following scheme:

1. Computation of the IHS components from the RGB channels 4/5/3.
2. Modification of the I component by adding the result from the SPOT high-pass filter.
3. Computation of the RGB components from the IHS channels Imod/H/S.

The multisensor image resulting from the colour transformation relates geometrically on the original SPOT scene. For an absolute rectification onto the geometry of to the 1:50 000 Topographic Map a transformation with a polynomial of the second degree has been executed which is based on the determination of coefficients by means of 13 control points. The mean error observed at the control points on the multisensor image amounted to 7.18 m at right angles to the flight direction (direction of pixel), and to 3.48 m in flight direction in case of a polynomial of the second degree. The resampling was here also made by a cubic convolution. The printing foils have been produced by a digital rasterization on the SCITEX high precision plotter and have been plotted with a resolution of 30 lines per millimetre.

4. Iconocarte thématique satellitaire: Rheinhessen au 1:50 000

La figure ci-jointe montre une section centrale de la carte thématique issue d'images satellitaires au 1:50 000 Rheinhessen qui a été produite en 1989 par la division "Recherches photogrammétriques" de l'Institut für Angewandte Geodäsie à Frankfurt am Main et qui a été publiée dans le cadre de la série I des *Nachrichten aus dem Karten- und Vermessungswesen*. On voulait démontrer par ce produit cartographique qu'il est possible de préparer en un minimum de temps des cartes thématiques nouvelles par un traitement adéquat des données des images satellitaires actuellement existantes. Par la combinaison des données des images de deux différents capteurs ayant une résolution géométrique différente (30 m pour le LANDSAT Thematic Mapper (TM) et 10 m pour le capteur SPOT High Resolution Visible (HRV)), on obtient une nouvelle image numérique qui englobe les avantages des deux systèmes (haute résolution tridimensionnelle pour SPOT et haute résolution spectrale pour TM) dans une forme optimale. On obtient une image muticolore englobant les couleurs des données TM et la grande résolution géométrique des données panchromatiques de SPOT, c'est-à-dire avec des dimensions de 10 × 10 m par pixel. Il est ainsi possible de créer des iconocartes au 1:50 000.

Pour produire par voie numérique les images multicapteurs à partir des données d'images-satellites, on utilise les méthodes et procédés du traitement d'image numérique. Le procédé central des différentes méthodes utilisées est la transformation de l'espace des couleurs selon l'intensité, la fréquence et la saturation des couleurs, transformation nommée IHS. Les séquences pour produire l'iconocarte multicapteurs à l'aide du système DIPIX ARIES III du traitement des images couvrant la carte topographique de Mainz (L6114) sont décrites comme suit.

Le contenu des images est basé sur de larges extraits de deux prises de vues par satellites. Les données du SPOT 1 HRV en mode panchromatique (0,51–0,73 μm) ont été acquises le 25 septembre 1986 (scène 050–249). Une composition de fausses couleurs des canaux 3 (0,63–0,69 μm), 4 (0,76–0,90 μm) et 5(1,55–1,75 μm) a été choisie parmi les données multispectrales de LANDSAT 5 TM du 17 août 1987 (scène 195–25). Une

reproduction chromatique RVB avec le canal 4 en tant que composante rouge, le canal 5 en tant que composante verte et le canal 3 en tant que composante bleue donne l'impression couleur comme elle se présente sur l'iconocarte multicapteurs. Les couleurs choisies se prêtent particulièrement à la différenciation fine de l'utilisation du sol dans les régions couvertes de végétation.

La rectification relative des données TM sur la géométrie de SPOT a été faite à l'aide de 32 points d'appui déterminés par voie interactive. Les paramètres de transformation d'un polynôme du troisième degré ont été utilisés pour réorganiser les données TM avec rééchantillonage par convolution cubique. Les erreurs moyennes sur points d'appui s'élevaient, avec cette disposition, à 1,44 pixel transversalement à la direction du vol et à 1,05 pixel dans la direction du vol.

Avant de commencer la transformation IHS, la distribution des valeurs de gris de chaque canal TM a été changée individuellement en utilisant l'histogramme correspondant, de manière à obtenir une image couleur avec un maximum de contraste. Les structures linéaires de l'image SPOT également élargie en ce qui concerne les valeurs de gris ont été extraites en utilisant un filtre passe-haut. Il s'ensuivit la production de l'image multicapteurs à partir de ces données d'entrée, selon le schéma suivant:

1. Calcul des composantes IHS à partir des canaux RVB 4/5/3.
2. Modification de la composante I en faisant l'addition avec le résultat du filtre passe-haut de l'image SPOT.
3. Calcul des composantes RVB à partir des canaux IHS Imod/H/S.

L'image multicapteurs résultant de la transformation des couleurs se réfère géométriquement à la scène originale de SPOT. Pour le calage absolu sur la géométrie de la carte topographique au 1:50 000 on a exécuté une transformation avec un polynôme du second degré, se basant sur la définition des coefficients à l'aide de 13 points d'appui. L'erreur moyenne observée sur l'image multicapteurs s'élevait à 7,18 m transversalement à la direction du vol (direction du pixel) et à 3,48 m dans la direction du vol, dans le cas d'un polynôme du second degré. Le

Reference

Pfeiffer, B. 1990. Die Multisensorbildkarte 1:50,000 Rheinhessen. *Nachrichten aus dem* *Karten- und Vermessungswesen*, Reihe I, Heft 104. Frankfurt am Main.

rééchantillonnage s'exécutait également avec une convolution cubique. On a restitué par voie numérique sur le phototraceur de précision SCITEX les images rectifiées, avec une résolution de 30 lignes par millimètre.

Référence

PFEIFFER, B. 1990. Die Multisensorbildkarte 1:50,000 Rheinhessen. *Nachrichten aus dem Karten- und Vermessungswesen*, Reihe I, Heft 104. Frankfurt am Main.

Result from the IHS transformation (multisensor image map HRV1/TM)

Résultat de la transformation IHS (Iconocarte multicapteurs HRV1/TM)

LAND USE/LAND COVER MAPPING

CARTOGRAPHIE DE L'OCCUPATION/UTILISATION DU SOL

Application Title
Land use/Cover maps of the Argungu Area, Northwest Nigeria, based on Landsat MSS data

Titre de l'application
Cartes de l'occupation/utilisation du sol de la région d'Argungu, Nord-Ouest du Nigeria, basées sur des données Landsat MSS

Localization/Localisation
Nigeria (Argungu)/Nigeria (Argungu)

RESPONSIBLE ORGANIZATIONS

Department of Geography
University of Waterloo
Isaiah Bowman Building
Waterloo, Ontario, Canada
N2L 3G1

Tel: 1/519 885 1211   Fax: 1/519 746 2031

ORGANISMES RESPONSABLES

Department of Geography
University of Lagos
Lagos
Nigeria

(Contact person: Prof. Andrzej Kesik)

5. Land use/Cover maps of the Argungu Area, Northwest Nigeria, based on Landsat MSS data

Land use/cover maps of the Argungu area are taken from an atlas I representing partial results of a joint research project of the Departments of Geography at the Universities of Waterloo, Canada, and Lagos, Nigeria. This project was designed to explore and demonstrate to Nigerian authorities the potential utility of Landsat MSS imagery for the extraction of information on agricultural resources for areas in which it is otherwise unavailable for use in development planning.

Black and white maps present the results of visual interpretation of the data and are based upon 1977 aerial photography and Landsat imagery acquired in December 1984, January 1986 and August 1986 (Figs 5.5.1, 5.5.2). Interpretation of the imagery was based upon enhancements produced at the University of Waterloo. The maps were compiled in the Department of Geography, University of Lagos, Nigeria, at a scale of 1:100 000, using a PROCOM-2 optical plotter. Results were checked during field work. The following major classes of land use/land cover are identified: Settlements, Water Bodies, Wetland, Agricultural Land, Shrubland/Grassland, Wooded Shrubland and Thicket, Grassland, Bare and Rocky Surfaces, Burnt Areas.

Colour maps present the results of digital interpretation of only the imagery acquired for December 1984 and August 1986 (Figs 5.b, 5.a). Digital analysis of MSS data was undertaken at the Department of Geography, University of Waterloo, Canada, using an ARIES-3 image analysis system. Colour printing was undertaken at the Ontario Centre for Remote Sensing, Toronto. Procedures were as follows:

1. Unsupervised classification of the Argungu sub-scene.
2. Field work in proposed training areas.
3. Digital analysis of training areas.
4. Supervised classification.
5. Post classification filtering.
6. Image registration to the Nigerian topographic map 1:100 000, using polynomial transformation equations.
7. Map reproduction: at the scale 1:100 000 by Applicon plotter, and at the scale 1:250 000 using a Versatec plotter.
8. Preparation for the atlas, using colour photocopy.

Colour land use/cover maps distinguish the following classes:

1. Grass/shrub
2. Fallow/upland
3. Bare surface
4. Settlements
5. Water bodies
6. Vegetated wetland
7. Non-vegetated wetland
8. Cultivated wetland
9. Standing crop
10. Cloud and Cloud shadow

Comparison of information derived from the visual and digital analysis of Landsat data (Table 5.5.1) suggests that each method has advantages for specific purposes.

Visual analysis was able in some cases to identify classes that have a high visual

Table 5.5.1. Comparison of Classification Results for Digital and Visual Analysis of Argungu Sub-area, 1984

Class	Dry season		Wet season	
	Hectares digital	Hectares visual	Hectares digital	Hectares visual
Grass/shrub	134 962	127 486	117 616	131 152
Fallow/upland	110 592	146 889	28 193	0
Bare surfaces	19 731	6 930	25 870	1 717
Settlements	869	2 025	869	1 551
Water bodies	0	305	986	2 382
Vegetated wetland	12 961	13 913	12 406	13 662
Non-vegetated wetland	0	237	264	0
Cultivated wetland	1 836	1 144	2 303	2 337
Standing crop	21 437	0	110 198	135 629
Cloud and cloud shadow	0	0	5 711	10 961

5. Cartes de l'occupation/utilisation du sol de la région d'Argungu, Nord-Ouest du Nigéria, basées sur des données Landsat MSS

Les cartes d'occupation/utilisation du sol de la région d'Argungu sont extraites d'un Atlas présentant les résultats partiels d'un projet commun de recherche des Départements de Géographie des Universités de Waterloo (Canada) et de Lagos (Nigéria). Ce projet était conçu pour explorer et démontrer, à l'intention des autorités nigérianes, l'utilité potentielle de l'imagerie Landsat MSS pour obtenir de l'information sur les ressources agricoles dans des zones où celle-ci n'est pas disponible, en vue du développement.

Les cartes monochromes présentent les résultats d'une interprétation visuelle des données, basés sur des photographies aériennes de 1977 et des images Landsat acquises en décembre 1984, janvier 1986 et août 1986 (figs 5.5.1, 5.5.2). L'interprétation des images a été réalisée à partir d'accentuations produites à l'Université de Waterloo. Les cartes ont été établies au Département de Géographie de l'Université de Lagos, à l'échelle de 1:100 000, à l'aide d'un phototraceur PROCOM-2. Les résultats furent vérifiés lors d'un contrôle terrain. Les classes principales suivantes d'occupation/utilisation du sol ont été identifiées: bâti, surfaces d'eau, zones humides, zones cultivées, prairies/broussailles, broussailles boisées et bosquet, prairies, sols nus et rocheux, zones brûlées.

Les cartes en couleur présentent les résultats de l'interprétation numérique des images acquises en décembre 1984 et août 1986 (figs 5.b, 5.a). L'analyse numérique des données MSS fut réalisée au Département de Géographie de l'Université de Waterloo, à l'aide d'un système d'analyse d'image ARIES-3. La restitution couleur fut réalisée au Centre de Télédétection de l'Ontario, à Toronto. Les procédures furent les suivantes:

1. Classification non supervisée de la sous-scène d'Argungu.
2. Levé terrain des zones d'entraînement proposées.
3. Analyse numérique des zones d'entraînement proposées.
4. Classification supervisée.
5. Filtrage post-classification.
6. Calage de l'image sur la carte topographique du Nigéria à 1:100 000, à l'aide d'équations de transformation polynômiale.
7. Restitution de la carte à l'échelle de 1:100 000, avec un traceur Applicon, et à 1:250 000 avec un traceur Versatee.
8. Préparation pour l'atlas, à l'aide de photocopie couleur.

Les cartes couleurs d'occupation/utilisation du sol distinguent les classes suivantes:

1. Herbe/buissons
2. Friches et hautes terres
3. Surface nue
4. Bâti
5. Surfaces d'eau
6. Végétation
7. zone sans végétation

Tableau 5.5.1. Comparaison des résultats de classification pour l'analyse numérique et l'analyse visuelle de la sous-zone d'Argungu, 1984

Classification	Saison sèche		Saison humide	
	Hectares numérique	Hectares visuelle	Hectares numérique	Hectares visuelle
Herbe/buisson	134 962	127 486	117 616	131 152
Friches et hautes terres	110 592	146 889	28 193	0
Surfaces nues	19 731	6 930	25 870	1 717
Bâti	869	2 025	869	1 551
Surfaces d'eau	0	305	986	2 382
Végétation	12 961	13 913	12 406	13 662
Zone sans végétation	0	237	264	0
Zone humide cultivée	1 836	1 144	2 303	2 337
Récolte sur pied	21 437	0	110 198	135 629
Nuage et ombre de nuage	0	0	5 711	10 961

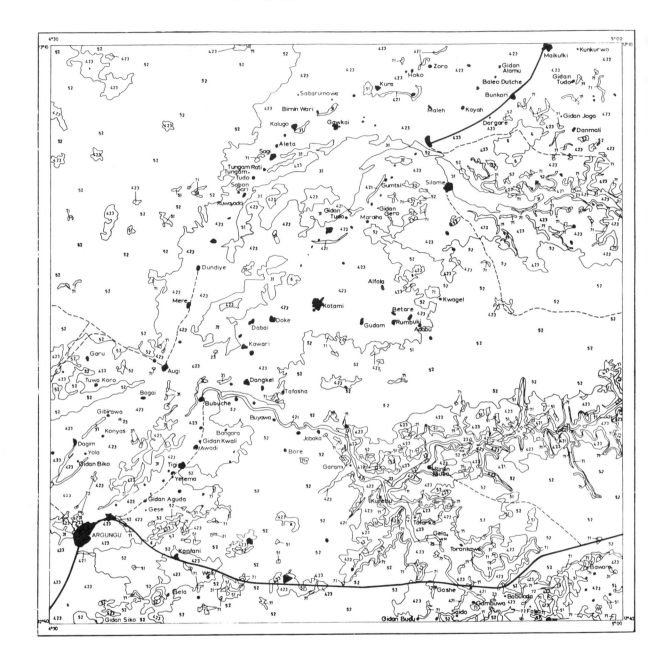

LAND-USE\COVER MAP OF ARGUNGU
(DRY SEASON 1984)

		HECTARES
■	SETTLEMENTS	2,025
2	WATER BODIES	305
21	Major Rivers	
22	Reservoirs	
23	Lakes/Ponds	305
3	WET LAND	13,913
31	Vegetated Wetland	13,913
32	Unvegetated Wetland	0
4	AGRICULTURAL LAND	148,725
41	Irrigated	
410	Rice	
411	Tree Crops	
412	Mixed Cropland	
413	Fallow/Just Harvested/Newly Planted	
42	Non Irrigated	148,725
420	Rice	
421	Fruit Trees/Sugar Cane	1,836
422	Mixed Cropland	
423	Fallow/Just Harvested/Newly Planted	146,889
5	SHRUBLAND/GRASSLAND	126,373
51	Grassland/Shrub	1,976
52	Shrubland/Grassland	124,397
53	Floodable Shrubland/Grass	0

		HECTARES
6	WOODED SHRUBLAND AND THICKET	1,113
7	BARE AND ROCKY SURFACES	6,930
71	Rocky/Stony/Lateritic	6,930
72	Sandy	0
8	BURNT AREAS	0
9	CLOUD COVER AND CLOUD SHADOW	0

Road on Base Apparent on Imagery
Road not on Base but Apparent on Imagery (New roads)
Road on Base but not Identifiable on Imagery
Forest Reserve
Dam Wall

Funded by the International Development Research Centre, Ottawa, Canada

Based on Visual Analysis of Landsat (FCC)
Data of Dec. 18, 1984. Compiled and Drafted under the direction of Dr. Peter
O Adeniyi at the Department of Geography, University of Lagos, Lagos, Nigeria
Reproduced at the Cartographic Unit, Faculty of Environmental Studies, University of
Waterloo, Waterloo, Ontario, Canada

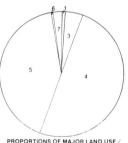

PROPORTIONS OF MAJOR LAND USE /
COVER CATEGORIES

INDEX MAP SHOWING THE PROJECT SUB-AREAS

Fig. 5.5.1.

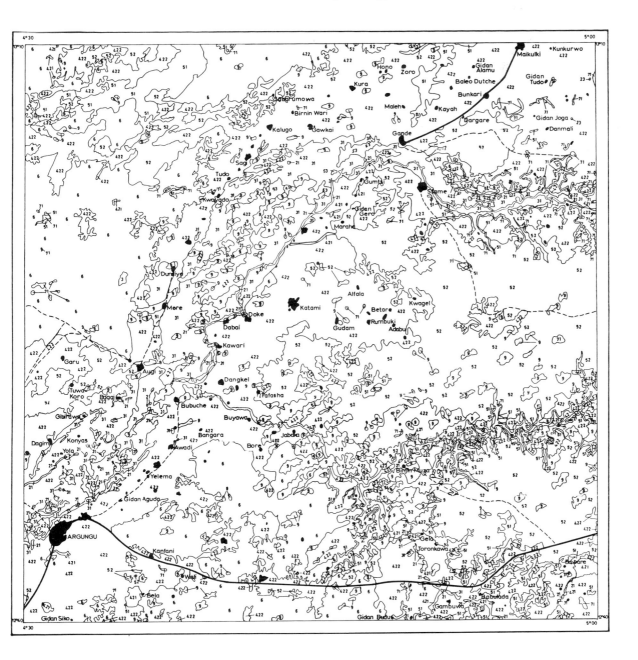

LAND-USE\COVER MAP OF ARGUNGU
(WET SEASON 1986)

		HECTARES
■	SETTLEMENTS	1,551
2	WATER BODIES	2,382
21	Major Rivers	2,379
22	Reservoirs	
23	Lakes/Ponds	3
3	WET LAND	13,662
31	Vegetated Wetland	13,662
32	Unvegetated Wetland	
4	AGRICULTURAL LAND	137,966
41	Irrigated	
410	Rice	
411	Tree Crops	
412	Mixed Cropland	
413	Fallow/Just Harvested/Newly Planted	
42	Non Irrigated	137,966
420	Rice	
421	Fruit Trees/Sugar Cane	2,337
422	Mixed Cropland	135,629
423	Fallow/Just Harvested/Newly Planted	
5	SHRUBLAND/GRASSLAND	77,876
51	Grassland/Shrub	9,914
52	Shrubland/Grassland	67,172
53	Floodable Shrubland/Grass	790

		HECTARES
6	WOODED SHRUBLAND AND THICKET	53,276
7	BARE AND ROCKY SURFACES	1,717
71	Rocky/Stony/Lateritic	1,717
72	Sandy	
8	BURNT AREAS	0
9	CLOUD COVER AND CLOUD SHADOW	10,961

▬▬	Road on Base Apparent on Imagery
────	Road not on Base but Apparent on Imagery (New roads)
- - - -	Road on Base but not Identifiable on Imagery
— — —	Forest Reserve
┼┼┼┼	Dam Wall

Funded by the International Development Research Centre, Ottawa, Canada

Based on Visual Analysis of Landsat (FCC)
Data of July 17, 1986. Compiled and Drafted under the direction of Dr. Peter
O Adeniyi at the Department of Geography, University of Lagos, Lagos, Nigeria.
Reproduced at the Cartographic Unit, Faculty of Environmental Studies, University of
Waterloo, Waterloo, Ontario, Canada

PROPORTIONS OF MAJOR LAND USE /
COVER CATEGORIES

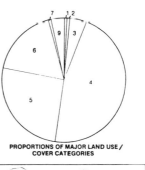

RIVER BASINS IN
NIGERIA

I. Sokoto - Rima
II. Hadejia - Jama'are
III. Lake Chad
IV. Upper Benue
V. Lower Benue
VI. Cross River
VII. Anambra - Imo
VIII. Niger
IX. Ogun - Oshun
X. Benin - Owena
XI. Niger Delta

INDEX MAP SHOWING THE PROJECT SUB-AREAS

Fig. 5.5.2.

discrimination but low spectral discrimination (e.g. settlements). Conversely, digital analysis was better for discrimination between areas which had been recently harvested, were newly planted or were lying fallow.

Both types of maps should be treated as complementary and necessary to the successful interpretation of land use and land cover.

Reference

ADENIYI, P. O. and BULLOCK, R. A. (Eds.). 1988. *Seasonal Land Use and Land Cover in Northwest Nigeria: An Atlas of the Central Sokoto-Rima Basin.* Waterloo: Department of Geography, University of Waterloo, Occasional Paper No. 8

8. Zone humide cultivée
9. Récolte sur pied
10. Nuage et ombre de nuage

La comparaison des informations dérivées des analyses visuelles et numériques des données Landsat (tableau 5.5.1) suggère que chaque méthode a ses avantages pour des buts spécifiques.

L'analyse visuelle a permis, dans certains cas, d'identifier des classes ayant une forte discrimination visuelle, mais une faible discrimination spectrale (par ex., le bâti). Inversement, l'analyse numérique discriminait mieux les zones récemment moissonnées, ou nouvellement plantées, ou laissées en friche.

Les deux types de cartes devraient être traités comme complémentaires et nécessaires à une interprétation satisfaisante de l'occupation et de l'utilisation du sol.

Reference

ADENIYI, P. O. and BULLOCK, R. A. (Eds.). 1988. *Seasonal Land Use and Land Cover in Northwest Nigeria: An Atlas of the Central Sokoto-Rima Basin.* Waterloo: Department of Geography, University of Waterloo, Occasional Paper No. 8

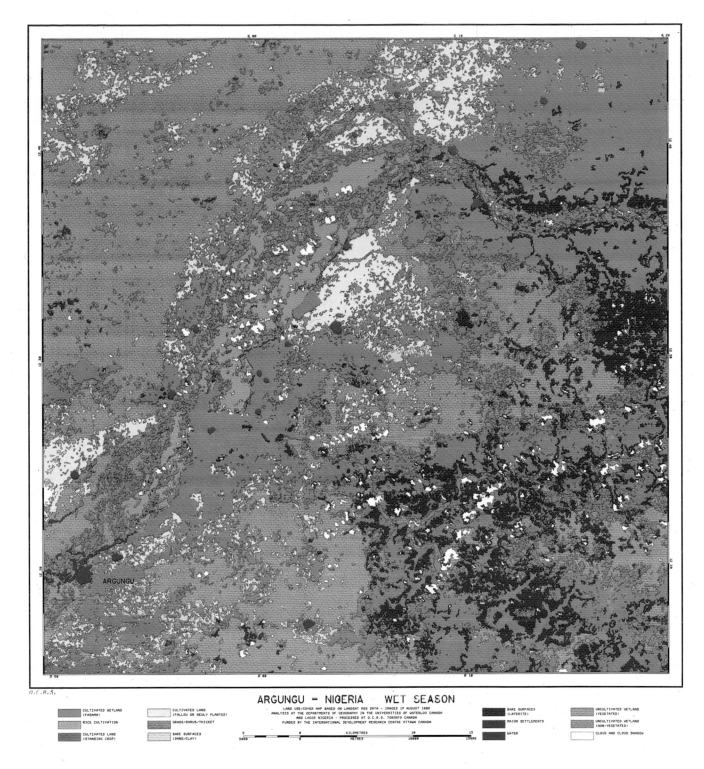

ARGUNGU — NIGERIA   WET SEASON

LAND USE/COVER MAP BASED ON LANDSAT MSS DATA - IMAGED 17 AUGUST 1986
ANALYSIS AT THE DEPARTMENTS OF GEOGRAPHY IN THE UNIVERSITIES OF WATERLOO CANADA
AND LAGOS NIGERIA - PROCESSED AT O.C.R.S. TORONTO CANADA
FUNDED BY THE INTERNATIONAL DEVELOPMENT RESEARCH CENTRE OTTAWA CANADA

CULTIVATED WETLAND (FADAMA)

RICE CULTIVATION

CULTIVATED LAND (STANDING CROP)

CULTIVATED LAND (FALLOW OR NEWLY PLANTED)

GRASS/SHRUB/THICKET

BARE SURFACES (SAND/CLAY)

BARE SURFACES (LATERITE)

MAJOR SETTLEMENTS

WATER

UNCULTIVATED WETLAND (VEGETATED)

UNCULTIVATED WETLAND (NON-VEGETATED)

CLOUD AND CLOUD SHADOW

Argungu - Nigeria - saison humide

Courtesy of the department of geography University of Waterloo, Ontario - Canada
Department of geography University of Lagos - Nigeria and Ontario Centre for Remote Sensing

Don gracieux du département de géographie de l'université de Waterloo (Ontario, Canada),
du département de géographie de l'université de Lagos - Nigeria et du Centre de télédétection
de l'Ontario

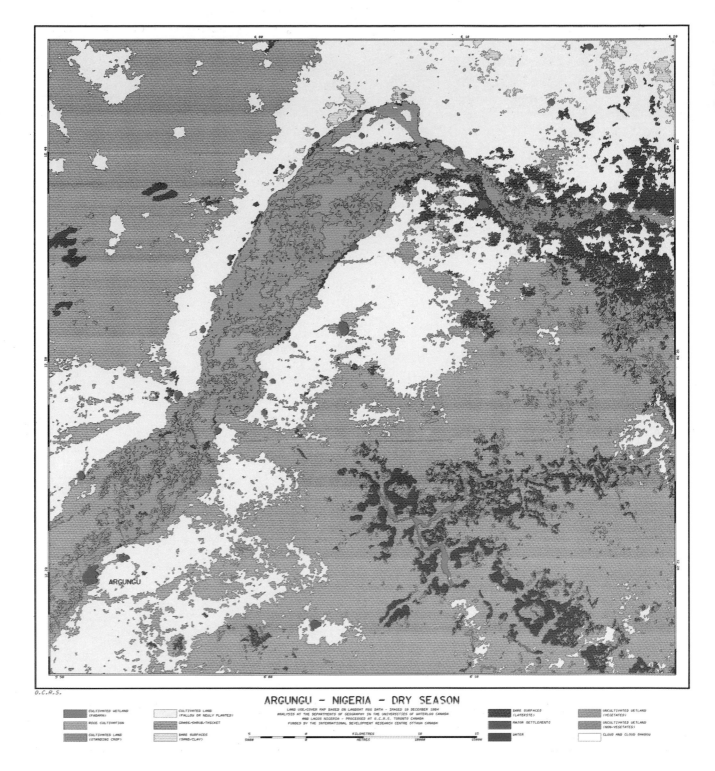

Argungu - Nigeria - saison sèche

*Courtesy of the department of geography University of Waterloo, Ontario - Canada
Department of geography University of Lagos - Nigeria and Ontario Centre for Remote Sensing*

Don gracieux du département de géographie de l'université de Waterloo (Ontario, Canada),
du département de géographie de l'université de Lagos - Nigeria et du Centre de télédétection
de l'Ontario

Application Title
Polygonal map compilation using remote sensing digital image processing techniques

Titre de l'application
Etablissement de cartes polygonales à l'aide de techniques de traitement d'images de télédétection

Localization/Localisation
Brazil (Formosa)/Brésil (Formosa)

RESPONSIBLE ORGANIZATIONS   ORGANISMES RESPONSABLES

Ministério da Aeronautica
Instituto de Cartografia Aeronautica
Praça Senador Salgado Filho, S/N
Aeroporto Santos Dumont
20021 Rio de Janeiro
Brazil

(Contact Person: Captain Cartograpic Engineer Carlos Alberto Gonçalves de Araujo)

6. Polygonal map compilation using remote sensing digital image processing techniques

1. Objectives

The research includes a methodology for the generation of polygonal maps from Landsat-TM data using computer graphic techniques. The polygonal maps represent different classes which can be used to update thematic maps. The proposed methodology provides a data file which can be used for the actualization of any type of thematic map, medium and small scale.

2. Methods used

See flow chart (Fig. 5.6.1).

3. Results

Type of documents
—polygonal maps,
—thematic maps.
Advantage obtained
—Utilization of satellite data for updating thematic maps.
Disadvantage
—The methodology can only be used for mapping or updating thematic maps at medium and small scales.

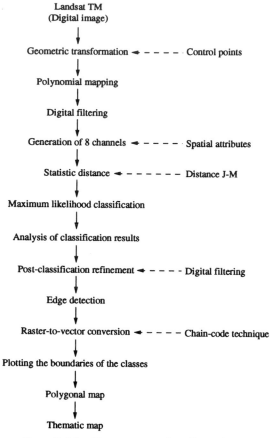

FIG. 5.6.1. Data processing flow chart.

6. Etablissement de cartes polygonales a l'aide de techniques de traitement d'images de télédétection

1. Objectifs

La recherche comporte une méthodologie pour créer des cartes polygonales issues de données Landsat-TM et mettant en oeuvre des techniques de dessin automatique. Les cartes polygonales doivent représenter une variété de classes qui peuvent être utilisées pour mettre à jour les cartes thématiques. La méthodologie proposée fournit un fichier de données qui peut être utilisé pour actualiser n'importe quel type de carte thématique à moyenne et petite échelle.

2. Méthode utilisée

Voir l'organigramme du traitement (fig. 5.6.1).

3. Résultats

Types de documents
—cartes polygonales,
—cartes thématiques.
Avantages obtenus
—Utilisation de données satellitaires pour la mise à jour des cartes thématiques.
Désavantages
—La méthode ne peut être utilisée que pour cartographier ou mettre à jour des cartes à moyenne et petite échelle.

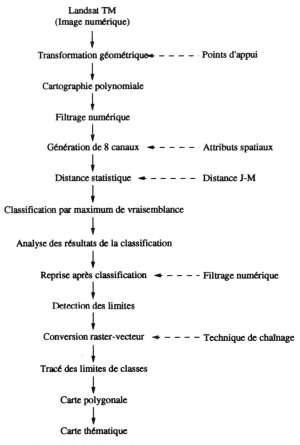

FIG. 5.6.1. Organigramme du traitement.

FORMOSA

CARTA TEMÁTICA
THEMATIC CHART
CARTE THÉMATIQUE

MINISTÉRIO DA AERONÁUTICA
BRASIL

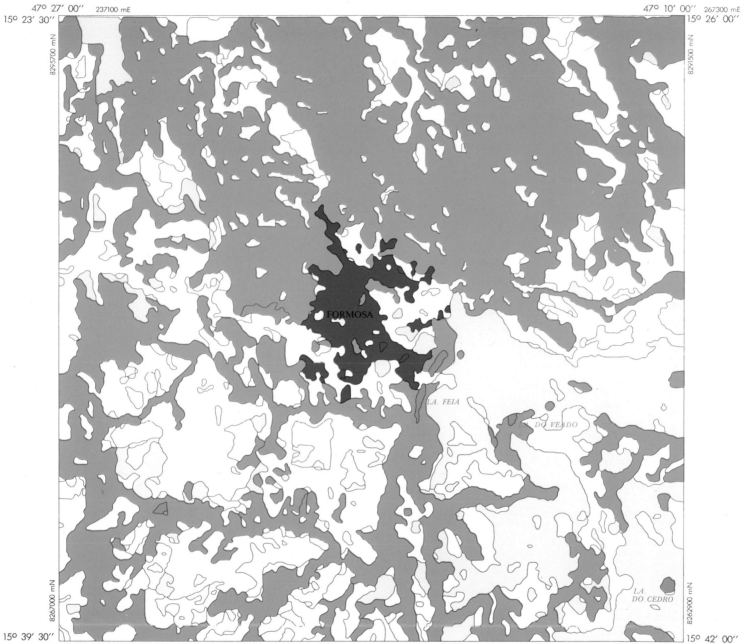

ESCALA / *SCALE* / *ÉCHELLE* 1:250.000

PROJEÇÃO UNIVERSAL TRANSVERSA DE MERCATOR
UNIVERSAL TRANSVERSE MERCATOR PROJECTION
PROJECTION TRANVERSE DE MERCATOR

1987

NG NQ

9° 30'

NORTE GEOGRÁFICO
TRUE NORTH
NORD GEOGRAPHIQUE

NORTE QUADRÍCULA
GRID NORTH
NORD DE LA GRILLE

LEGENDA
LEGEND / LEGENDE

FLORESTA / *FOREST*
FORÊT

CERRADO / *CERRADO*
SAVANE

SUPERFÍCIE LÍQUIDA / *BODY OF WATER*
SURFACE EN EAU

ÁREA URBANA / *URBAN AREA*
ZONE URBAINE

ÁREAS NÃO CLASSIFICADAS / *UNCLASSIFIED AREAS*
ZONES SANS CLASSIFICATION

Elaborado a partir de dados do
**LANDSAT - TM DISSERTAÇÃO
DE MESTRADO**
CAP. ENG. CARLOS ALBERTO GONÇALVES DE ARAÚJO

*Prepared from LANDSAT - TM data
imagery   MASTER DISSERTATION*

*Produit à partir de données
LANDSAT - TM   THÈSE DE MÉTRICE*

Application Title
Land cover mapping in Spain

Titre de l'application
Cartes de l'occupation du sol en Espagne

Localization/Localisation
Spain (Corunna)/Espagne (La Corogne)

RESPONSIBLE ORGANIZATIONS   ORGANISMES RESPONSABLES

Instituto Geografico Nacional
General Ibanez de Ibero, 3
28003 Madrid, Spain

Tel: 34/1 533 38 00   Fax: 34/1 254 67 43

Compiled by
José M. Lopez-Vizoso
Instituto de Economia y Geografia Aplicadas del
Consejo Superior de Investigaciones Cientificas,
Pinar, 25
28006 Madrid, Spain

Tel: 34/1 411 24 63   Fax: 34/1 411 30 77

7. Land cover mapping in Spain

1. Objectives

Land use/cover inventories are just one facet among many others which put together the territorial information set that a country must collect. Consequently, a good land cover inventory, its evolution and cartographic representation are always basic building blocks of such a set.

There have been many pilot projects in Spain to survey land cover, but only a few had led to cartographic outputs. The point has been the lack of a National Land Cover Programme. Only the 1:50 000 crops and vegetation map series, maintained by the Ministry of Agriculture, is worth a mention as a valuable but incomplete cartographic effort.

In January 1988, the Instituto Geografico Nacional (IGN) prepared a report about the need to complete a land use/cover map of Spain in 1:100 000 scale by using visual analysis techniques on satellite imagery. The Programme was meant to be the Spanish chapter of the CORINE Land Cover Project.

Regardless of CORINE's methodology, we experimented in a specific site in the north-west of Spain (Galicia) automatic classification procedures with a cartographic target, and an automatic method to prepare the printing colour separates (1:100 000 scale).

2. Methods used

Our study area, a 1:100 000 UTM sheet in the Spanish Topographic Map Series, is covered by the first quarter of the 204-30 TM scene that was taken on during 9 July 1985.

A subscene of 2048×2048 pixels (3775 km^2, approximately) was selected, including a large littoral area: "As Marinas" county, together with an inland mountainous area. The geographic differences between these two zones are important. These physical aspects must be taken into account because of their influence on spectral signatures of the land cover classes.

The two methods employed for image enhancement were linear stretch and filtering. The last method was designed to reinforce visual boundaries between spectral classes (edge enhancement).

Several colour composites were employed to select the most suitable training areas. Associ-

ARTIFICIAL SURFACES

Urban fabric
Mineral extraction sites

AGRICULTURAL AREAS

Complex cultivation patterns
Complex cultivation patterns,
with significant areas

FOREST AREAS

Coniferous forest
Broad leaved forest
Mixed forest

SHRUB AND/OR HERBACEOUS
VEGETATION ASSOCIATION

Transitional woodland-shrub
Moors and heathland
Burnt areas

OPEN SPACES WITH LITTLE
OR NO VEGETATION

Beaches, dunes, sands
Bare rocks

WETLANDS

Marshes and intertidal flats

WATER BODIES

Inland and maritime waters

OTHERS

Clouds and unclassified areas

FIG. 5.7.1.

ation species, density, importance of undergrowth, lithology and slopes were taken into account. This approach led us to consider the image spectral variation. Some statistical parameters from the training fields were extracted to characterize each spectral class.

We employed the maximum likelihood procedure to classify the image. This method evaluates both the variance and correlation of the spectral response patterns of the categories when classifying an unknown pixel.

We obtained a classified image within 53 spectral classes; four of them were different types of water and wetlands. Thereafter we grouped the spectral classes into land cover thematic classes. This was done using a land use/cover classification system: the CORINE land cover nomenclature.

In order to present the results it was necessary to correct the classified file from random distortions by analysing ground control

7. Carte de l'occupation du sol en Espagne

1. Objectifs

L'inventaire de l'occupation et de l'utilisation du sol est un aspect parmi beaucoup d'autres qui regroupent l'ensemble des informations territoriales qu'un pays doit recueillir. C'est ainsi qu'un bon inventaire de l'occupation du sol, son évolution et sa représentation cartographique sont toujours des éléments de base d'un tel ensemble.

Il y a eu beaucoup de projets pilotes en Espagne pour décrire l'utilisation du sol, mais très peu ont débouché sur des sorties cartographiques. Le motif a été le manque d'un programme national d'utilisation du sol. Seule une carte au 1:50 000 des récoltes et de la végétation, maintenue par le ministère de l'Agriculture, peut être mentionnée comme un effort cartographique valable mais incomplet.

En janvier 1988, l'Institut Géographique National (IGN) prépara un rapport sur la nécessité de réaliser une carte d'Espagne de l'occupation et de l'utilisation du sol, à l'échelle du 1:100 000, en utilisant les techniques d'analyse visuelle de l'imagerie spatiale. Le programme devait être la participation espagnole au projet CORINE Land Cover.

Indépendamment de la méthodologie de CORINE, nous avons expérimenté, dans un site spécifique du Nord-Ouest de l'Espagne (en Galice) des procédures de classification automatique avec un but cartographique, et une méthode automatique pour sortir les planches de couleurs séparées à l'échelle du 1:100 000.

2. Méthodes utilisées

Notre zone d'étude, une feuille UTM au 1:100 000 de la série des cartes topographiques espagnoles, est couverte par le premier quart d'une scène Landsat TM (204-30) prise le 9 juillet 1985.

Un extrait de 2048×2048 pixels (environ 3775 km^2) fut choisi, incluant une large zone littorale "As Marinas", contiguë à une zone intérieure montagneuse. Les différences géographiques entre ces deux zones sont importantes. Ces aspects physiques doivent être pris en compte par leur influence dans les signatures spectrales des classes d'utilisation du sol.

SURFACES ARTIFICIELLES

Zones urbaines, industrielles
Carrières, mines

SURFACES AGRICOLES

Formes de culture complexes
Formes de culture complexes dont
une part significative de pâturages

SURFACES FORESTIERES

Forêt de conifères
Forêt de feuillues
Forêt de mixte

**BROUSSAILLES ET/OU ASSOCIATION DE
VEGETATIONS HERBACEES**

Transition broussailles/bois
Landes et bruyères
Surfaces brûlées

**ESPACES OUVERTS AVEC PEU OU
PAS DE VEGETATION**

Plages, dunes, sables
Roches nues

ZONES HUMIDES

Marais, estran

SURFACES D'EAU

Eaux maritimes et continentales

AUTRES

Nuages et zones non classées

FIG. 5.7.1.

Les deux méthodes employées pour l'accentuation de l'image étaient l'étalement linéaire et le filtrage. La dernière méthode devait renforcer les frontières visuelles entre les classes spectrales (accentuation des limites).

Plusieurs compositions colorées furent employées pour sélectionner les zones d'entraînement les plus adéquates. Nous prîmes en compte des associations d'espèces, de densité, l'importance du sous-bois, la lithologie et les pentes. Cette approche nous conduit à considérer la variation de l'image spectrale. Des paramètres statistiques des zones d'entraînement furent extraits pour caractériser chaque classe spectrale.

La procédure du maximum de vraisemblance a été utilisée pour classer l'image. Cette méthode évalue tant la variance que la corrélation des types de réponse spectrale des catégories pour classer les pixels inconnus.

Nous avons retenu une répartition en 53

points. During this process we resampled the pixel size from original 30×30 m to 20×20 m to eliminate jags in the cartographic output. Many of the isolated pixels were eliminated by using a majority filter, resulting in a final smoothed map.

Another important issue was the selection of the map background. We selected a simplified version of main roads and urban structure. The same criteria were applied on the other subjects we considered (hydrology, place-names, contour lines, UTM grid and some plots of coast-line).

The digital classified file was transformed into three new files, according to the three colours employed in printing, which were taken to obtain directly the printing colour separates, using a very high resolution digital integrated film recorder: SCITEX-Response 350. The final output was drawn by a laser plotter RAYSTAR (60 lines per centimeter resolution). So we got the different separates before litography printing.

3. Results

In order to assess the accuracy of the map, some test areas were selected using systematic stratified non-aligned sampling.

The class which scored most mistakes was mixed forest. Some other forest areas and urban areas had a high confusion ratio as well. The correctness for the overall map was 87.1 percent.

In our digital classification of a map sheet from the north-west of Spain $1:100\,000$, a complete system from digital tapes to map was developed, trying to explain different steps in an operational way.

For the moment we can conclude that digital classification is more accurate in tracing the limits between classes. We can also obtain definitive cartography and it is possible to store directly the data in raster mode. But we need good knowledge of the study area. It is quite difficult to develop a common strategy of this type for large countries, like Spain, covered by many Landsat scenes.

Nowadays the number of classes that we can identify is very limited (no more than 10 or 15, normally). Perhaps, in the wide context of a country, we could point out up to 20 classes. We can assert that, in most cases, it is impossible to reach more than the third Corine nomenclature level using digital classification.

A land cover map using remote sensing and digital classification schemes has been completed, but the aim of the Corine Land Cover Project, managed in Spain by the IGN, is to develop a land cover multipurpose database. Finally, we must emphasize the need to update the information every five or ten years, depending on the dynamics of the area.

classes spectrales; quatre d'entre elles étaient des variétés d'eau et de zones humides. Ensuite on a regroupé les classes spectrales dans des classes thématiques d'occupation du sol. Ceci fut fait en utilisant un système de classification de l'occupation/utilisation du sol, la nomenclature CORINE.

Dans le but de présenter les résultats, il était nécessaire de corriger le fichier classé des distortions aléatoires en analysant des points de contrôle au sol. Durant cette procédure nous avons rééchantillonné la taille des pixels de 30×30 m à 20×20 m pour éliminer les dentures sur le document cartographique. Beaucoup de pixels isolés furent éliminés en utilisant un filtre majoritaire, conduisant à un lissage final.

Une autre sortie importante est la sélection du fond de carte. Nous avons choisi une version simplifiée des routes principales et des structures urbaines. On a appliqué les mêmes critères sur les autres objets considérés (hydrologie, toponymie, courbes de niveau, grille UTM et traits de côte).

Le fichier numérique classé a été transformé en trois nouveaux fichiers, suivant les trois couleurs d'impression, et qui donnent directement les trois films de couleur séparée, en utilisant un enregistreur sur film à très haute résolution: le SCITEX-Response 350. La sortie finale fut réalisée par un traceur à laser RAYSTAR (résolution de 60 lignes au centimètre). Ainsi furent obtenues les planches séparées avant impression offset.

3. Résultats

Dans le but d'estimer la précision de la carte, quelques zones-tests furent choisies en utilisant un échantillonnage systématique, stratifié et non aligné.

La classe qui comptait le plus d'erreurs fut celle de la forêt mixte. D'autres zones forestières et urbaines avaient également un fort pourcentage de confusion. L'exactitude pour l'ensemble de la carte fut de 87,1%.

Dans notre classification numérique sur une carte du Nord-Ouest de l'Espagne au $1:100\,000$, un système complet de bandes à carte fut élaboré tâchant de définir les étapes nécessaires à un processus opérationnel.

Pour le moment nous pouvons conclure que la classification numérique est plus précise pour tracer les limites entre classes. Nous pouvons également obtenir une cartographie définitive et il est possible de garder les données en mode maillé. Mais nous avons besoin d'une bonne connaissance de la zone étudiée. Il est très difficile de développer une stratégie commune de ce type pour de vastes pays comme l'Espagne, couverts par de nombreuses scènes Landsat.

Malgré tout le nombre de classes qui peuvent être identifiées reste très limité (pas plus de 10 à 15, normalement). Peut-être dans le vaste contexte d'un pays pouvons-nous atteindre 20 classes. Nous pouvons affirmer que, dans la majorité des cas, il est impossible d'atteindre mieux que le 3e niveau de la nomenclature CORINE, en utilisant une classification numérique.

Une carte d'occupation du sol utilisant des processus de télédétection et de classification numérique a été achevée, mais le but du projet CORINE Land Cover, piloté en Espagne par l'IGN, est de développer une base de données d'occupation du sol multi-applications. Enfin, nous devons souligner la nécessité de mettre à jour les informations tous les cinq ou dix ans, suivant le dynamisme de la région.

MOPU
INSTITUTO GEOGRÁFICO NACIONAL

A CORUÑA

MAPA DE OCUPACIÓN DEL SUELO

LEYENDA

SUPERFICIES EDIFICADAS E INFRAESTRUCTURAS
- Zonas urbanizadas
- Canteras y minas a cielo abierto (*)

SUPERFICES AGRICOLAS
- Mosaico de cultivos anuales con praderas y/o pastizales
- Mosaico de cultivos anuales con praderas y/o pastizales (predominio de praderías)

SUPERFICES ARBOLADAS
- Coníferas
- Frondosas
- Poblaciones mixtas

SUPERFICES DE VEGETACIÓN ARBUSTIVA Y/O HERBÁCEA
- Matorral boscoso de transición
- Landas y matorrales
- Zonas quemadas

SUPERFICES DESNUDAS O CON POCA VEGETACIÓN
- Playas, dunas arenales y zonas de vegetación psamófila (*)
- Roquedo y suelo destruido

ZONAS HÚMEDAS
- Marismas, áreas de inundación periódica y zonas de vegetación halófila

SUPERFICES DE AGUA
- Aguas marinas y continentales

OTROS
- Nubes y zonas sin clasificar

(*) Elaborado mediante técnicas de fotointerpretación convencional

DESCRIPCIÓN DEL PROYECTO

INFORMACIÓN TEMÁTICA.
Satélite Landsat 5, Sensor TM. Imagen utilizada 204-33 (1)
Fecha 9-7-1985. Tamaño del píxel 20 metros.
TRATAMIENTO DE LA IMAGEN.
Clasificación automática por el método de máxima probabilidad
Corrección geométrica. Filtro de suavizado.
SISTEMA DE CLASIFICACIÓN.
Adaptación del sistema propuesto para el Mapa de Ocupación
del Suelo de la Comunidad Europea, Proyecto CORINE (Coordination, information, Environment)
POSITIVOS BASE TEMÁTICA.
Obtenidos directamente para la Incoma en el plotter Raystar,
sistema Response 350, SCITEX
BASE TOPOGRÁFICA.
Generalización del Mapa Topográfico Nacional 1:50.000 (I.G.N.)

DIVISIÓN ADMINISTRATIVA

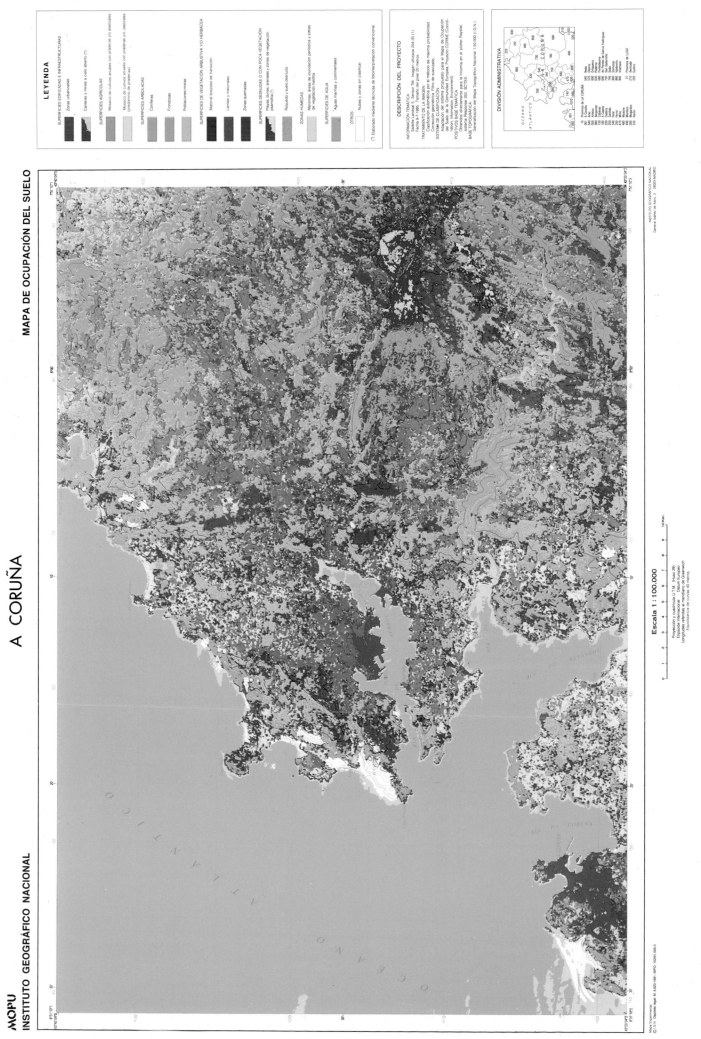

Escala 1:100.000

Proyección y cuadrícula U.T.M. (Huso 29)
Elipsoide Internacional   Datum Europeo
Longitudes referidas al meridiano de Greenwich
Equidistancia de curvas 40 metros

INSTITUTO GEOGRÁFICO NACIONAL
General Ibáñez de Íbero, 3   28003 MADRID

Application Title
Land use and cover map 1 : 250 000 of Catalonia (32 000 km^2) obtained by digital treatment of multitemporal TM data

Titre de l'application
Carte au 1 : 250 000 de l'occupation et de l'utilisation du sol de Catalogne (32 000 km^2) obtenue par traitement numérique de données TM multidates

Localization/Localisation
Spain (Catalonia)/Espagne (Catalogne)

RESPONSIBLE ORGANIZATIONS   ORGANISMES RESPONSABLES

Institut Cartografic de Catalunya
Balmes 209-211
08006 Barcelona
Spain

Tel: 34/93 218 87 58   Fax: 34/93 218 89 59

(Contact person: Sr Xavier Baulies Bochaca)

8. Land use and cover map 1:250 000 of Catalonia (32 000 km²) obtained by digital treatment of multitemporal TM data

Catalonia is a nation located on the Mediterranean north-east coast of Spain. It covers over 32 000 km² and has a complex topography with a 16 percent of areas above 1100 m and 11 percent below 100 m.

1. Objectives

The images used were recorded in late spring and late summer of 1987, in order to pick up dynamic phenological aspects of the land cover in the process of multispectral classification. This information was completed with images from the early summer in the mountainous regions in order to diminish the phenomenon of shadows. To cover the whole area of Catalonia it was necessary to treat 10 TM quarter images for each period of the year chosen.

2. Methods used

The territory of Catalonia (32 000 km²) was segmented into 27 sections corresponding to the 27 sheets of the series at a scale of 1:100 000. For each sheet a multitemporal file was constructed and the number of variables was reduced by selecting the 5, 6 or 7 most effective principal components. A new segmentation of these areas into four sections corresponding to the four 1:50 000 sheets gave us the basic classification units.

Each of these units was submitted to a unsupervised multispectral classification. The labelling of the classes obtained was established automatically from their correlation with training samples. These samples covered a 3 percent of the total area, and their cover type was determined by photointerpretation of stereoscopic pairs of aerial photographs at a scale of 1:22 000. The visual supervision of the result made it possible to establish the definitive product after carrying out a certain amount of interaction with the process, such as reclassification of the impure classes, change of assignation and the making of masks in order to avoid spectral confusion.

The discrimination between different uses corresponding to a similar cover type, as in the case of urban cover, residential estates, industrial areas, transport infrastructures and barren areas, was performed thanks to a previous stratification based on photointerpretation, including a margin around the urban uses in question. In this way it was possible to carry out the process of classification in an independent way inside and outside the stratum. Then a modal filter was applied to the classified image to eliminate the small heterogeneities; however, the urban uses were never subjected to this filter in order to conserve the maximum detail. This whole process, applied to each segment corresponding to the 1:50 000 sheets, allowed us to obtain these definitively classified areas. Then a mosaic was carried out in order to reconstruct the whole area of Catalonia, and in the printing stage a laser restoration system was

	A	B	C	D	E	F	G	H	I	J	K	L	M	N	O	P	Q	R	S
1	%	1.	3.	5.	6.	7.	8.	9.	10.	11.	12.	13.	14.	15.	16.	17.	18.	19.	20.
2	1	83.7	10.9	0.2	0.2	0.1	0.1	0.3	0.0	0.0	0.0	0.0	0.1	0.0	0.0	2.1	0.7	0.0	0.0
3	3	6.1	77.4	0.0	0.0	0.1	0.0	0.0	0.0	0.0	0.0	0.0	0.0	0.0	0.0	1.2	1.1	0.0	0.1
4	5	1.1	0.6	82.9	11.7	1.8	0.1	1.0	0.7	0.2	0.1	0.1	0.3	0.0	0.4	0.1	0.3	0.0	0.2
5	6	0.4	0.3	5.1	79.8	0.1	0.3	1.2	0.4	0.3	0.0	0.0	0.0	0.0	2.6	0.2	0.2	0.0	0.1
6	7	1.7	2.1	5.7	1.5	89.9	7.2	4.6	1.9	1.1	0.8	0.8	2.2	0.0	0.0	1.3	0.0	0.0	0.0
7	8	0.4	0.1	1.1	3.4	1.3	81.7	6.8	0.0	0.0	0.1	0.1	2.3	0.0	0.3	0.7	0.0	0.0	0.0
8	9	0.2	0.1	0.2	0.5	0.3	2.1	81.8	0.0	0.0	0.0	0.0	0.2	0.0	0.0	0.2	0.0	0.0	0.0
9	10	0.1	0.1	0.5	0.5	0.5	0.1	0.1	80.8	3.0	1.4	1.4	1.4	0.1	7.5	0.3	0.0	0.0	0.1
10	11	0.0	0.0	0.0	0.0	0.3	0.1	0.0	5.7	82.9	2.9	2.9	2.0	0.0	0.1	0.2	0.0	0.0	0.2
11	12	0.1	0.0	0.0	0.0	1.0	0.2	0.1	3.5	6.3	90.2	2.9	4.1	0.1	2.1	0.5	0.0	0.0	1.3
12	13	0.0	0.0	0.0	0.0	0.0	0.0	0.1	0.0	0.0	0.1	89.9	0.2	0.0	0.0	0.7	0.0	0.0	0.0
13	14	0.2	1.0	1.6	1.7	3.5	6.9	2.8	6.5	5.8	4.1	4.1	83.8	5.7	3.7	5.2	0.0	0.0	0.9
14	15	0.0	0.0	0.0	0.0	0.0	0.0	0.0	0.0	0.0	0.0	0.0	1.3	84.8	0.0	6.7	0.0	16.9	0.0
15	16	0.0	0.0	0.3	0.0	0.0	0.0	0.0	0.0	0.0	0.0	0.0	0.0	0.0	79.8	0.0	1.2	0.0	0.0
16	17	3.9	5.8	1.8	0.4	1.0	0.9	1.1	1.3	0.2	0.1	0.1	1.8	9.3	1.2	76.5	1.2	10.3	0.4
17	18	0.0	0.0	0.0	0.0	0.0	0.0	0.0	0.0	0.0	0.0	0.0	0.0	0.0	0.6	0.1	92.6	0.0	0.0
18	19	0.0	0.0	0.0	0.0	0.0	0.0	0.0	0.0	0.0	0.0	0.0	0.0	0.2	0.0	0.8	0.0	72.8	0.0
19	20	0.0	0.0	0.3	0.1	0.0	0.0	0.0	0.1	0.1	0.1	0.1	0.2	0.4	1.7	0.4	1.2	2.9	96.5
20																			
21	NPC	30139.	7930.	61045.	21275.	158009.	62423.	16151.	49623.	38998.	167938.	12553.	185916.	46102.	3022.	35770.	3768.	261.	20806.
22																			
23	NPC: number pixels-control — general accuracy 85.6%																		
24																			

FIG. 5.8.1.

8. Carte au 1:250 000 de l'occupation et de l'utilisation du sol de Catalogne (32 000 km²) obtenue par traitement numérique de données TM multidates

La Catalogne est une nation située sur la côte méditerranéenne, au Nord-Est de l'Espagne. Elle couvre environ 32 000 km² et a une topographie complexe avec 16% des terres au-dessus de 1100 m et 11% au-dessous de 100 m.

1. Objectifs

Les images utilisées furent enregistrées à la fin du printemps et à la fin de l'été de 1987, dans le but de prendre en compte des aspects phénologiques ponctuels de l'occupation du sol lors des procédés de classification de données multispectrales. Ces informations furent complétées par des images du début de l'été dans des régions montagneuses dans le but de diminuer les phénomènes d'ombre. Pour couvrir la totalité de la Catalogne il fut nécessaire de traiter 10 quarts d'images TM pour chaque période de l'année choisie.

2. Méthodes utilisées

Le territoire de la Catalogne (32 000 km²) fut divisé en 27 secteurs suivant le découpage en 27 feuilles de la série au 1:100 000. Pour chaque feuille un fichier multidates fut élaboré et le nombre des variables fut réduit en sélectionnant 5, 6 ou 7 des composantes principales les plus importantes. Un nouveau découpage de ces zones en 4 sections correspondant aux 4 feuilles au 1:50 000 nous donna les unités de base de la classification.

Chacune de ces unités était soumise à une classification multispectrale non supervisée. L'identification des diverses classes fut réalisée automatiquement par corrélation avec des zones d'entraînement. Ces zones couvraient 3% de la zone totale, et leur affectation était déterminée par photo-interprétation de couples stéréoscopiques à l'échelle du 1:22 000. Un contrôle visuel du résultat permit d'établir définitivement le produit après avoir éliminé un certain nombre d'interactions dans le procédé, telles qu'une reclassification des classes équivoques, un changement de répartition et la confection de masques dans le but d'éviter des confusions spectrales.

La discrimination entre les diverses parties d'un couvert identique, comme dans le cas des étendues urbaines avec quartiers résidentiels, zones industrielles, infrastructures de transport et terrains vagues est réalisée grâce à une répartition préalablement traitée par photo-interprétation incluant une marge autour de ces zones urbaines en question. Dans ce sens il est possible de transposer le procédé de classification de façon indépendante à l'intérieur ou à l'extérieur de la strate. Ensuite un filtre fut appliqué à l'image classée pour éliminer de petites hétérogénéités, à l'exception des étendues urbaines, dans le but de conserver le maximum de détails. L'ensemble du procédé, appliqué à chaque section correspondant aux feuilles au 1:50 000, nous a permis d'obtenir la classification définitive. Ensuite une mosaïque

	A	B	C	D	E	F	G	H	I	J	K	L	M	N	O	P	Q	R	S
1	%	1.	3.	5.	6.	7.	8.	9.	10.	11.	12.	13.	14.	15.	16.	17.	18.	19.	20.
2	1	83.7	10.9	0.2	0.2	0.1	0.1	0.3	0.0	0.0	0.0	0.0	0.1	0.0	0.0	2.1	0.7	0.0	0.0
3	3	6.1	77.4	0.0	0.0	0.1	0.0	0.0	0.0	0.0	0.0	0.0	0.0	0.0	0.0	1.2	1.1	0.0	0.1
4	5	1.1	0.6	82.9	11.7	1.8	0.1	1.0	0.7	0.2	0.1	0.1	0.3	0.0	0.4	0.1	0.3	0.0	0.2
5	6	0.4	0.3	5.1	79.8	0.1	0.3	1.2	0.4	0.3	0.0	0.0	0.0	0.0	2.6	0.2	0.2	0.0	0.1
6	7	1.7	2.1	5.7	1.5	89.9	7.2	4.6	1.9	1.1	0.8	0.8	2.2	0.0	0.0	1.3	0.0	0.0	0.0
7	8	0.4	0.1	1.1	3.4	1.3	81.7	6.8	0.0	0.0	0.1	0.1	2.3	0.0	0.3	0.7	0.0	0.0	0.0
8	9	0.2	0.1	0.2	0.5	0.3	2.1	81.8	0.0	0.0	0.0	0.0	0.2	0.0	0.0	0.2	0.0	0.0	0.0
9	10	0.1	0.1	0.5	0.5	0.5	0.1	0.1	80.8	3.0	1.4	1.4	1.4	0.1	7.5	0.3	0.0	0.0	0.1
10	11	0.0	0.0	0.0	0.0	0.3	0.1	0.0	5.7	82.9	2.9	2.9	2.0	0.0	0.1	0.2	0.0	0.0	0.2
11	12	0.1	0.0	0.0	0.0	1.0	0.2	0.1	3.5	6.3	90.2	2.9	4.1	0.1	2.1	0.5	0.0	0.0	1.3
12	13	0.0	0.0	0.0	0.0	0.0	0.0	0.1	0.0	0.0	0.1	89.9	0.2	0.0	0.0	0.7	0.0	0.0	0.0
13	14	0.2	1.0	1.6	1.7	3.5	6.9	2.8	6.5	5.8	4.1	4.1	83.8	5.7	3.7	5.2	0.0	0.0	0.9
14	15	0.0	0.0	0.0	0.0	0.0	0.0	0.0	0.0	0.0	0.0	0.0	1.3	84.8	0.0	6.7	0.0	16.9	0.0
15	16	0.0	0.0	0.3	0.0	0.0	0.0	0.0	0.0	0.0	0.0	0.0	0.0	0.0	79.8	0.0	1.2	0.0	0.0
16	17	3.9	5.8	1.8	0.4	1.0	0.9	1.1	1.3	0.2	0.1	0.1	1.8	9.3	1.2	76.5	1.2	10.3	0.4
17	18	0.0	0.0	0.0	0.0	0.0	0.0	0.0	0.0	0.0	0.0	0.0	0.0	0.0	0.6	0.1	92.6	0.0	0.0
18	19	0.0	0.0	0.0	0.0	0.0	0.0	0.0	0.0	0.0	0.0	0.0	0.2	0.0	0.8	0.0	72.8	0.0	
19	20	0.0	0.0	0.3	0.1	0.0	0.0	0.0	0.1	0.1	0.1	0.1	0.2	0.4	1.7	0.4	1.2	2.9	96.5
20																				
21	NPC	30139.	7930.	61045.	21275.	158009.	62423.	16151.	49623.	38998.	167938.	12553.	185916.	46102.	3022.	35770.	3768.	261.	20806.
22																				
23		NPC : number pixels-control																		
24		general accuracy 85.6%																		

Fig. 5.8.1.

used to create the appropriate grid films. A background added to the lands adjoining Catalonia was obtained by the use of TM 4 channel of the satellite image.

At the publishing stage, toponymical information and the borders of Catalonia and its comarques (administrative division) were added.

3. Results

The evaluation was made through the construction of a table of omissions–commissions. This table was established automatically by comparison between the code of the pixels of the test sites and that of the same pixels situated in the classified image, and counting the coincidences and confusions. These test sites correspond to a statistical sample different from that of the training sites and representative of the cover type and uses to be distinguished, equivalent to 3 percent of Catalonia.

The general accuracy obtained as a weighted average by percent presence of each cover type was 85.6 percent and the matrix corresponding to the whole Catalonia is indicated at the bottom.

It must be said that the cover corresponding to low-density urban areas and residential areas (2) was not included in this evaluation in order to maintain the real heterogeneities in the classified image; nor was the cover of road infrastructure included (4), as it was incorporated from photointerpretation.

fut réalisée pour reconstituer l'ensemble du territoire de la Catalogne. Au stade de l'impression, un système de restitution à laser fut mis en oeuvre pour réaliser les films tramés appropriés. Un fond a été ajouté sur les territoires limitrophes de la Catalogne en utilisant le canal 4 de l'image satellitaire TM.

A la publication nous avons ajouté des informations toponymiques, les frontières et les divisions administratives (comarques).

3. Résultats

L'évaluation a été faite via l'établissement d'une matrice de confusion, réalisée automatiquement en comparant les réponses de chaque pixel entre le test et l'image classée et en comptant les coïncidences et les confusions. Ces zones-tests correspondent à un échantillonnage statistique différent de celui des zones d'entraînement et représentatifs des types d'occupation/utilisation du sol à distinguer, et elles couvrent 3% de la Catalogne. La précision générale obtenue comme moyenne pondérée en pourcentage de la présence de chaque couvert est de 85,6% et la matrice qui en découle pour l'ensemble de la Catalogne figure dans les pages précédentes (fig. 5.8.1).

Il est à noter que les couverts correspondants aux zones urbaines de faible densité et aux quartiers résidentiels (2) ne sont pas pris en compte dans cette évaluation pour maintenir une réelle hétérogénéité de l'image classée, de même pour l'infrastructure routière (4) qui a été extraite par photointerprétation.

Generalitat de Catalunya
Departament de Política Territorial
i Obres Públiques
Institut Cartogràfic de Catalunya

Mapa d'usos del sòl de Catalunya 1:250 000

Photographical reduction of Land use and cover map of Catalonia; approximate scale 1:1 600 000

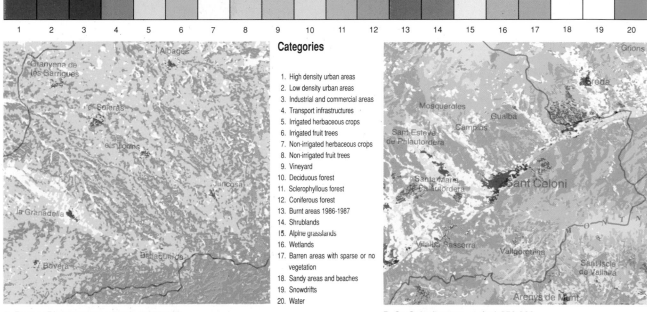

| 1 | 2 | 3 | 4 | 5 | 6 | 7 | 8 | 9 | 10 | 11 | 12 | 13 | 14 | 15 | 16 | 17 | 18 | 19 | 20 |

Categories

1. High density urban areas
2. Low density urban areas
3. Industrial and commercial areas
4. Transport infrastructures
5. Irrigated herbaceous crops
6. Irrigated fruit trees
7. Non-irrigated herbaceous crops
8. Non-irrigated fruit trees
9. Vineyard
10. Deciduous forest
11. Sclerophyllous forest
12. Coniferous forest
13. Burnt areas 1986-1987
14. Shrublands
15. Alpine grasslands
16. Wetlands
17. Barren areas with sparse or no vegetation
18. Sandy areas and beaches
19. Snowdrifts
20. Water

A. Borges Blanques' area, scale 1:250 000

B. St. Celoni's area, scale 1:250 000

Application Title
Utilization of the East Slovakian lowland landscape

Titre de l'application
Utilisation des basses terres de Slovaquie Orientale

Localization/Localisation
Slovakia/Slovaquie

RESPONSIBLE ORGANIZATIONS   ORGANISMES RESPONSABLES

Geograficky ustav SAV
Obrancov mieru 49
814 73 Bratislava

Geographical Institute
Slovak Academy of Sciences
Bratislava
Czechoslovakia

(Contact person: J. Feranec, J. Otahel)

9. Utilization of the East Slovakian lowland landscape
(An interpretation scheme set up applying colour infrared space photographs)

To create the interpretation scheme we used two-colour infrared space photographs—positives on transparent base, made by Kozmos–Priroda in August 1985, original scale approximately 1:275 000 and size 30 × 30 cm.

A valuable source of information about forms of landscape utilization were colour infrared space photographs and also interpretation schemes in scale 1:10 000 obtained by interpretation of multispectral aerial photographs. They included approximately 80 percent of lowland areas. The subject information came from the frontier and northern lowland parts, from where multi-spectral space photographs were obtained from sowing plans and by field mapping.

The collections of data formed the original basis of specific variants of analogue interpretation of colour infrared space photographs. It used 192 interpretation form schemes of landscape utilization of the East-Slovakian lowlands, scale 1:10 000, sowing plans scales 1:10 000 and 1:5000 from frontier and northern lowland parts and further field mapping results scale 1:10 000.

Individual objects of the Earth surface identified on colour infrared space photographs scale 1:275 000 were visually compared on a lighting desk with interpretation schemes, sowing plans and field mapping results. After comparison of each area/object, identified on colour infrared photographs with its content documented in some of the auxiliary materials, the interpreter recorded contours and colour signatures, indicating relevance to the defined forms of landscape utilization on FOLEX foil. Debatable cases were marked onto the map, scale 1:10 000, and verified directly in the field.

The result of application of the methodical process was the setting up of an interpretation scheme of landscape utilization in the East-Slovakian lowlands, in scale approximately 1:200 000.

Used colour infrared space photographs, which are noted for highly distinguishing ability, represented a valuable source of actual information about present landscape utilization forms.

(The compilation, drawing and printing of the original interpretation scheme was carried out at the Geographical Institute of the Slovak Academy of Sciences, Brastislava.)

9. Utilisation des basses terres de Slovaquie Orientale
(Schéma d'interprétation établi à partir de photographies spatiales en infrarouge couleur)

Pour réaliser le document d'interprétation nous avons utilisé deux photographies spatiales positives en infrarouge couleur, sur support transparent, prises par Kosmos-Priroda en août 1985 à l'échelle approximative du 1:275 000, avec un format de 30 × 30 cm.

Une source valable d'informations sur les formes de l'occupation du sol était fournie par les photographies spatiales en infrarouge couleur ainsi que par des schémas d'interprétation à l'échelle du 1:10 000 obtenus par interprétation de photographies aériennes multispectrales. Elles concernent environ 80% des basses terres. Les informations sur les zones frontières et les basses terres du Nord, objet des photographies spatiales multispectrales, furent obtenues à partir de cartes agricoles et par des relevés de terrain.

Ces ensembles de données ont servie de base de départ pour l'interprétation analogique de photographies spatiales en infrarouge couleur. Celle-ci a résulté en 192 schémas d'interprétation de l'occupation du sol des basses terres de Slovaquie au 1:10 000, des cartes agricoles au 1:10 000 et au 1:5000 des zones frontières et des terres du Nord, et des levés de terrain cartographiés à l'échelle du 1:10 000.

Les objets identifiés sur les photographies en infrarouge couleur à l'échelle du 1:275 000 étaient comparés sur table lumineuse, avec les schémas d'interprétation, les cartes agricoles et les levés de terrain. Après comparaison de chaque objet ou zone, identifié sur la photographie infrarouge couleur avec son contenu documenté sur quelques matériels auxiliaires, l'interprète identifiait les contours et les signatures en couleur (corrélées aux types d'utilisation du sol) sur un support Folex. Les cas douteux étaient mentionnés sur une carte au 1:10 000 et vérifiés directement sur le terrain.

Le résultat de l'application de la méthode mentionnée plus haut de l'interprétation de l'occupation du sol des basses terres de l'Est de la Slovaquie fut présenté à l'échelle approximative du 1:200 000.

L'utilisation de photographies spatiales infrarouges couleur, qui sont d'unc utilisation très aisée, représente une source valable d'informations sur les formes réelles de l'utilisation du sol.

(La rédaction, le tracé et l'impression du schéma d'interprétation original ont été réalisés à l'Institut de Géographie de l'Académie Slovaque des Sciences, Bratislava.)

Paysages urbains ou industriels
    à fonction résidentielle
    à fonction industrielle
    à fonction minière
    à zones de loisir

Paysages agricoles
    avec vergers et vignobles
    avec prairies
    avec des terres arables
        —avec végétation
        —sans végétation

Paysages forestiers
    avec forêts de feuillus prédominants

Surfaces d'eau
    Surfaces d'eau artificielles
    Surfaces d'eau naturelles, eaux stagnantes
    Cours d'eau

Frontière d'Etat
Limite des basses terres
Limite des formes d'utilisation des paysages, des parcelles de terres arables, des vergers et vignobles

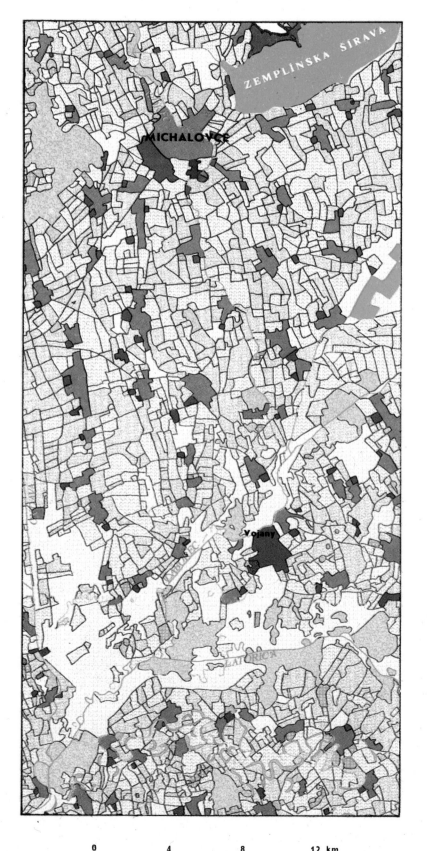

LEGENDA

URBANIZOVANÁ A TECHNIZOVANÁ KRAJINA
Urbanized and technicalized landscape

 s obytnou funkciou
with residential function

 s výrobnou funkciou
with productional function

 s ťažobnou funkciou
with mining function

 s rekreačnou funkciou
with recreational function

POĽNOHOSPODÁRSKA KRAJINA
Agricultural landscape

 so sadmi a vinicami
with orchards and vineyards

 s trávnatými porastami
with grasslands

s ornou pôdou
with arable land

 s vegetáciou
with vegetation

 bez vegetácie
without vegetation

LESNÁ KRAJINA
Forest landscape

 s prevahou listnatých stromov
with leafy forests prevailing

VODY
Waters

 umelé vodné plochy
artificial water surfaces

prirodzené vodné plochy a močiare
natural water surfaces and
dead waters

vodné toky
water courses

štátna hranica
State boundary

hranica nížiny
the lowland's boundary

**hranice foriem využitia krajiny,
parciel ornej pôdy, sadov a viníc**
boundaries of utilization forms
of the landscape, of the plots
of arable land, of orchards and
vineyards

0   4   8   12 km

URBAN CARTOGRAPHY

CARTOGRAPHIE URBAINE

Application Title
Housing sample survey using Landsat data

Titre de l'application
Inventaire de l'habitat à l'aide de données Landsat

Localization/Localisation
Saudi Arabia/Arabie Saoudite

RESPONSIBLE ORGANIZATIONS   ORGANISMES RESPONSABLES

Ministry of Housing
Riyadh
Saudi Arabia

Institut Géographique National
2 Avenue Pasteur
94160 Saint-Mandé
France

Tel: 33/1 43 98 80 00   Fax: 33/1 43 65 69 54

(Contact person: M. Charles Etienne-Jaouen)

10. Housing sample survey using Landsat data

The aim of the housing sample survey was to provide the government of Saudi Arabia with the basic comprehensive data on the housing situation in the kingdom, such as size and quality of the housing on which to base its policies and programmes in the future.

106 Landsat images were analysed by computer to make the first overall location and area estimates of all sedentary human settlements in Saudi Arabia, as well as to make a simple classification of land uses in those settlements.

Type of documents delivered:

—Two full sets of computer compatible tapes containing four band Landsat MSS data, geometrically rectified to UTM projection, with RMS errors in position less than 240 m and on a fifth band all ground control points and administrative boundaries.
—Two sets of computer compatible tapes containing for each settlement to which land-use category the corresponding pixels have been classified. Each matrix is accompanied by an alphanumeric string which identified each settlement by name, code, administrative area, UTM coordinates and surface of each land-use category.
—A national atlas is produced to demonstrate the spatial configuration of land-use in all settlements of the kingdom.
—Mosaics in IR colour using 106 Landsat data scenes, scale 1 : 1 000 000.

The main advantage of this method is to get a speedy land-use map of each settlement and the area of each class of buildings.

Disadvantages: in some regions, especially in the mountains, the pixel size of Landsat data MSS was insufficient, so interactive methods were occasionally used.

10. Inventaire de l'habitat à l'aide de données Landsat

Le but de l'inventaire de l'habitat était de fournir au gouvernement de l'Arabie Saoudite des données de base exhaustives sur la situation de l'habitat dans le royaume, en importance et qualité, afin d'orienter sa politique et ses programmes dans le futur.

On a analysé par ordinateur 106 images Landsat pour faire la première localisation et estimation d'ensemble de toutes les agglomérations permanentes d'Arabie Saoudite, ainsi qu'un classement élémentaire de l'utilisation du sol dans ces agglomérations.

Types de documents fournis:

—Deux jeux complets de bandes magnétiques compatibles-ordinateurs comprenant les données des 4 bandes Landsat MSS, rectifiées géométriquement en projection UTM, avec des écarts-types en position inférieurs à 240 m, et sur une cinquième bande tous les points d'appui au sol et les limites administratives.
—Deux jeux de bandes magnétiques compatibles-ordinateurs contenant pour chaque agglomération une matrice de pixels classés par type d'usage du sol. Chaque matrice est accompagnée d'une chaîne de caractères alphanumériques, identifiant l'agglomération par son nom, son code, sa zone administrative, ses coordonnées UTM et la surface de chaque catégorie d'usage du sol.
—Un atlas national montre la configuration spatiale de l'usage du sol dans toutes les agglomérations du royaume.
—Des mosaïques en infra-rouge couleur utilisant les 106 scènes Landsat à l'échelle du 1 : 1 000 000.

Le principal avantage de cette méthode est d'obtenir rapidement une carte d'usage du sol de chaque agglomération et la surface de chaque classe de bâti.

Le désavantage est que dans quelques régions, principalement en montagne, la taille des pixels Landsat MSS était inadaptée, aussi des méthodes interactives ont-elles été parfois utilisées.

	المسطح بالهكتار Area in hectares		الطرق الرئيسية Major roads	٣٠٦ 306
مناطق سكنية عالية الكثافة Residential areas high density	٨٠٥ 805		المسطح الإجمالي للمنطقة اص ٤ Total settlement area * (p. 4)	٦ ٨٢٠ 6 820
مناطق سكنية منخفضة الكثافة Residential areas low density	٢ ٨٧٨ 2 878	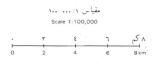	مطارات Airports	
مباني طينية Mud buildings	٠ 0			
مناطق صناعية و تجارية Industrial and commercial areas	١٠٥٩ 1 059		مسقط مركاتور المستعرض العالمي Universal Transverse Mercator Projection	
مواقع بناية عالية الكثافة استخدامات متعددة High intensity building sites Mixed use	٦٢٢ 622			
مواقع بناية منخفضة الكثافة استخدامات متعددة Low intensity building sites Mixed use	٠ 0		مقياس ١: ١٠٠ ٠٠٠ Scale 1:100,000	
اسطح مائية Water	٦١ 61		٠ ٢ ٤ ٦ ٨ كم 0 2 4 6 8 km	
مناطق خضراء Green areas	١٦٦ 166			
مناطق تجميع الفضلات و القمامة Waste dumps	٥٠٣ 503			

Application Title
Compiling of urban area thematic maps from multispectral image data

Titre de l'application
Etablissement de cartes thématiques du milieu urbain de Prague à partir de données d'images en mode multibande

Localization/Localisation
Czechoslovakia (Prague)/Tchécoslovaquie (Prague)

RESPONSIBLE ORGANIZATIONS   ORGANISMES RESPONSABLES

Geodetic and Cartographic Enterprise
Remote Sensing Centre
Kostelni 42
170 30 Praha 7
Tchécoslovaquie

Tel: (42) 2 37 14 41-9

(Contact persons: K. Charvat, V. Cervenka, J. Uhlir, E. Milesova)

11. Compiling of urban area thematic maps from multispectral image data

Gathering of information on land use are the main goals of remote sensing methods. This task is of special importance in regions with complicated structural zoning, e.g. in urban agglomerations and their surroundings. It is advantageous to use the Thematic Mapper data for this purpose. One of the possible methods of image data preprocessing will be outlined in this contribution. It is aimed to pointing up the urban vegetation and its fundamental categories.

The method described applies modern principles of artificial intelligence, especially from the branch of neural networks. It is based on the so-called "back propagation algorithm" which enables one to compress effectively the whole information, contained in the original Thematic Mapper data, into three new synthetic channels, as well as its decompression back into the original channels. The algorithm uses a three-layer unidirectional network, where each neuron (node) from one layer is connected with all neurons of the previous layer (see Fig. 5.11.1).

Any connection between two neurons i and j has a certain weight $w(ij)$. In the course of data processing, spectral values of processed image elements are assigned to individual neurons of the input layer. Neuron values in higher layers are computed from the expression

$$x(i) = S [\text{sum above } j \text{ of } w(ij) \cdot x(j)]$$

where j is a set of neurons from the previous layer, and S is a certain (usually sigmoid) function.

Data processing by means of such neural network runs in two steps: adaptation, and evaluation. In the course of adaptation the weights of connections are changed until required results are obtained in the output layer. If the goal of this network is to perform data compression, it is necessary to require the equality of input and output values. If such configuration is obtained, then the values of the middle layer (having lower capacity than input and output layers) can be considered as the effective compression of the original information. The evaluation of the image data consists in the introducing of all pixel values to the input of the "instructed" network.

The values obtained in the middle layer (having three neurons in the case of Thematic Mapper data processing) may be used after their registration on the film as the RGB components for colour composite production. Assignment of individual components is chosen empirically. The colour composites contain almost all information from the original image data sets.

The photomap, experimentally compiled, comprehends the part of Landsat Thematic Mapper image (14 August 1988, track 192, frame 025—EURIMAGE) at the scale of 1:100 000 approximately. It shows the main types of land use in the City of Prague and its surroundings:

1. Various types of built-up area:
   —blocks of buildings in the town centre and industrial zones,
   —family houses with gardens on larger areas,
   —new settlements at marginal parts of the city.
2. Significant transport surfaces (particularly highways, railway stations, airports).
3. Water surfaces (rivers and artificial reservoirs).
4. Vegetation and land cover:
   —deciduous forest, coniferous forest,
   —town greenery, including orchards (larger areas),
   —meadows and arable land with green crops (e.g. fodder crops, sugar beet),
   —arable land without crops,
   —bare soil (large building sites, waste deposits and other territories under devastation).

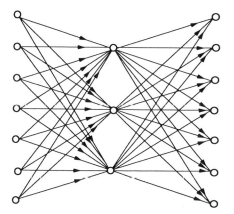

Layer 1 (input)   Layer 2   Layer 3 (output)

Fig. 5.11.1.

11. Etablissement de cartes thématiques du milieu urbain de Prague à partir de données d'images en mode multibande

L'acquisition d'informations sur l'usage du sol est un des objectifs principaux des méthodes de télédétection. Cette tâche est spécialement importante dans les régions ayant une structure particulièrement compliquée, par ex. les agglomérations urbaines et leurs alentours. Il est judicieux d'utiliser les images Landsat Thematic Mapper (TM) à cet effet. Une des méthodes possibles de prétraitement des données d'image est présentée dans le texte ci-dessous. Elle a pour but de mettre en évidence la végétation urbaine et ses catégories fondamentales.

La méthode en question utilise les principes modernes de l'intelligence artificielle, notamment ceux des réseaux de neurones. Elle est basée sur l'algorithme appelé "back propagation", qui permet de comprimer toutes les informations contenues dans les données d'images originales du TM, en trois nouveaux canaux synthétiques, et inversement de les décomprimer dans leurs canaux d'origine. L'algorithme utilise un réseau unidirectionnel à 3 niveaux, où chaque neurone est relié à tous les neurones du niveau précédent (voir fig. 5.11.1).

Toute liaison entre deux neurones i et j a un certain poids $w(ij)$. Au cours du traitement des données les valeurs spectrales des éléments de l'image traitée sont attribuées aux neurones individuels du niveau d'entrée. Les valeurs des neurones aux niveaux supérieurs sont calculées d'après la formule:

$$x(i) = S[\text{sum above } j \text{ of } w(ij) \cdot x(j)]$$

où j est un ensemble des neurones du niveau précédent et S est une certaine fonction, généralement sigmoïde.

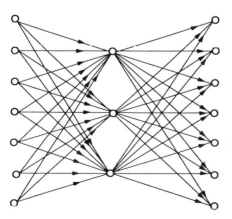

Fig. 5.11.1.

Le traitement des données à l'aide de ce réseau de neurones comprend deux phases: l'adaptation et l'évaluation. Pendant la phase d'adaptation les poids des liaisons sont changés jusqu'à obtention des résultats attendus au niveau de sortie. Si ce réseau est utilisé pour la compression de données, on doit exiger l'égalité des valeurs à l'entrée et à la sortie. Si ce schéma est obtenu, on peut considérer que les valeurs au niveau intermédiaire (dont la capacité est plus petite que celle des niveaux d'entrée et de sortie) représentent effectivement les informations originales comprimées. L'évaluation des données d'image consiste en l'introduction des valeurs de tous les pixels à l'entrée du réseau "instruit".

Les valeurs obtenues au niveau intermédiaire (qui contient trois neurones dans le cas du Landsat-TM) peuvent être utilisées après leur enregistrement sur le film comme les composantes RVB pour l'établissement des compositions colorées. L'attribution des composantes individuelles est choisie empiriquement. Les compositions colorées comportent presque toute l'information provenant des données de l'image d'origine.

La photocarte, établie expérimentalement comporte la partie d'image Lansat-TM (du 14 août 1988, track 192, frame 025, EURIMAGE) à l'échelle du 1:100 000 environ. Elle indique les types fondamentaux d'utilisation du sol dans Prague et ses environs:

1. Différents types de zones bâties:
   —le centre ville avec blocs d'immeubles et zones industrielles,
   —les maisons individuelles avec jardins sur des zones plus vastes,
   —les nouveaux quartiers à la périphérie de la ville.
2. Surfaces importantes destinées au transport (en particulier les gares, les autoroutes, les aéroports).
3. Plans d'eau (rivières principales, réservoirs artificiels),
4. Végétation et occupation du sol:
   —forêts de feuillus et de conifères,
   —espaces verts, y compris les vergers (zones plus vastes),
   —prés et sol arable avec des cultures

A key for the thematic interpretation has been compiled by means of comparing the results with the colour composites of aerial multispectral photographs acquired at almost the same time as TM images and with available thematic maps.

vertes (plantes fourragères, betterave à sucre),

—sol arable sans culture,

—sol nu (surfaces en construction, dépôts d'ordures, et autres terrains vagues).

Une clé pour l'interprétation thématique a été obtenue par comparaison des résultats avec les compositions colorées de photographies aériennes multibandes acquises presque en même temps que les images TM et à l'aide de cartes thématiques existantes.

Types of built-up areas

blocks of buildings in the town centre and industrial zones

family houses with gardens on larger areas

new settlements at marginal parts of the city

Transport surfaces

Water surfaces

Vegetation and land cover

deciduous forest

coniferous forest

town greenery including orchards (larger areas)

meadows and arable land with green crops

arable land without crops

bare soil and territories under devastation

1:100 000

N

PRAHA

Vltava

Vltava

Application Title
Representation of a large urban area

Titre de l'application
Représentation d'une grande agglomération urbaine

***Localization*/Localisation**
***Brazil (Brasilia)*/Brésil (Brasilia)**

RESPONSIBLE ORGANIZATIONS

Direction du Service Géographique
de l'Armée de Terre brésilienne
(DSG)/5a DL
Rua Major Daemon, 81
20081 Rio de Janeiro
Brasil

(Contact person:
M. Luis de Andrade)

ORGANISMES RESPONSABLES

Institut Géographique National
2, avenue Pasteur
94160 Saint-Mandé
France
Tél: 33/1 43 98 80 00
Fax: 33/1 43 98 84 45

(Contact person:
M. Patrice Foin)

12. Representation of a large urban area

1. Objectives

To evaluate the possibilities of obtaining cartographic products using SPOT satellite imagery.

2. Methods used

(a) Geometric corrections using existing topographic maps.

(b) Contrast enhancement using filter weighting factors.

(c) Radiometric enhancement of bands by a histogram calculation for each band. Pixel analysis between the maximum and minimum levels, as well as the percentage of pixels in each level.

(d) Production of a colour composite in false colours by associating channel 3 (red), channel 2 (green) and channel 1 (blue).

(e) Production of a colour composite in pseudo-natural colours.

(f) Map production using photo-plotter.

(g) Production of a Cromalin proof print.

(h) Preparation of data obtained from other documents. Lettering, conventional signs, a few spot heights, coordinate origin, marginal information, etc., were transferred to the proof print.

(i) Cartographic operations and visual interpretation.

(j) Printing of the final document.

3. Results

(a) Space map at 1:50 000 covering the Brasilia region.

(b) Advantages:
—Rapid production at relatively low costs.
—Reveals very recent developments.
—Expressive document.

(c) Defects:
—Insufficient resolution (20 m) for distinguishing the street network (10 m resolution will be necessary) and the urban built-up area data.
—Scale too small.

12. Représentation d'une grande agglomération urbaine

1. Objectifs

Evaluer les possibilités d'obtention de produits cartographiques avec l'utilisation d'images du satellite SPOT.

2. Méthodes utilisées

(a) Corrections géométriques à l'aide de cartes topographiques existantes.

(b) Accentuation des contrastes utilisant des facteurs pondérés de filtrage.

(c) Amélioration radiométrique des bandes par le calcul de l'histogramme de chaque bande. Analyse des pixels entre les niveaux minimum et maximum ainsi que le pourcentage des pixels dans chaque niveau.

(d) Sortie d'une composition colorée en fausse couleur en associant le canal 3 (rouge), le canal 2 (vert) et le canal 1 (bleu).

(e) Sortie d'une composition colorée en pseudo-couleur naturelle.

(f) Sortie cartographique avec utilisation de phototraceur.

(g) Réalisation d'une épreuve d'essai cromalin.

(h) Préparation des données issues d'autres documents. Sur l'épreuve d'essai ont été portés des écritures, des signes conventionnels, quelques points cotés, l'origine des coordonnées, des données d'habillage, etc.

(i) Travaux de cartographie et d'interprétation visuelle.

(j) Impression du document cartographique.

3. Résultats

(a) Une spatiocarte au 1:50 000 couvrant la zone de Brasilia.

(b) Avantages:
—Obtention rapide et coût relativement bas.
—Identification des quartiers les plus récents.
—Document expressif.

(c) Défauts:
—Résolution de 20 m insuffisante pour pouvoir distinguer le réseau des rues (une résolution de 10 m serait nécessaire) et les constructions d'une zone bâtie.
—Echelle trop petite.

SOIL MAPPING

CARTOGRAPHIE DES SOLS

Application Title
Interpretation scheme of forms of the surface water-logging intensity of the soils in the southern part of the Danube Lowland at the scale 1:50 000

Titre de l'application
Schéma d'interprétation des formes de l'intensité des teneurs en eau des sols du Bas-Danube à l'échelle de 1:50 000

Localization/Localisation
Slovakia/Slovaquie

RESPONSIBLE ORGANIZATIONS   ORGANISMES RESPONSABLES

Geograficky ustav SAV
Obrancov mieru 49
814 73 Bratislava

Fakulta stavebni CVUT
Laborator DPZ
Thakurova 7
166 29 Praha 6
Czechoslovakia

(Contact persons: J. Feranec, J. Kolar)

13. Interpretation scheme of forms of the surface water-logging intensity of the soils in the southern part of the Danube Lowland at the scale 1:50 000

The interpretation scheme documents possibilities in the utilization of Landsat 5 TM data with indirect identification of the surface water-logging intensity of soils without vegetation. Forms of the surface water-logging intensity of soils are identified by means of their relevant physiognomic signs.

With interpretation, Landsat 5 TM data from 12 April 1988 were used. There were also used field mapping results from training and test areas using the Pericolor system. The Bayesian classifier was used for classification.

Water-logging of soil is defined as a temporary or permanent flooding of all its pores (in surface layer) with precipitation or groundwater coming to the surface particularly in depressive positions. According to this definition it is possible by applying space images and data obtained by synchronous field measurements and mapping and by means of physiognomic indications of water-logging (they represent significant appearance characteristics of water-logged soils without vegetation with different intensity, manifesting themselves in coloured images, particularly in near infrared parts of the spectrum, with marked change of pixel values, or as the case may be, with a characteristic structure), to identify forms of surface water-logging of soils without vegetation in 4 degrees (see the legend of the interpretation scheme):

Form "V"—Water concentrated on the soil surface. Its area ranges from some tens of square metres up to hundreds of them.

Form "I"—Intensively water-logged soils. This covers sporadic and smaller areas of water concentrated on the surface. Distribution of these areas forms a characteristic structure, formed particularly by ploughed land and by water concentrated on the surface, or as the case may be, by intensively water-logged soil in the surface layer.

Form "II"—Less intensively water-logged soils. It is represented by areas with very sporadic (or none) occurrence of water concentrated at the surface. Within this form, areas with soils partially water-logged at the surface layer are dominant by area.

Form "O"—Other areas (soils without vegetation occurring within them, are relatively dry; arable soil with vegetation, forms of forest landscape, urbanized and industrialized landscape).

The interpretation scheme presents valuable information which can be used in practice, i.e. with projecting hydro-improvement arrangements of areas of interest. Furthermore, with observing the effectiveness of existing drainage systems, etc.

13. Schéma d'interprétation des formes de l'intensité des teneurs en eau des sols du bas-Danube

Le schéma d'interprétation présente les possibilités de l'utilisation des données de Landsat 5 TM pour une identification indirecte des intensités de de teneur en eau de surface dans les sols nus. Les formes de l'intensité des teneurs en eau de surface dans les sols sont identifiées par les caractères physionomiques corespondants (apparences).

Pour cette interprétation on a utilisé des données Landsat 5 TM du 12 avril 1988, ainsi que des cartes de terrain, issues de zones d'entraînement et de contrôle traitées sur système Pericolor. Pour la classification on a utilisé la méthode de Bayes.

La teneur en eau d'un sol est définie comme une irrigation temporaire ou permanente de tous les pores (de la couche de surface), avec précipitation ou en nappe, venant à la surface particulièrement dans des parties dépressives aussi bien qu'en zone plane. Selon cette définition il est possible, en utilisant des images spatiales et des données obtenues par des mesures simultanées sur le terrain puis leur report sur carte au moyen d'indicateurs physionomiques des teneurs en eau, (ces indicateurs représentent les caractéristiques apparentes significatives de la présence d'eau dans les sols nus, qui ont des réponses différentes dans les images couleurs, principalement dans la partie proche infrarouge du spectre, avec des changements marqués des valeurs de pixel et, le cas échéant, une structure caractéristique), afin d'identifier 4 degrés dans le classement de présence d'eau dans les sols nus, comme indiqués dans la légende du document joint:

Forme "V"—Concentration d'eau en surface du sol. Les surfaces sont de quelques dizaines de mètres carrés jusqu'à quelques centaines.

Forme "I"—Sols intensivement imprégnés d'eau. Cette forme est caractérisée par des zones sporadiques et plus petites de concentration d'eau en surface. La distribution de ces zones forme une structure caractéristique constituée principalement de terres labourées et d'eau concentrée en surface, ou le cas échéant par un sol intensivement imprégné d'eau sur la couche de surface.

Forme "II"—Sols moins intensément imprégnés d'eau. Cette forme est caractérisée par des zones avec une occurrence sporadique ou inexistante de concentration d'eau de surface. Dans cette catégorie, des zones avec des sols partiellement imprégnés d'eau en surface dominent par plaques.

Forme "O"—Concerne les autres sols, ou zones sans végétation apparaissant parmi eux, et qui sont relativement secs; terres arables avec végétation, paysages de forêts, étendues urbanisées ou industrialisées.

Le schéma d'interprétation fournit des informations valables qui peuvent être utilisées en pratique, soit pour des projets d'aménagement hydraulique de zones intéressantes, soit en observant l'efficacité de systèmes de drainage existants, etc.

SURFACE WATER-LOGGING INTENSITY OF THE SOILS
Danube Lowland (Czechoslovakia)

INTENSITE DE LA TENEUR EN EAU DES SOLS
Région du Bas-Danube (Tchécoslovaquie)

Form V : water concentrated on soil surface / eau concentrée en surface
Form I : intensively waterlogged soils / sols à forte teneur en eau
Form II : less intensively waterlogged soils / sols à moindre teneur en eau
Form O : other areas / autres zones

AGRICULTURAL MAPPING

CARTOGRAPHIE AGRICOLE

Application Title
Evaluation of the arable-land area on SPOT imagery of Al Fayyum (Egypt)

Titre de l'application
Evaluation de la surface agricole utile du Fayoum (Egypte)

Localization/**Localisation**
Egypt (Fayum)/**Egypte (Le Fayoum)**

<table>
<tr><td>*RESPONSIBLE ORGANIZATIONS*</td><td>ORGANISMES RESPONSABLES</td></tr>
<tr><td>Egyptian Survey Authority
Orman Giza
Cairo, Egypt</td><td>Institut Géographique National
2 avenue Pasteur
94160 Saint-Mandé
Tél: (33) 1 43 98 80 00
Fax: (33) 1 43 98 84 00</td></tr>
</table>

(Contact person: M. Rantrua)

14. Evaluation of the arable-land area on SPOT imagery of Al Fayyum (Egypt)

1. Objective of the application

The aim of this pilot investigation was to evaluate the arable-land area, in the Al Fayyum region, using a SPOT XS image. That research was undertaken jointly by the Egyptian Survey Authority and the Institut Géographique National (France) in 1987.

2. Method used

The determination of the arable-land area was undertaken by a computer-assisted supervised classification. The principle is to select on the image samples corresponding to homogeneous zones and then to identify them in the field. Digital procedures were then used to classify the pixels of the image in accordance with their radiometric resemblance to the pixels of the samples depending on their "distance" from the class centres in the radiometric space. The field enquiry and the selection of the themes are here fundamental. It was also important to verify a posteriori the validity and the quality of the classification using check areas.

Arable land area	Non-agricultural area
1. Crops	5. Open water
2. Tree crops	6. Natural vegetation
3. Fallow land	7. Uncultivated land
4. Flooded land	8. Towns

3. Results

The results of a first classification showed that processing a scene as a single block produced inevitable confusions between certain classes.

A stratification of the image was then undertaken (four strata) as well as a classification by stratum.

The distribution matrix for the complete group of four strata based on pixels selected at random gave the following results:

Theme	Arable land area
1	394 well classified on 398 checked
2	45 well classified on 68 checked
3	191 well classified on 209 checked
4	19 well classified on 27 checked

Theme	Non-agricultural areas
5	185 well classified on 187 checked
6	2 well classified on 3 checked
7	28 well classified on 63 checked
8	36 well classified on 43 checked

Confusion between arable-land areas and non-agricultural areas is as follows:

Arable-land area   96% (well classified)
Non-agricultural area   91.2% (well-classified)

14. Evaluation de la surface agricole utile du Fayoum (Egypte) sur image SPOT

1. Objectif de l'application

Le but de cette étude pilote était d'évaluer la surface agricole utile de la région du Fayoum à partir d'une image SPOT XS. Cette recherche a été menée conjointement par l'Egyptian Survey Authority et l'Institut Géographique National (France) en 1987.

2. Méthode utilisée

La détermination de la surface agricole utile a été effectuée par classification supervisée assistée par ordinateur. Le principe est de sélectionner sur l'image des échantillons correspondant à des zones homogènes, et de les identifier ensuite sur le terrain. Des traitements numériques classent alors les pixels de l'image suivant leur ressemblance radiométrique avec les pixels des échantillons, en fonction de leur "distance" aux centres de classe dans l'espace radiométrique. L'enquête de terrain et le choix des thèmes sont ici fondamentaux. Il importe également de vérifier a posteriori la validité et la qualité de la classification sur les zones de contrôle.

Surface agricole	Surface non-agricole
1. Cultures	5. Eau libre
2. Cultures arborées	6. Végétation naturelle
3. Terre en attente de culture	7. Terre non cultivée
4. Terre inondée	8. Villes

3. Résultats

Les résultats d'une première classification ont montré qu'en traitant d'un bloc la scène, des confusions entre certaines classes étaient inévitables.

Une stratification de l'image a alors été effectuée (4 niveaux) ainsi qu'une classification par strate.

La matrice de répartition sur l'ensemble des 4 niveaux, établie à partir de pixels tirés au hasard donne les résultats suivants:

Thème	Surface agricole
1	394 bien classés sur 398 vérifiés
2	45 bien classés sur 68 vérifiés
3	191 bien classés sur 209 vérifiés
4	19 bien classés sur 27 vérifiés

Thème	Surface non agricole
5	185 bien classés sur 187 vérifiés
6	2 bien classés sur 3 vérifiés
7	28 bien classés sur 63 vérifiés
8	36 bien classés sur 43 vérifiés

Confusion entre surface agricole et surface cultivée:

Surface agricole	96% (bien classés)
Surface non agricole	91,2% (bien classés)

Organizational Flow Chart

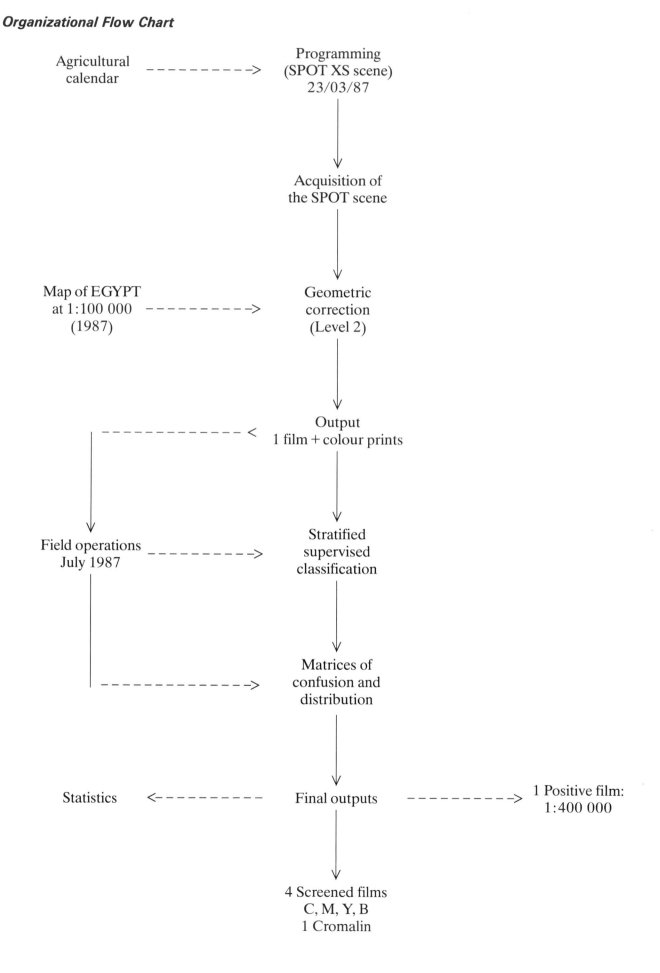

Organigramme du traitement

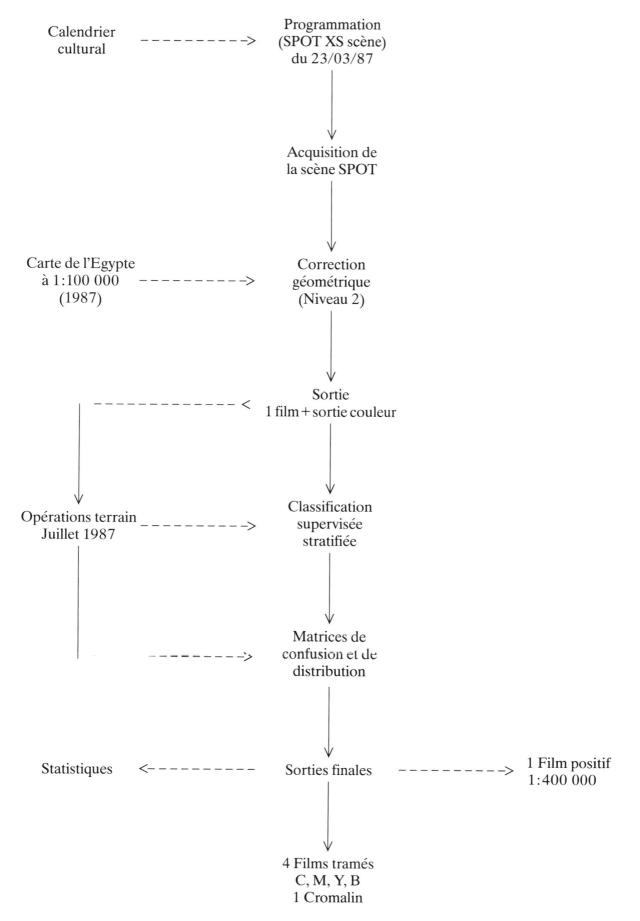

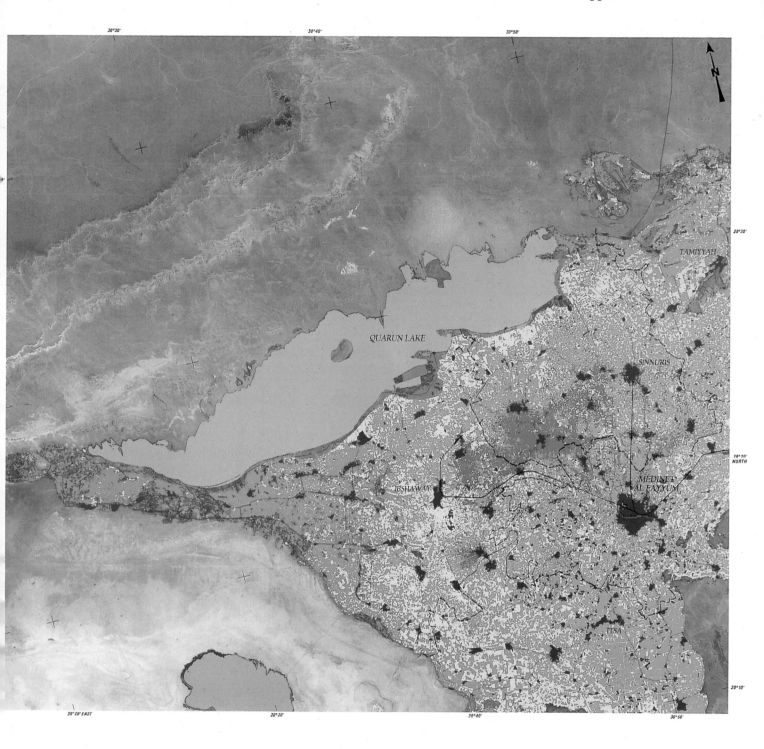

QUARUN LAKE

TAMIYYAH

SINNURIS

IBSHAWAY

MEDINET
AL FAYYUM

ITSA

ARABLE LAND AREA

CROPS
147 000 Fed.

TREE-CROPS
18 000 Fed.

FALLOWLAND
74 800 Fed.

FLOODED LAND
7 200 Fed.

TOTAL AREA ARABLE LAND ON IMAGE
247 000 Feddans

NON AGRICULTURAL AREA

OPEN WATER
66 800 Fed.

NATURAL VEGETATION
760 Fed.

BARE GROUND
UNCULTIVATED
11 450 Fed.

HOUSING
12 900 Fed.

TOTAL AREA NON AGRICULTURAL LAND ON IMAGE
91 910 Feddans

SAND

AREA IN FEDDANS
1 Fed. = 0,42 Hect. = 10,5 SPOT pixels

PLANIMETRY

ROADS

RAILWAYS

CANALS

- - - - - DISTRICT BOUNDARIES

© IGN PARIS 1988   © ESA LE CAIRE 1988
SPOT ® PRODUCT   CNES-IGN © CNES 1987

MAPPING OF OCEAN AND COASTAL ZONES

CARTOGRAPHIE OCEANIQUE ET LITTORALE

Application Title
Evaluation of potential sea-water farming zones in New Caledonia

Titre de l'application
Evaluation de zones à potentiel aquacole en Nouvelle Caledonie

Localization/Localisation
France (New Caledonia)/France (Nlle Calédonie)

RESPONSIBLE ORGANIZATIONS   ORGANISMES RESPONSABLES

IFREMER
Centre de Brest
Pêche aquaculture
B.P 70
29280 PLOUZANE
France

Tel: 33/ 98 22 43 18   Fax: 33/ 98 22 45 33

(Contact person: M. Jacques Populus)

15. Evaluation of potential sea-water farming zones in New Caledonia

Production of general-inventory maps at 1:150 000

The main advantage, for regional developers, is to be able to locate rapidly the size and positions of sites, and also the mutual positions of sites or groups of sites. That synoptic aspect is supplied by the small scale. The 1:150 000 scale selected here is compatible with the size of SPOT scenes. It makes it possible to safeguard the detailed aspect whilst at the same time obtaining a good overall view.

The western coast of New Caledonia is covered by six sheets, from the Belep Islands (sheet 1) to St Vincent's Bay (sheet 6).

Spot imagery, produced by Spot Image Co., in the Universal Transverse Mercator (UTM) projection are thus directly superposable on IGN-F maps of New Caledonia. After dynamic adaptations giving the best contrast, the images were transferred to an automatic cartography station for the following purposes:

(a) Computer-assisted photo-interpretation of potential sites, area calculations and production of closed loops surrounding groups of sites.

(b) Transfer of digitized main roads and communal boundaries on the IGN-F maps. Positioning of place-names.

(c) Production of the so-called cartographic attributes: geographic framework, references, grid, etc.

(d) Inset-data editing:
—On the right-hand side there are a certain number of meteorological and hydrological data, which are criteria of direct interest to shrimp aquaculture. Some are local (flow, rainfall) whilst others cover the entire territory.
—The lower part includes the legend for the superposed items, the sheet-index and the positions of the detailed maps at 1:25 000 scale relating to the sheet.

15. Evaluation des zones à potentiel aquacole en Nouvelle Calédonie

Confection des cartes d'inventaire général au 1:150 000

L'intérêt premier, pour l'aménageur, est de pouvoir situer rapidement la taille et la position des sites, et aussi les positions mutuelles des sites ou groupes de sites. Cet aspect synoptique est apporté par la petite échelle. Le 1:150 000 choisi ici est compatible avec la taille des scènes SPOT. Il permet de ne pas porter préjudice au détail tout en donnant une bonne vision d'ensemble.

La côte ouest de la Nouvelle Calédonie est couverte par six feuilles allant des îles Belep (feuille 1) à la Baie St Vincent (feuille 6).

Les images SPOT sont produites par la société SPOT Image en projection Mercator Transverse Universelle, donc directement superposables aux cartes IGN de la Nouvelle Calédonie. Après des adaptations de dynamique donnant le meilleur contraste, les images sont transférées sur un poste de cartographie automatique où sont effectuées plusieurs tâches:

(a) Photo-interprétation assistée par ordinateur des sites potentiels, calcul des surfaces et confection des "bulles" enveloppant les groupes de sites.

(b) Report des voies de communication et limites communales digitalisées sur les cartes IGN. Positionnement des toponymes.

(c) Confection des attributs cartographiques proprement dits: cadre géographique, repères, grille, etc.

(d) Rédaction de cartouche:
   —Sur la partie droite se trouvent un certain nombre de données de météorologie et d'hydrologie, critères d'intérêt direct pour l'aquaculture de crevettes. Certaines sont locales (débits, pluviométrie) d'autres valables pour tout le Territoire.
   —La partie inférieure comprend la légende des éléments en superposition, le tableau d'assemblage des feuilles et la position des cartes de détail au 1:25 000 relatifs à la feuille.

MAPPING OF ICE AND SNOW FIELDS
CARTE DES COUVERTS DE GLACE ET NEIGE

Application Title
Alpine snow cover mapping with Metsat data, USA

Titre de l'application
Carte de la couverture neigeuse alpine avec des données Metsat, EUA

Localization/Localisation
USA (Durango, Colorado)/EUA (Durango, Colorado)

RESPONSIBLE ORGANIZATIONS   ORGANISMES RESPONSABLES

U.S. Geological Survey,
National Mapping Division,
EROS Data Center,
Sioux Falls, South Dakota 57198 USA

Tel: 605/594-6114   Telex 910-668-0301

(Contact person: Joseph V. Baglio)

16. Alpine snow cover mapping with Metsat data, USA

1. Objectives

Seasonal alpine snowpacks provide a large percentage of fresh water supplies in the western United States and are primary sources of water for domestic supply, irrigation and electrical power generation, among other uses. Seasonal snowpacks are also causes of seasonal flooding in many areas and, as well, are fundamental elements in climate systems (Kukla *et al.*, 1977; Kukla, 1981). Periodic mapping for monitoring seasonal snowpack extents is, consequently, an important task (Rango and Martinec, 1986). Snow cover area mapping can be conducted utilizing Advanced Very High Resolution Radiometer (AVHRR) image data. Near real-time AVHRR data are available through the AVHRR Data Acquisition and Processing System (ADAPS) at the US Geological Survey, Earth Resources Observation Systems (EROS) Data Center in Sioux Falls, South Dakota, USA. ADAPS receives image data from AVHRR sensors on the NOAA-9 and NOAA-10 Sun-synchronous, polar-orbiting, Advanced TIROS-N series satellites. Up to eight daytime passes per day are provided from the two satellites for a receiving footprint covering the Coterminous United States.

2. Methods used

Multispectral (multichannel) AVHRR image data comprise radiances measured in the visible, reflected-infrared and thermal-infrared regions of the electromagnetic spectrum, with nominal image resolutions of 1.1 km at nadir. Digital image processing techniques have been developed for using these data in concert with a georegistered digital elevation model (MNT) for mapping snow cover area. The snow mapping process (see accompanying schematic) involves four distinct steps: cloud/snow discrimination, image normalization, snow classification and area calculation.

Channel 1 (0.58–0.68 μm wavelength), channel 3 (3.55–3.93 μm wavelength) and channel 4 (10.3–11.3 μm wavelength) image data are used to discriminate snow cover from cloud cover. Differences between the calibrated,

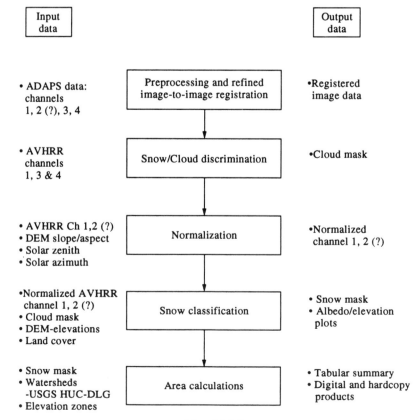

Fig. 5.16.1.

16. Carte de la couverture neigeuse alpine avec des enregistrements Metsat, EUA

1. Objectifs

Les chutes de neige alpine saisonnières procurent un vaste pourcentage de réserve en eau dans l'ouest des Etats-Unis et sont la principale source de fourniture en eau domestique, d'irrigation, de création d'énergie électrique, entre autres usages. Les chutes saisonnières sont également source d'inondations saisonnières dans beaucoup de régions tout comme elles sont des éléments fondamentaux du système climatique (Kukla *et al.*, 1977; Kukla, 1981). Une carte périodique pour suivre l'étendue des chutes de neige saisonnières est par conséquent une tâche importante (Rango et Martinec, 1986). La carte des zones neigeuses peut être réalisée en utilisant les données du radiomètre avancé à très haute résolution (AVHRR). Les données AVHRR, proches du temps réel, sont disponibles par l'intermédiaire du système d'observation des ressources de la terre (ADAPS) auprès du service géologique des US

au centre des ressources de la terre, Sioux Falls, Sud Dakota, USA. ADAPS reçoit les données du capteur AVHRR des satellites hélio-synchrone NOAA-9 et NOAA-10, d'orbite polaire, satellites des séries avancées TIROS-N. Près de huit passages par jour sont réalisés par les deux satellites pouvant capter l'ensemble des Etats-Unis.

2. Méthode utilisée

Les données AVHRR multispectrales (plusieurs canaux) dont les radiances sont mesurées dans le visible, l'infrarouge réfléchi et l'infrarouge thermique du spectre electromagnétique, ont une résolution nominale au nadir de 1,1 km. Les techniques de traitement d'images numériques ont été développées en utilisant des données couplées à un modèle numérique de terrain (MNT) géoenregistré pour cartographier les zones neigeuses. Le procédé de cartographie de la neige (voir schéma d'accompagnement)

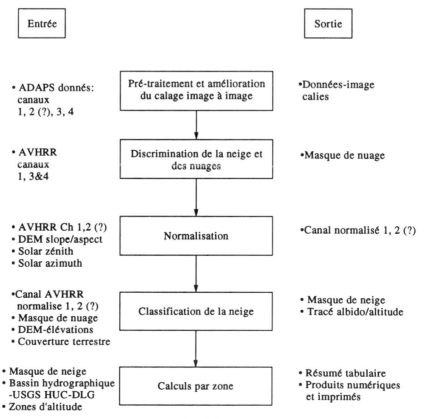

Fig. 5.16.1.

daytime, channel 3 apparent temperatures of clouds and most terrestrial surfaces, specifically snow, occur as a result of differences in the reflectivity of these surfaces at the channel 3 wavelength (Saunders, 1986). Digital analysis of the relationships between channel 1 reflectance and the difference of channel 3 and channel 4 apparent temperatures is completed to produce a cloud cover image.

Channel 1 and channel 2 data are normalized by correcting image reflectances for a common solar-illumination condition, for example, to a vertical Sun. The correction enables multidate comparison of the image data. A normalization method that incorporates terrain effects (Hoben and Justice, 1980) is currently being investigated.

A normalized reflectance image is analysed in an interactive procedure utilizing the cloud cover image, the MNT, and a classified forest canopy image to identify a minimum snowpack reflectance-elevation threshold. The minimum snowpack reflectance-elevation thresholds are used in a snowpack classification algorithm to identify:

—open, continuous snow cover,
—discontinuous snow cover or snow cover potentially obscured by forest canopy,
—snow cover inferred beneath cloud cover.

The classified snowpack image is overlayed with an image of selected surface water drainage basins for calculation of basin snow-cover percentages. Future research will examine inclusion of real-time point measurements of snowpack properties (Shafer, 1985) into the snow mapping process and will focus more closely on quantifying forest canopy effects.

3. Results

(a) Image processing techniques described

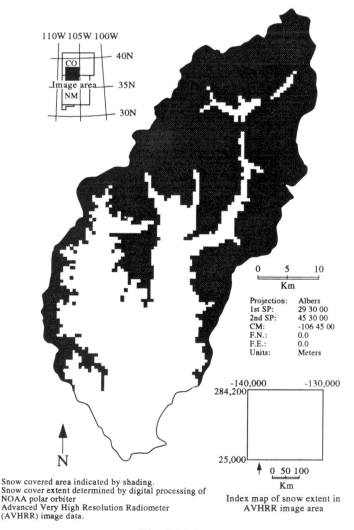

Snow covered area indicated by shading.
Snow cover extent determined by digital processing of NOAA polar orbiter
Advanced Very High Resolution Radiometer (AVHRR) image data.

Index map of snow extent in AVHRR image area

Fig. 5.16.2.

englobe 4 niveaux distincts: discrimination nuage-neige, image de normalisation, classification de la neige, zone de calcul.

Le canal 1 (0,58 à 0,68 μm de longueur d'onde), le canal 3 (3,55 à 3,93 μm de longueur d'onde) et le canal 4 (10,3 à 11,3 μm de longueur d'onde) fournissent les éléments de discrimination entre les couvertures neigeuses et les couvertures nuageuses. Les différences entre les calibrages des canaux 1 et 3, de jour, fournissent les températures des nuages et de la plupart des surfaces terrestres, en particulier la neige, apparaissent comme le résultat des différences de réflectivité de ces surfaces au canal 3 (Saunders, 1986). Les analyses numériques des relations entre la réflectance du canal 1 et les différences des températures apparentes des canaux 1–3 et des canaux 1–4, sont jointes pour fournir l'image de la couverture nuageuse.

Les données des canaux 1 et 2, sont normalisées par correction des réflectances selon des conditions données d'illumination solaire par exemple pour un soleil vertical. La correction permet des comparaisons multiples des données-images. Une méthode de normalisation incorporant des effets de terrain (Holben et Justice, 1980) est en cours d'investigation.

Une image de réflectance normalisée est analysée par une procédure interactive utilisant l'image du couvert nuageux, le MNT, et une image témoin des forêts pour identifier un minimum d'élévation du seuil de réflectance de la couverture neigeuse. Les minimums d'élévation du seuil de réflectance de la couverture neigeuse sont utiles dans l'algorithme de classement des couches neigeuses:

—couverture ouverte ou continue,
—couverture discontinue ou potentiellement masquée par couvert forestier,
—neige identifiée sous une couverture nuageuse.

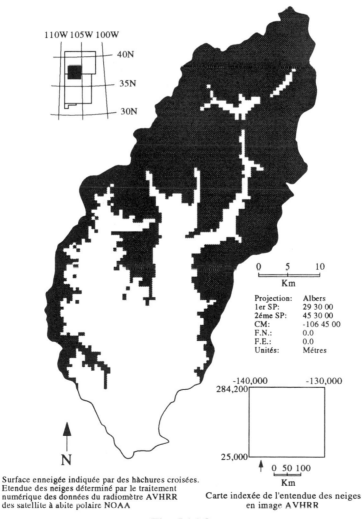

Surface enneigée indiquée par des hàchures croisées.
Etendue des neiges déterminé par le traitement numérique des données du radiomètre AVHRR des satellite à abite polaire NOAA

Carte indexée de l'entendue des neiges en image AVHRR

Fig. 5.16.2.

above are being adapted for use in a National Oceanic and Atmospheric Administration (NOAA), National Weather Service operational snow cover monitoring programme.

(b) Digital AVHRR image data, resampled to 540 m pixel resolution, were utilized for mapping snow cover over selected basins of the San Juan Mountains, Colorado, USA. Experimental snow cover maps were generated for 15 different dates during the 1987 and 1988 snow cover mapping seasons. Processing of digital Landsat Thematic Mapper data, co-incident with two AVHRR image dates during the 1987 mapping season, verified the results.

(c) Snow-covered area within the Animas River Basin above Durango, Colorado mapped for 12 April 1988 (see accompanying map) illustrates mapping capabilities. Note that the snow classification scheme described above has been generalized into a snow/no-snow mask. Also note that the accompanying map was produced using a computer based geographic information system (see Figs 5.16.1 and 5.16.2).

4. References

HOLBEN, B. N. and JUSTICE, C. O. 1980. The topographic effect on spectral response from nadir-pointing sensors. *Photogrammetric Engineering and Remote Sensing*, vol. 46, No. 9, pp. 1191–1200.

KUKLA, G. J., ANGELL, J. K., KORSHOVER, J., DRONIA, H., HORSHIAI, M., NAMIAS, J., RODWALD, M., YAMAMOTO, R. and IWASHIMA, T. 1977. New data on climatic trends. *Nature* No. 270, pp. 573–580.

KUKLA, G. J. 1981. *Snow Covers and Climate: Glaciological Data.* Report GD-11 (Snow Watch 1980), World Data Center a for Glaciology (Snow and Ice), Boulder, Colorado, pp. 27–39.

RANGO, A. and MARTINEC, J. 1986. The need for improved snow-cover monitoring techniques, hydrologic applications of space technology. *Proceedings of the Cocoa Beach Workshop*, Florida, August 1985. IAHS Publication No. 160, pp. 173–179.

SAUNDERS, R. W. 1986. An automated scheme for removal of cloud contamination from AVHRR radiances over western Europe. *Int. J. of Remote Sensing*, vol. 7, No. 7, pp. 867–886.

SHAFER, B. A. 1985. Integration of SNOTEL and remotely sensed snow cover area in water supply forecasting. *Proceedings of the Nineteenth International Symposium on Remote Sensing of Environment.* Ann Arbor, Michigan, October 1985, Environmental Research Institute of Michigan, pp. 1045–1056.

L'image des zones enneigées classées est superposée à une image des surfaces des bassins versants pour calculer les pourcentages de bassins recouverts de neige. De futures recherches examineront l'inclusion en temps réel des mesures ponctuelles des propriétés du couvert neigeux (Shafer, 1985) dans le processus de cartographie de la neige et permettra de quantifier au plus près les effets du couvert forestier.

3. Résultats

(a) Les techniques de traitement d'image décrites ci-dessus vont être adoptées pour être utilisées par la NOAA, au service national météorologique, pilotant un programme de suivi du couvert neigeux.

(b) Les données numériques de AVHRR, rééchantillonnées à une dimension de pixel de 540 m, furent utilisées pour cartographier le couvert neigeux de bassins sélectionnés des montagnes de San Juan—Colorado—USA. Des cartes expérimentales furent générées à 15 dates différentes, en 1987 et 1988 durant les saisons neigeuses. Un traitement de données Landsat TM, concomitantes avec 2 données AVHRR durant la saison de 1987 a vérifié les résultats.

(c) La zone neigeuse du bassin "Animas River", au dessus de Durango, Colorado, cartographiée le 12 avril 1988 (voir document ci-joint) illustre les possibilités cartographiques. Notons que le schéma de classification de la neige décrit plus haut a été généralisé par un masque neige/non-neige. Notons aussi que la carte d'accompagnement a été réalisée avec un système d'information géographique (voir figs 5.16.1 et 5.16.2).

4. Références

HOLBEN, B. N. and JUSTICE, C. O. 1980. The topographic effect on spectral response from nadir-pointing sensors. *Photogrammetric Engineering and Remote Sensing*, vol. 46, No. 9, pp. 1191–1200.

KUKLA, G. J., ANGELL, J. K., KORSHOVER, J., DRONIA, H., HORSHIAI, M., NAMIAS, J., RODWALD, M., YAMAMOTO, R. and IWASHIMA, T. 1977. New data on climatic trends. *Nature* No. 270, pp. 573–580.

KUKLA, G. J. 1981. *Snow Covers and Climate: Glaciological Data*. Report GD-11 (Snow Watch 1980), World Data Center a for Glaciology (Snow and Ice), Boulder, Colorado, pp. 27–39.

RANGO, A. and MARTINEC, J. 1986. The need for improved snow-cover monitoring techniques, hydrologic applications of space technology. *Proceedings of the Cocoa Beach Workshop*, Florida, August 1985. IAHS Publication No. 160, pp. 173–179.

SAUNDERS, R. W. 1986. An automated scheme for removal of cloud contamination from AVHRR radiances over western Europe. *Int. J. of Remote Sensing*, vol. 7, No. 7, pp. 867–886.

SHAFER, B. A. 1985. Integration of SNOTEL and remotely sensed snow cover area in water supply forecasting. *Proceedings of the Nineteenth International Symposium on Remote Sensing of Environment*. Ann Arbor, Michigan, October 1985, Environmental Research Institute of Michigan, pp. 1045–1056.

MAPPING OF VEGETATION

CARTOGRAPHIE DE LA VEGETATION

Application Title
Seasonal vegetation greenness mapping in Africa for grasshopper and locust control

Titre de l'application
Carte de "verdeur" de la végétation saisonnière en Afrique pour le contrôle des sauterelles et criquets

Localization/Localisation
West Africa/Afrique de l'Ouest
Mali/Mali
Senegal/Sénégal

RESPONSIBLE ORGANIZATIONS   ORGANISMES RESPONSABLES

U.S. Geological Survey,
National Mapping Division,
EROS Data Center,
Sioux Falls, South Dakota 57198 USA

Tel: 605/594-6114   Telex 910–668–0301

(Contact person: G. Gray Tappan)

17. Seasonal vegetation greenness mapping in Africa for grasshopper and locust control

1. Objectives

Large grasshopper populations plague West Africa with the return of near normal rainfall in the past several years. In 1985 and 1986, the Senegalese grasshopper, *Oedaleus senegalensis*, threatened cropland in a number of Sahelian countries. A major upsurge in the population of the desert locust, *Schistocerca gregaria*, occurred in late 1987, resulting in the present plague situation which threatens agriculture over vast areas in the northern half of Africa. Major international donors have mounted emergency assistance programmes for grasshopper and locust control. In early 1987, the US Agency for International Development (USAID) prepared for a major campaign in West Africa. One goal of the programme was to improve grasshopper prediction and survey techniques. The ability to monitor the distribution and seasonal changes of natural and agricultural vegetation in the Sahel was considered as an integral aspect of grasshopper and locust control efforts. For this reason, the Bureau for Africa of USAID requested that the EROS Data Center of the US Geological Survey (USGS) conduct a pilot project to develop, test and evaluate a near-real-time monitoring procedure using satellite data and geographic information system technologies in support of their grasshopper and locust control programmes. With the successful completion of the pilot project, the programme continued in an operational mode in 1988–89 in support of continent-wide efforts to combat the present locust plague.

2. Methods used

The approach to monitoring and mapping the distribution and growth of seasonal vegetation over large regions involves the use of satellite data from the NOAA TIROS-N series of polar-orbiting satellites, carrying the AVHRR (Advanced Very High Resolution Radiometer) sensor. Image data recorded over West Africa are purchased in digital form. Three to five images per week over a given region are ingested into the AVHRR Data Reception and Processing System (ADAPS) where they are screened for cloud cover, radiometrically calibrated and registered to a map base. Using the visible (channel 1) and near-infrared (channel 2) spectral data, the images are then transformed into the normalized difference vegetation index (NDVI), which is strongly correlated to amounts of green vegetation cover and biomass. The individual NDVI images are cloud-masked and combined over a two-week period to generate a single, maximum value NDVI composite scene. The range of digital values is subdivided into 19 different colours based on a natural colour sequence for ease of interpretation. The scaling is designed to be most sensitive to emerging and low amounts of vegetation growth. Finally, the composite scenes are combined with a cartographic database containing locational information (including roads, place names, administrative boundaries) to produce the final "greenness" maps. The maps are field-oriented hard copy products at scales ranging from 1:1 000 000 to 1:2 500 000.

3. Results

The vegetation index or greenness maps are the primary products of the programme. The maps are produced at 14-day intervals through the rainy season. Each series of country-wide maps depicts the complex seasonal green-up and dry-down vegetation patterns and the relative amount of green vegetation. The one-kilometre spatial resolution of the image data is considered particularly useful and critical for detailed monitoring of vegetation within such important locust breeding areas as wadis in arid areas. In 1987, maps were produced for Senegal, the Gambia, Mauritania, Niger and Chad. In 1988, maps were being produced to monitor the summer seasonal vegetation patterns of Mauritania, Mali, Niger, Chad and the Sudan, while winter season maps were being prepared for Morocco, Algeria and Tunisia.

The maps are used primarily by the crop protection services and co-operating international donor organizations for monitoring favourable grasshopper and locust habitats. Pest control teams in each country use the maps for

17. Carte de "verdeur" de la végétation saisonnière en Afrique pour le contrôle des sauterelles et criquets

1. Objectifs

De nombreuses nuées de sauterelles ont dévasté l'Afrique de l'Ouest avec le retour des chutes de pluie quasi-normales durant les dernières années. En 1985 et 1986, les sauterelles du Sénégal, *Oedaleus senegalensis*, menacèrent les récoltes dans bon nombre de contrées sahéliennes. Une grande vague de criquets du désert, *Schistocerca gregaria*, déferla au début de 1987; le résultat en est la situation préoccupante actuelle qui menace l'agriculture sur de vastes étendues dans la moitié nord de l'Afrique. Beaucoup d'instances internationales ont mis sur pied des programmes d'urgence pour lutter contre les sauterelles et criquets. Au début de 1987, l'agence américaine pour le développement international (USAID) a préparé une importante campagne en Afrique de l'Ouest. Le but du programme était de tester certaines techniques de surveillance et de prédiction des sauterelles. L'aptitude à suivre la distribution et les changements saisonniers de la végétation naturelle et agricole du Sahel était considérée comme un aspect primordial des efforts de lutte contre les sauterelles et criquets. Pour cette raison, le bureau USAID pour l'Afrique demanda au centre de données EROS du service géologique des US (USGS) d'élaborer un projet-pilote pour développer, tester, et évaluer une procédure de suivi en temps réel, à partir de données satellitaires et de systèmes d'information géographique, en appui à leur programme de contrôle des sauterelles et criquets. Avec l'achèvement réussi du projet-pilote, le programme a été poursuivi en mode opérationnel en 1988–89 servant de base aux efforts déployés à l'échelle continentale pour combattre le fléau des criquets.

2. Méthodes utilisées

Mener à bien un projet cartographique de distribution et d'évolution de la végétation saisonnière de vastes régions implique l'utilisation de données-satellites NOAA TIROS-N, d'orbite polaire, disposant du capteur AVHRR (radiomètre à très haute résolution). Les images enregistrées sur l'Afrique de l'Ouest sont acquises sous forme numérique. De trois à cinq images par semaine sur une région donnée sont saisies par le système de réception et de traitement de données (ADAPS) du capteur AVHRR, où elles sont traitées pour le couvert nuageux, la calibration radiométrique puis calées sur une base cartographique. Avec les données spectrales du visible (canal 1) et du proche infrarouge (canal 2) les images fournissent un index normalisé des différentes végétations (NDVI), qui est fortement corrélé à la biomasse et à la verdeur du couvert végétal. Les images NDVI sont corrigées des couvertures nuageuses et combinées, sur une période de deux semaines, en une seule avec les valeurs les plus fortes des NDVI. L'éventail des valeurs numériques est divisé en 19 plages colorées en une séquence de couleurs naturelles pour faciliter l'interprétation. La palette est choisie pour être sensible aux fortes et faibles variations du couvert végétal. Puis les scènes composées sont combinées à une base cartographique comportant les informations locales (réseau routier, toponymie, limites administratives) pour produire les cartes finales de la végétation saisonnière. Les cartes sont éditées à des échelles comprises entre le 1:1 000 000 et le 1:2 500 000.

3. Résultats

Les cartes d'index de végétation ou de "verdeur" sont les premiers produits à intervalles de 14 jours durant la saison des pluies. Chaque série de cartes à couverture nationale présente les modèles complexes de croissance/décroissance de la végétation et le taux relatif de végétation verte. La résolution de 1 km de l'image spatiale peut être considérée comme particulièrement appropriée pour suivre en détail la végétation des zones où sévissent les criquets, comme les oueds en contrées arides. En 1987, des cartes furent produites pour le Sénégal, la Gambie, la Mauritanie, le Niger et le Tchad. En 1988, des cartes ont été produites pour suivre les modèles de végétation saisonnière d'été en Mauritanie, Mali, Niger, Tchad et Soudan tandis que des cartes pour la saison d'hiver ont été réalisées pour le Maroc, l'Algérie et la Tunisie.

planning and conducting field and aerial surveys in areas with potential infestations. For grasshoppers, and in particular the Senegalese grasshopper (*Oedaleus senegalensis*) which is known to cause severe crop damage, map use is based on the principle that initial seasonal rainfall triggers both the growth of herbaceous vegetation and, if present, the hatching of grasshopper eggs. For locusts, and in particular the desert locust, the maps serve as indicators of areas favourable for reproduction and development (source areas) as well as sites likely to be invaded by migrating locust swarms (invasion areas). By focusing on land areas that are greening or are presently green, survey teams can significantly narrow down the areas to be covered by both land and aerial survey efforts, avoiding areas that are dry. This is a major consideration given the vastness of the Sahelian and Saharan regions, the inadequacy or unavailability of environmental data and the high cost of conducting aerial surveys.

4. References

HOLBEN, B. N. 1986. Characteristics of maximum-value composite images from temporal AVHRR data. *International Journal of Remote Sensing*, vol. 7, No. 11, pp. 1417–1434.

LAUNOIS, M. 1979. An ecological model for the study of the grasshopper *Oedaleus senegalensis* in West Africa. *Trans. Roy. Soc. Lond.* vol. B 287, pp. 345–355.

STANCIOFF, A., STALJANSSENS, S. and TAPPAN, G. 1986. *Mapping and Remote Sensing of the Resources of the Republic of Senegal: A Study of the Geology, Hydrology, Vegetation and Land Use Potential.* Remote Sensing Institute, South Dakota State University, Brookings, South Dakota. SDSU-RSI-86-01.

TAPPAN, G., LOVELAND, T., ORR, D., MOORE, D., HOWARD, S. and TYLER, D. 1988. *Pilot Project for Seasonal Vegetation Monitoring in Support of Grasshopper and Locust Control in West Africa.* US Geological Survey, International Technical Report, EROS Data Center, USGS, Sioux Falls, South Dakota.

Les documents sont utilisés principalement par les services de protection des récoltes et les organisations d'aide et coopération internationales pour une détection efficace des habitats de sauterelles et de criquets. Des équipes de contrôle de la peste, dans chaque pays, utilisent ces documents pour planifier et organiser une surveillance terrestre et aérienne des zones à haut risque d'infection. Pour les sauterelles, et en particulier celles du Sénégal (*Oedaleus senegalensis*) qui sont réputées pour les sévères dommages causés aux récoltes, l'utilisation de la carte est basée sur le principe que le début de la saison des pluies indique la croissance de la végétation herbacée et l'incubation des oeufs de sauterelles si elles sont présentes. Pour les criquets, et en particulier ceux du désert, les cartes peuvent indiquer les zones favorables à la reproduction et au développement (zones de croissance) ainsi que les sites pouvant être envahis par des essaims (zones d'invasion). En repérant les aires qui verdoient ou qui sont déjà vertes, les équipes de surveillance peuvent facilement y porter leurs efforts tant terrestres qu'aériens en évitant celles qui sont sèches. C'est une considération importante étant donnée l'étendue des régions sahéliennes et sahariennes, l'inadéquation ou l'absence de données sur l'environnement et le coût élevé d'une surveillance aérienne.

4. References

HOLBEN, B. N. 1986. Characteristics of maximum-value composite images from temporal AVHRR data. *International Journal of Remote Sensing,* vol. 7, No. 11, pp. 1417–1434.

LAUNOIS, M. 1979. An ecological model for the study of the grasshopper *Oedaleus senegalensis* in West Africa. *Trans. Roy. Soc. Lond.* vol. B 287, pp. 345–355.

STANCIOFF, A., STALJANSSENS, S. and TAPPAN, G. 1986. *Mapping and Remote Sensing of the Resources of the Republic of Senegal: A Study of the Geology, Hydrology, Vegetation and Land Use Potential.* Remote Sensing Institute, South Dakota State University, Brookings, South Dakota. SDSU-RSI-86-01.

TAPPAN, G., LOVELAND, T., ORR, D., MOORE, D., HOWARD, S. and TYLER, D. 1988. *Pilot Project for Seasonal Vegetation Monitoring in Support of Grasshopper and Locust Control in West Africa.* US Geological Survey, International Technical Report, EROS Data Center, USGS, Sioux Falls, South Dakota.

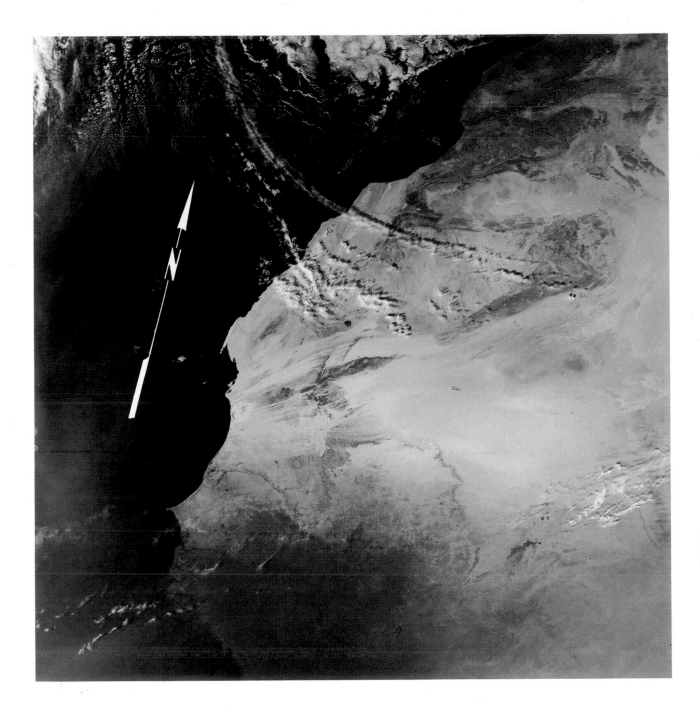

Figure 1. A NOAA-9 AVHRR color composite image of West Africa acquired on January 2, 1987. Digital data from images like this one were used to compute the Normalized Difference Vegetation Index for the Vegetation Index Maps. Image scale : 1:12,960,000.

Figure 1. Image en composition colorée, du capteur AVHRR du satellite NOAA-9, sur l'Afrique de l'ouest, prise le 2 janvier 1987. Les données numériques d'images semblables ont été utilisées pour calculer l'index normalisé de différence de végétation pour les cartes d'index de végétation. L'échelle est de 1:12 960 000

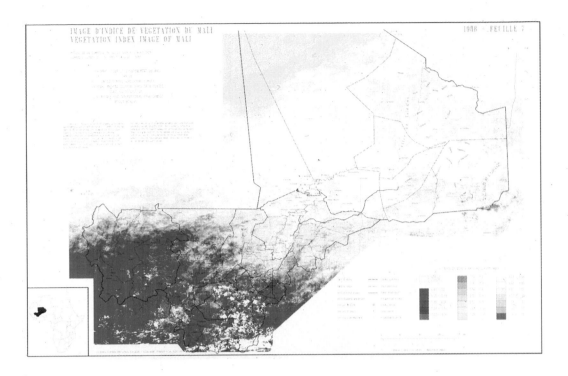

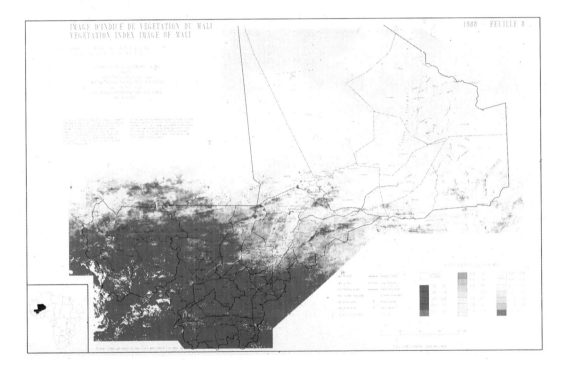

Figure 2.An example of two consecutive Vegetation Index or "greenness" maps produced for Mali. The top map represents vegetation greenness conditions between July 19 and August 1, 1988, while the bottom one covers the period of August 2 to August 15, 1988. Notice the striking increase in vegetation growth, and the development of greenness in the wadis (linear drainage features) of northeastern Mali. Greenness in the wadis indicates very favorable breeding conditions for the Desert Locust. Original map scale is 1:2,500,000.

Figure 2. Un exemple de deux index de végétation consécutifs ou cartes de "verdoyance", produites pour le Mali. La carte du haut représente les conditions de la végétation verdoyante entre le 19 juillet et le 1er août 1988, tandis que celle du bas concerne la période du 2 au 15 août 1988. Noter la nette croissance végétale et le développement de la verdoyance dans les oueds (traînées linéaires) du Nord-Est du Mali. La verdoyance dans les oueds indique des conditions favorables à la reproduction des criquets du désert.
Carte originale à l'échelle du 1:2 500 000.

Figure 3. A series of Vegetation Index Maps of Senegal indicating the development of vegetation cover as the rains progress northwards. The adjacent ground photos were taken near Tiel, Senegal. The photo dates of June 22, July 21, and August 3, 1987 fall within the three periods represented by the maps. A qualitative comparison between the map greenness values and ground conditions can be made.

Figure 3. Un ensemble de cartes d'index de végétation du Sénégal, indiquant le développement de la couverture végétale durant la progression des pluies vers le nord. Les photos adjacentes ont été prises près de Tiel (Sénégal). Les dates de prises de vues : 22 juin, 21 juillet et 3 août 1987 coïncident avec les périodes représentées par les cartes. Une comparaison qualitative peut être faite entre les valeurs de "verdoyance" des cartes et les conditions au sol.

LANDSCAPE MAPPING

CARTOGRAPHIE DES PAYSAGES

Application Title
Three-dimensional remote sensing data representation for regional planning

Titre de l'application
Représentation tridimensionnelle de données de télédétection pour l'aménagement régional

Localization/Localisation
Austria (Styria)/Autriche (Styrie)

RESPONSIBLE ORGANIZATIONS   ORGANISMES RESPONSABLES

Institute for Image Processing and Computer Graphics
Joanneum Research
Wastiangasse 6
A-8010 Graz, Austria

Tel: 43/316 8021   Fax: 43/316 8021/20

(Contact person: Dr Manfred Buchroithner)

18. Three-dimensional remote sensing data representation for regional planning

1. Objectives

Within the last few years the environmental consciousness in regional planning has been increasing tremendously. For certain planning projects, in particular in alpine terrain, remote sensing data application seems to be a most appropriate tool. The superposition of space-borne multispectral remote sensing data over a high-resolution digital elevation model (MNT) and its oblique perspective representation was one aim of preparatory activities for a regional development project in the Styrian Alps, Eastern Austria. The objective was to present most natural views of the Erzberg area displaying the actual recent conditions. This area contains an abandoned surface iron mining site with all the problems induced by the closing of the mine. The local authorities are looking for alternative uses, and perspective views based on a MNT and spaceborne satellite data and simulations of planning variants should assist in decision making.

2. Methods used

In order to achieve a most nature-like depiction of the project area, the remote sensing data were geocoded using a high-precision geocoding approach based on a 10-m resolution MNT.

The digital elevation model was created by means of the software package GTM (Graz Terrain Model) developed at the Institute for Image Processing and Computer Graphics in Graz. The input data for GTM were map contour lines and some special information, e.g. hilltops, drain and ridge lines. The contour lines were scanned from the official Austrian Topographic Map 1:50 000, and the additional information was digitized. This data was then used to compute the MNT with algorithms that interpolate raster heights from the digitized contour lines. The output of GTM is a raster of heights, slope gradients and slope directions. Slope gradient and slope direction data were used to determine the normal vectors for later use in rendering the landscape.

The MNT grid used for the Erzberg area had a size of 1290×1040 cells. As the resolution was 10 m, an area of $12.9 \times 10.4 \ \text{km}^2$ located around a centre point at $47°32'30''$ latitude and $14°53'20''$ longitude had to be treated.

As for the satellite data used, a hybrid approach was taken. Impressive results could be obtained by combining the following two types of satellite image data to compute a texture: Landsat Thematic Mapper (TM) and SPOT HRV panchromatic data. The TM satellite image contains colour information on the basis of 30×30 m pixels, and the SPOT image data contain contrast information with a resolution of 10 m. Using a colour look-up table, the spectral TM bands 3, 2 and 1 can be mapped into the RGB channels of a "true colour" image.

The two satellite images were then combined to get a true colour texture with a resolution of 10 m. However, the geometric rectification using the MNT had to be performed before the data could be used for texture generation.

After the geocoding process, the two images were combined. First, the TM image has been enlarged by a factor of 3 and transformed from the RGB into the IHS colour space. Then the intensity band of the SPOT image was added to the TM image using a look-up table. Finally, the result was retransformed into the RGB colour space.

The texture in the Erzberg area has been generated using TM scene 190–127 and SPOT scene 67–254. The selection of the TM bands (band 1 in blue, band 2 in green, band 3 in red), the colour adjustment using a LUT, and the combination of the TM and SPOT image data was performed with the Institute's Context Vision GOP 302 image processing system.

For the rendering of the perspective scene, a particular method adapted at the Institute for Image Processing and Computer Graphics in Graz was used.

The MNT and the surface information given by the satellite data can be represented by a regular quadrilateral mesh with height, colour and normal vector in each grid point. An alternative is the use of a triangular mesh. However, by subdividing the quadrilateral mesh into a triangular mesh, a higher amount of shading artefacts can be observed than when using Gouraud shading. As the rendering had to be performed at a graphic workstation which

18. Représentation tridimensionnelle de données de télédétection pour l'aménagement régional

1. Objectifs

Durant les dernières années l'importance de l'environnement dans l'aménagement régional s'est considérablement développée. Pour certains projets d'aménagement, en particulier en zone alpine, les applications à partir de données satellitaires semblent être l'instrument le mieux approprié. La superposition de données de télédétection satellitaire multispectrale et d'un modèle numérique de terrain de haute résolution (MNT) puis sa représentation en perspective oblique furent l'objet d'activités préparatoires pour un projet de développement régional dans les Alpes de Styrie, en Autriche de l'Est. L'objectif était de montrer des vues les plus naturelles de la région d'Erzberg décrivant les récentes conditions réelles. Cette région contient un site métallifère minier abandonné, avec tous les problèmes induits par la fermeture de la mine. Les autorités locales sont à la recherche de solutions de rechange et les vues perspectives appuyées sur un MNT et les données satellitaires ainsi que les diverses simulations de développement peuvent aider à la décision.

2. Méthodes utilisées

Afin d'atteindre à une représentation la plus naturelle de la zone, les données de télé-détection s'appuyaient sur un MNT défini avec une résolution de 10 m.

Le modèle numérique de terrain (MNT) fut réalisé au moyen du progiciel GTM (Graz Terrain Model) développé à l'Institut de traitement d'image et de cartographie auto-matique de Graz. Les données d'entrée pour GTM furent les courbes de niveau et des informations spéciales comme les sommets, les lignes de crête et les talwegs. Les courbes de niveau ont été scannées à partir de la carte topographique officielle d'Autriche au 1:50 000 et les informations complémentaires furent numérisées. Ces informations furent alors utilisées pour réaliser le MNT avec des algorithmes interpolant les altitudes aux noeuds d'une grille, à partir de courbes de niveau. Le résultat du GTM est une grille de points cotés,

de gradients et directions de pente. Gradients et direction de pente servent à déterminer la normale au terrain pour ensuite représenter le paysage.

La grille MNT utilisée pour la zone d'Erzberg avait une dimension de 1290×1040 mailles. Avec une résolution de 10 m, une zone de 12,9 km sur 10,4 km autour du point central de 47°32′30″ de latitude et 14°53′20″ de longitude a été traitée.

Pour les données satellitaires utilisées, une approche mixte a été adoptée. Des résultats spectaculaires peuvent être obtenus en combinant les deux types de données satellitaires suivants pour composer la texture: Landsat Thematic Mapper (TM) et SPOT-HRV panchromatique. Les images satellitaires TM contiennent des informations colorées sur la base d'un pixel de 30 m $\times$ 30 m alors que l'image SPOT fournit des informations contrastées avec une résolution de 10 m. En utilisant une table de couleurs, les bandes spectrales 3, 2 et 1 de TM peuvent être cartographiées dans les 3 canaux d'une image en "vraie couleur" (RVB).

Les deux images satellitaires sont alors combinées pour obtenir une texture en vraie couleur d'une résolution de 10 m. Cependant, une rectification géométrique utilisant le MNT a du être réalisée avant que les données puissent être utilisées pour générer la texture.

Après le procédé de géocodage, les deux images furent combinées; tout d'abord l'image TM a été agrandie d'un facteur 3 et transformée de l'espace coloré RVB dans celui dénommé IHS. Ensuite la bande d'intensité de l'image SPOT était ajoutée à l'image TM en utilisant une table de correspondance. Finalement le résultat était retransformé dans l'espace de couleur RVB.

La texture de la zone à Erzberg a été réalisée en utilisant les scènes TM 190-127 et SPOT 67-254. La sélection des bandes TM (bande 1 en bleu, bande 2 en vert, bande 3 en rouge), l'ajustement des couleurs utilisant un LUT, la combinaison des images TM et SPOT ont été réalisés par un système de traitement d'image de l'Institut, le "Context Vision GOP 302".

Pour le rendu de la perspective on a utilisé une méthode particulière mise au point à

creates Gouraud shading, a quadrilateral mesh was used.

A perspective transformation of a MNT causes a varying data density within the picture. The perspective distortion will be specially visible if the viewpoint is near the ground level of the MNT, which results in a higher amount of information at the horizon than near the viewpoint. In this case, the patches far from the viewpoint result in an inadequate picture quality in the respective region of the landscape.

This problem was solved by using non-rectangular meshes, e.g. radial meshes with different resolutions. The use of non-rectangular meshes only works well when one creates a still picture, because the geometry of the mesh depends on the camera position. In order to save valuable computing time, the problem was solved by using a pyramid structure of the data.

An image pyramid keeps the same data in different resolutions. Different filter algorithms can be used to reduce the resolution of data. An appropriate filter for the MNT data is a 2×2 or 3×3 average filter, and for the texture a Gaussian filter. The Gaussian filter minimized the aliasing effects due to textures. A sequence of filter operations generated the MNT and the texture information as required for efficient rendering.

Optimal rendering, which provides constant data density within the picture, could be achieved by computing the adequate resolution for each patch of the regular quadrilateral mesh depending on the camera parameter. Adjacent patches of the mesh have then been used at the same resolution. A division of the landscape into subareas helped to enhance the rendering in the following way: the optimal resolution was computed only once for all patches of an area, and a quick software clipping was performed for

this area. Therefore the landscape was divided in equal-sized areas of the x–y plane. The information stored for each area is its bounding volume as well as the MNT and the texture data which are available in different resolutions in the pyramid. The bounding volume is stored to speed up the software clipping of whole areas in the landscape.

A special procedure first checks whether an area's bounding volume is visible or not. If it was visible, the optimal resolution for the area was computed. The variable "number of patches" gave the maximal number of patches in x- or y-direction within an area. Another parameter specified the quality of the rendering. The calculated maximal edge length of the frustum was then compared with the side length, which is an estimation of an area's side length in picture coordinates, to choose a pyramid level.

By using the above pyramid data structure, an approximately constant data density could be achieved after the perspective transformation.

3. Results

Based on the above described method of high-quality remote sensing, a secondary image product has been generated. The advantages of this map-like product, which provides an excellent planning tool, is that it consumes comparatively low computing costs due to the image pyramid approach and represents a very natural view of the landscape due to the combination of Landsat TM and SPOT HRV panchromatic data.

In addition, based on the introduction of camera parameters varying from scene to scene, the production of an overflight simulation can be achieved. (Which, by the way, has been performed in this project.)

l'Institut de traitement d'image et de cartographie automatique de Graz.

Le MNT et les informations satellitaires de surface peuvent être représentés par un réseau quadrilatère régulier avec altitude, couleur et vecteur normal en chaque point de la grille. Une autre possibilité est l'utilisation d'un réseau triangulaire. Cependant, en subdivisant le réseau quadrilatère en un réseau triangulaire, un plus grand nombre d'artefacts d'ombre peut être observé quand on utilise l'estompage Gouraud. Comme le rendu devait être traité à l'atelier graphique qui a créé l'estompage Gouraud, on a utilisé un réseau quadrilatère.

Une transformation perspective du MNT entraîne une variation de la densité d'information au sein de l'image. La distorsion de perspective sera particulièrement visible si le point de vue est proche du niveau de base du MNT, avec comme résultat une plus grande densité d'informations à l'horizon que près du point de vue. Dans ce cas, les mailles éloignées du point de vue peuvent être inférieures à un pixel, ce qui donne souvent un mauvais rendu. En plus, les mailles élémentaires proches du point de vue donnent une qualité d'image inappropriée dans cette partie du paysage.

Ce problème fut résolu en utilisant un réseau non rectangulaire, à savoir un réseau radial avec différentes résolutions. L'usage d'un réseau non-rectangulaire donne de bons résultats quand on crée une image fixe, car la géométrie du réseau dépend de la position de la caméra. Dans le but d'économiser du temps d'ordinateur, le problème fut résolu en utilisant une structure pyramidale des données.

Une image pyramidale garde les mêmes données dans différentes résolutions. Divers algorithmes filtrants peuvent être utilisés pour réduire la résolution. Un filtre approprié aux données MNT est un filtre de moyenne sur des données 2×2 ou de 3×3, et pour la texture un filtre gaussien. Le filtre gaussien minimise les effets déformants dus à la texture. Une suite d'opérations de filtrage a généré le MNT puis l'information de texture, de façon à obtenir un rendu de qualité.

Un rendu optimal qui fournit une densité d'information constante dans l'image peut être atteint en calculant la résolution adéquate pour chaque maille élémentaire du réseau quadrilatère régulier en fonction des paramètres de la prise de vues. Les mailles adjacentes du réseau sont affectées alors de la même résolution. Un partage du paysage en sous-zones permet d'améliorer le rendu de la façon suivante: la résolution optimale était calculée une seule fois pour toutes les taches élémentaires d'une zone, et un rapide découpage par programme fut réalisé pour cette zone. Le paysage était donc divisé en aires égales du plan x–y. Les informations stockées pour chaque aire est son volume englobant ainsi que le MNT et les données de texture qui sont disponibles à différentes résolutions dans la pyramide. Le volume englobant est stocké pour accélérer le découpage par logiciel pour toutes les aires du paysage.

Une procédure spéciale contrôle tout d'abord si le volume englobant de chaque aire est visible ou non. S'il est visible, la résolution optimale pour cette aire est définie. Le paramètre "nombre de taches" donne le nombre maximum de taches élémentaires dans les directions x et y de l'aire choisie. Un autre paramètre spécifie la qualité du rendu. La longueur maximale du tranchant de la pyramide est alors comparée avec la longueur du côté, qui est une estimation de la longueur du côté, de la zone en coordonnées-image, pour choisir un niveau de pyramide.

En utilisant la structure de la pyramide ci-dessus une valeur de la densité approximativement constante peut être obtenue après transformation en perspective.

3. Résultats

Basé sur la méthode décrite ci-dessus de télédétection à haute qualité, un produit-image secondaire a pu être développé. L'avantage de ce document "pseudo-carte", qui est un excellent outil d'aménagement, est son faible coût relatif en calcul ordinateur grâce à l'approche d'image pyramidale et aussi parce qu'il fournit une vue très naturelle du paysage par la combinaison des données Landsat TM et SPOT-HRV pan-chromatique.

De plus, en introduisant des paramètres de la prise de vues, variables de scène à scène, on peut déboucher sur la production d'une simulation de vol. (Qui en fait a été réalisée dans ce projet.)

*Three-Dimensional Remote Sensing
Data Representation for Regional Planning*

Représentation tridimensionnelle de données
de télédétection pour l'aménagement régional

Styria (Austria) / Styrie (Autriche)

Application Title
Use of SPOT for studying a highway project in the French Rhone-Alps region

Titre de l'application
Utilisation de SPOT pour l'étude d'un projet d'autoroute en région Rhône-Alpes

Localization/Localisation
France (Rhone-Alps region)/France (région Rhône-Alpes)

RESPONSIBLE ORGANIZATIONS   ORGANISMES RESPONSABLES

Centre d'études techniques de l'équipement
B.P. 128
38081 L'ISLE D'ABEAU CEDEX

Tel: 33/1 74 27 28 50   Fax: 33/1 74 27 09 45   Telex: 900 427 F

(Contact person: Martine Chatain)

19. Use of SPOT for studying a highway project in the French Rhone-Alps region

1 Objectives

The study of a highway project 50 km long (150 km of alternative plans) required the use of a Geographic Information System for the analysis of environmental constraints, the selection of alternative plans and the definition of their impacts on the environment.

The data supplied by the SPOT satellite were used in accordance with two consideration axes:

—What type of use and what information quality are obtainable depending on the different project study phases?
—What cost (comparison with photo-interpretation and its integration into the GIS)?

2. Methods used

—Acquisition of a stereoscopic pair of images in the multispectral mode (20 m resolution) and a panchromatic image (10 m resolution), available in the catalogue.
—Radiometric and geometric pre-processing: the geometric registration of the images in the same cartographic reference system (Lambert coordinates) implies the production of a digital terrain model by automatic stereoscopy. The geometric rectifications are based on an analytic deformation model (without ground-control points) developed by the GEOIMAGE Company which combines all the mechanical data of the satellite flight, a modelling of the optics and the entire group of parameters and files of ephemerides included on the SPOT bands.
—Multisensor fusion: three new spectral bands P + XS1, P + XS2 and P + XS3 containing spectral information from the three associated XS bands (20 m resolution) and the information with 10 m resolution from the panchromatic, supplied improved-quality images for the interpretation of certain sectors of the study.
—Resolution gain: visual improvement of the SPOT data were obtained locally by bicubic over-sampling in order to increase the resolution (without expanding the pixels)

and by rectification of the local contrasts (to eliminate fuzzy effects).
—Multispectral classification: land use mapping was based on multispectral classification.
—Textural analysis: an analysis of the texture was carried out for the panchromatic data, by using mathematical morphology methods, in order to extract the urban zones. That technique made it possible to extract the different sizes of buildings, roads, old centres and industrial buildings. That information enriched the preceding mapping.

3. Results

—Land-use: Forestry, agriculture, town growth.

Land-use mapping covering the study zone, obtained from the preceding processing, can be integrated into the GIS after a generalization operation.

—Three-dimensional representation (attached illustration).

One of the advantages of using the digital terrain model is the possibility of producing perspective views which can accept information digitized in other ways: strip zones, routes, administrative boundaries, etc.

Those tools not only have the technical advantage of supplying experts with a new analysis tool for a given geographic zone, but more especially represent a very important support for presenting projects to a public of non-technicians (elected representatives, general public, etc.).

4. Advantages obtained

The automatic processing carried out with SPOT data supplies information with an accuracy which is quite compatible with the study of the major options for selecting the route of a new highway.

(a) In the framework of a computer-processing of the totality of the data, the resulting cost of using SPOT is less than

19. Utilisation de SPOT pour l'étude d'un projet d'autoroute en région Rhône-Alpes

1. Objectifs

L'étude d'un projet autoroutier de 50 km (150 km de variantes) a nécessité la mise en oeuvre d'un Système d'Information Géographique utilisé pour l'analyse des contraintes d'environnement, le choix des variantes et la définition de leurs impacts sur l'environnement.

Les données fournies par le satellite SPOT ont été utilisées selon deux axes de réflexion:

— Quel type d'utilisation et quelle qualité d'information en fonction des différents stades d'études du projet?
— Quel coût, (comparaison avec la photo-interprétation et son intégration dans les SIG)?

2. Méthodes utilisées

— Acquisition d'un couple stéréoscopique d'images en mode multispectral (résolution 20 m) et d'une image panchromatique (résolution 10 m), disponibles en catalogue.
— Prétraitements radiométriques et géométriques: Le recalage géométrique des images dans un même référentiel cartographique (coordonnées Lambert) implique la réalisation d'un modèle numérique de terrain par stéréoscopie automatique. Ces rectifications géométriques s'appuient sur un modèle de déformation analytique (sans amer) développé par la société GEOIMAGE, et intégrant l'ensemble des éléments de mécanique de vol du satellite, une modélisation de l'optique, et l'ensemble des paramètres et fichiers d'éphémérides inclus sur les bandes SPOT.
— Fusion multicapteur: Trois nouvelles bandes spectrales $P + XS1$, $P + XS2$ et $P + XS3$ contenant l'information spectrale des trois bandes XS (à 20 m) associées et l'information de résolution (à 10 m) du panchromatique, ont fourni des images de meilleure qualité pour l'interprétation sur certains secteurs d'étude.
— Gain en résolution: Localement, l'amélioration visuelle des données SPOT a été obtenue par suréchantillon bicubique pour augmenter la résolution (sans grossir les pixels), et par redressement des contrastes locaux (pour éliminer les effets de flou).
— Classification multispectrale: Une cartographie de l'occupation du sol résulte de la classification multispectrale.
— Analyse texturale: Une analyse de texture a été réalisée sur les données panchromatiques en utilisant des méthodes de morphologie mathématique afin d'extraire les zones urbaines. Cette technique a permis l'extraction de différentes tailles de bâti, des voiries, des centres anciens, et des bâtiments industriels. Ces informations enrichissent la cartographie précédente.

3. Résultats

— Occupation du sol: Forêt, agriculture, urbanisation.

La cartographie de l'occupation du sol recouvrant la zone d'étude, résulte des traitements précédents, elle est intégrable au SIG à la suite d'une opération de généralisation.

— Représentation en trois dimensions (illustration ci-jointe).

Un des intérêts de l'utilisation du modèle numérique de terrain est la possibilité de réaliser des vues en perspective qui peuvent intégrer des informations numérisées par ailleurs: fuseaux, tracés, limites administratives, etc.

Ces outils ont à la fois un intérêt technique en fournissant aux spécialistes un nouvel outil d'analyse d'une zone géographique mais aussi et surtout représentent un formidable support de présentation des projets à un public de non-techniciens (élus, grand public).

4. Avantages obtenus

Les traitements automatiques effectués sur SPOT fournissent des informations d'une précision tout à fait compatible avec l'étude des grandes options de choix de passage d'une autoroute.

(a) Dans le cadre d'un traitement informatique de la totalité des données, le coût résultant de l'utilisation de SPOT est

that for normal photo-interpretation methods. In addition, computer-processing, when properly used enables one to make decisions within fairly short time-periods.

(b) The production of a Digital Terrain Model based on SPOT makes it possible to print three-dimensional displays to which one can easily add other data such as road development zones, as well as administrative boundaries. They are communication tools particularly well adapted for informing elected representatives and the general public during the different stages of consultation stipulated by law.

The new tools make possible a much greater availability of technical teams, for better project reliability and for improved communication quality.

inférieur aux méthodes traditionnelles de photo-interprétation. De plus, les traitements informatiques, bien maîtrisés, permettent des interventions dans de meilleurs délais.

(b) La réalisation d'un Modèle Numérique de Terrain issu de SPOT permet l'édition de visualisation en trois dimensions dans lesquelles on intègre facilement des informations telles que les zones de développement des tracés, ainsi que des limites administratives. Ce sont des outils de communication particulièrement bien adaptés à l'information des élus et du public lors des différentes étapes de concertation prévues dans les procédures.

Ces nouveaux outils permettent une plus grande disponibilité des équipes techniques pour une meilleure fiabilité des projets et une meilleure qualité de la communication.

Exploitation en trois dimensions d'une image SPOT panchromatique avec incrustation de l'urbanisation (extraction automatique) et des zones d'insertion possible d'un projet d'autoroute.

FLOODED-ZONE MAPPING

CARTOGRAPHIE DES ZONES INONDEES

Application Title
Mapping of damage caused by floods

Titre de l'application
Cartographie des dégâts causés par les inondations

Localization/Localisation
Tunisia (Sidi Bou Zid)/Tunisie (Sidi Bou Zid)

RESPONSIBLE ORGANIZATIONS   ORGANISMES RESPONSABLES

Centre National de Télédétection
B.P. 200
1080 TUNIS—CEDEX

Tel : 216 1 760 900   Fax: 216 1 760 890   Telex: 14580 DEFNAT

(Contact person: Mr H. Ben Moussa)

20. Mapping of damage caused by flooding

Spacemap of Sidi Bou Zid, Tunisia, 1:100 000

1. Objectives

Comparative study for evaluating damage (mainly on agricultural land and major infrastructures) caused by the January 1990 floods in the centre and southern regions of the country.

2. Methods used

(a) Geometric correction of image obtained after the floods (February 1990 Level 1B) with respect to the June 1988 image (Level 2S).

(b) Comparative study:
   —Calculation of the normalized vegetation index and brightness index on two dates (88–90), application of logic operators.
   —Thresholding on channel XS3 for extracting watery zones.
   —Visual interpretation of the hydrographic network and road-system infrastructures.

(c) Map-editing: Superposition of information plans produced by the comparative study, marginal information (Lambert grid, place-names, index-sheets, etc.).

(d) Cartographic output:
   —film plotting (Vizir),
   —flashing and offset.

3. Résults

(a) Space map at 1:100 000 covering the Sidi Bou Zid region.

(b) —Calculation of damaged agricultural areas (in hectares).
   —Calculation of the lengths of damaged road infrastructures (in linear metres).

4. Advantages and drawbacks

(a) Advantages:
   —Availability and accessibility of instantaneous information on the damage (image recorded one week after the disaster).
   —Rapid map-production (2 months, including methodology development and map output).
   —Possibility of updating documents linked with the project (hydrographic network, road network).

(b) Drawbacks:
   —Relatively high cost of image-acquisition when numerous request.
   —Insufficient resolution for interpreting the damage, particularly in urban zones and on small land parcels.

20. Cartographie des dégâts causés par les inondations

Spatiocarte sur Sidi Bou Zid, Tunisie, 1 : 100 000

1. Objectifs

Etude comparative pour l'évaluation des dégâts (principalement sur les terres agricoles et les grandes infrastructures) causés par les inondations de janvier 1990 dans le centre et le sud du pays.

2. Méthodes utilisées

(a) Correction géométrique de l'image prise après les inondations (février 1990, niveau 1B) par rapport à l'image de juin 88 (niveau 2S).

(b) Etude comparative:
  —Calcul de l'indice de végétation normalisé et l'indice de brillance sur 2 dates (88–90), application d'opérateurs logiques.
  —Seuillage sur canal XS3 pour extraction des zones humides.
  —Interprétation visuelle du réseau hydrographique et des infrastructures routières.

(c) Rédaction cartographique: superposition des plans d'informations issus de l'étude comparative, habillage cartographique (grille Lambert, toponymie, tableau d'assemblage, etc.).

(d) Sortie cartographique:
  —restituteur film (Vizir),
  —flashage et impression offset.

3. Résultats

(a) Spatiocarte au 1 : 100 000 couvrant la région de Sidi Bou Zid.

(b) —Calcul des surfaces agricoles endommagées (en ha).
  —Calcul des longueurs d'infrastructures routières endommagées (en mètre linéaire).

4. Avantages et inconvénients

(a) Avantages:
  —Disponibilité et accessibilité d'une information instantanée sur les dégâts (image enregistrée une semaine après la catastrophe).
  —Rapidité de l'établissement de la carte (2 mois y compris développement de la méthodologie et sortie cartographique).
  —Possibilité de mise à jour des documents annexes au projet (réseau hydrographique, réseau routier).

(b) Inconvénients:
  —Coûts relativement élevés de l'acquisition des images en programmation rouge.
  —Résolution insuffisante pour interprétation des dégâts notamment en milieux urbains et sur petits parcellaires.

CARTE DES DEGATS CAUSES PAR LES INNONDATIONS AU 1/100 000

EXTRAIT DE LA CARTE DE SIDI BOUZID

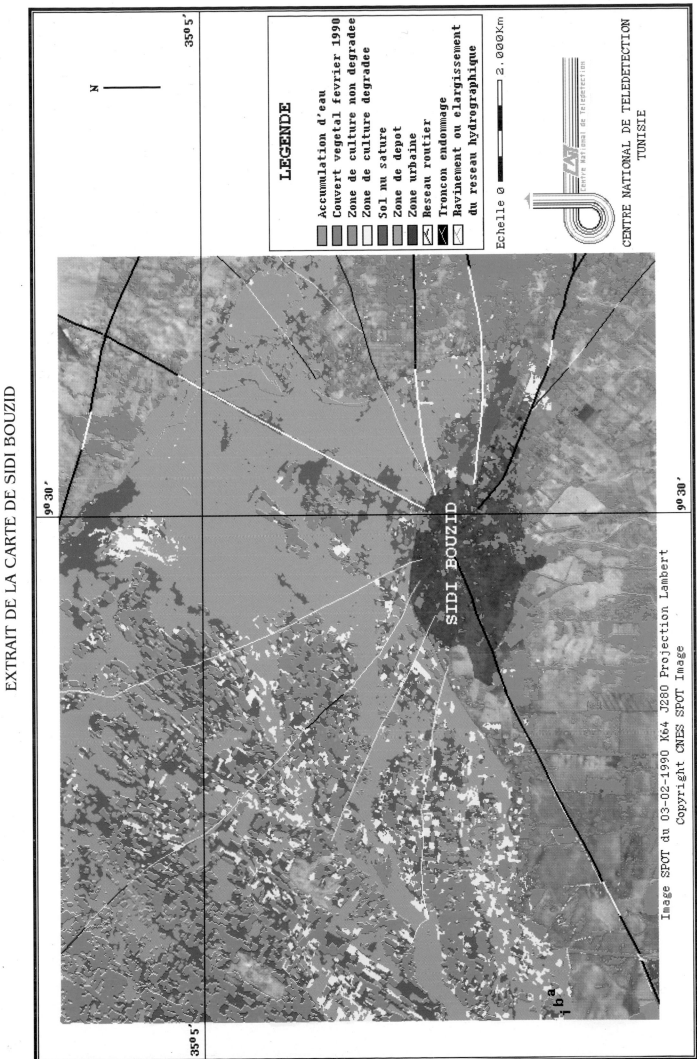

LEGENDE

- Accumulation d'eau
- Couvert vegetal fevrier 1990
- Zone de culture non degradee
- Zone de culture degradee
- Sol nu sature
- Zone de depot
- Zone urbaine
- Reseau routier
- Troncon endommage
- Ravinement ou elargissement du reseau hydrographique

Echelle 0   2.000Km

N

35⁰5'

9⁰30'

SIDI BOUZID

CENTRE NATIONAL DE TELEDETECTION
TUNISIE

Image SPOT du 03-02-1990 K64 J280 Projection Lambert
Copyright CNES SPOT Image

Chapitre 1

L'IMAGERIE DE TELEDETECTION SPATIALE ET SON IMPORTANCE EN CARTOGRAPHIE

Auteurs
Andrzej Ciolkosz* et Andrzej B. Kesik†

*Institute of Geodesy and Cartography, Warsaw, Poland
†Department of Geography, University of Waterloo, Waterloo, Ontario, Canada

TABLE DES MATIÈRES

1.1 Introduction

Les trois dernières décennies de développement en cartographie, télédétection (TD) et systèmes d'informations géographiques (SIG), ont été caractérisées par l'émergence d'une nouvelle interface entre ces trois disciplines. Beaucoup d'efforts ont été faits pour une meilleure intégration des différents types de données cartographiques, et pour une insertion plus réussie des données de télédétection dans les banques de données d'environnement mondiales ou régionales. L'importance de la télédétection, en tant que sous-système d'information sur l'environnement, s'est accrue sensiblement avec l'arrivée de nouveaux satellites dotés de capteurs photographiques et électroniques capables de fournir des données et des images liées aux éléments physiques et humains de la géosphère.

La représentation cartographique des variations spatiales et temporelles des éléments de la géosphère constitue le principal objectif de la cartographie thématique. Cet objectif est diversement atteint, selon différents programmes cartographiques réalisés à des niveaux mondiaux, régionaux et locaux. La cartographie thématique est reconnue par beaucoup comme une activité importante, vitale même pour l'inventaire des ressources, la gestion de l'environnement, et l'aménagement.

Le développement rapide des systèmes satellitaires a fait croître l'acquisition des données de TD et leur application aux travaux de production cartographique. L'interface cartographie/TD/SIG peut être présentée selon différents modèles (Fisher et Lindenberg, 1989). Le modèle de "triple interaction" (fig. 1.1) paraît refléter les relations actuelles entre les trois disciplines sans prépondérance particulière de l'une d'elles. Les zones de recouvrement représentent les domaines de recherche interdisciplinaire, avec potentiel de développement de nouveaux outils, méthodologies et stratégies pour une meilleure compréhension du Système Terre (NASA, 1986).

L'utilisation des images satellites comme source d'information en cartographie thématique survient dans l'un des trois cas suivants:

—Le recours aux données de TD spatiale est nécessaire parce que celles-ci représentent la source de l'information désirée.

—Les données de TD spatiale sont requises en tant que source supplémentaire d'information.
—Les données de TD spatiale sont souhaitées en tant que subsitut à d'autres sources d'information pouvant être considérées comme moins fiables, moins économiques et non satisfaisantes en matière de délais.

Les images satellites ou les données numériques sont acquises, traitées et reproduites, en vue de leur exploitation, par les organisations internationales ou nationales responsables du fonctionnement des systèmes spatiaux, expérimentaux ou opérationnels. La diffusion et l'application des images de satellites à des fins cartographiques, que ce soit à un niveau mondial, régional ou local, dépend de la disponibilité des données de TD, de l'accès à celles-ci, du budget des projets envisagés, et de la méthodologie cartographique.

Les informations de TD spatiale se présentent et sont exploitées sous deux formes différentes:

—*Images:* générées par des sytèmes photographiques et électroniques de TD, et restituées selon les techniques de tirage photo et de reproduction d'image, conduisant à une sortie reprographique (tableau 1.1).
—*Données:* ensemble de valeurs numériques représentant les signaux électriques émis par les capteurs. Les données sont normalement enregistrées sur des bandes magnétiques "compatibles-ordinateur" (Computer-Compatible Tapes—CCT) et ainsi aptes à l'analyse assistée par ordinateur, qui peut impliquer accentuation et classification d'image.

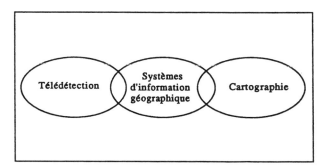

Fig. 1.1 Modèle de "triple interaction" entre TD, cartographie et SIG (d'après Fisher et Lindenberg, 1989)

Tableau 1.1 Appareils de sortie reprographique (d'après Ferns, 1984)

Hard-copy device	Quality	Cost
Colour film writer/plotter (e.g. Optronics, Dicomed)	****	*****
Black and white film writer/ plotter	*****	****
Colour camera (e.g. Matrix, Dunn, Modgraph)	***	***
Ink-jet printers	**	***
Colour dot matrix printers	*	**
Camera (35 mm) screen photography	**	*

Les images satellites reproduites sous forme de tirages noir et blanc ou de compositions colorées sont directement exploitables par photo- interprétation et analyse visuelle/ manuelle. Ce processus comporte habituellement l'identification et la délimitation d'objets, ainsi que leur association, utilisant en priorité le système oeil-cerveau d'interprétation d'image, appuyé sur des outils optico-mécaniques. Classification et "catégorisation" des objets analysés sont effectuées sur la base du processus subjectif d'interprétation, guidé par l'expérience de l'imagiste et de sa connaissance du sujet étudié (voir chapitre 2). Les classifications taxonomiques d'objets analysés et cartographiés constituent les produits dérivés de l'interprétation d'image. Ils témoignent fréquemment de leur dépendance des systèmes existants de classification d'objets élaborés par telle ou telle science de la Terre. Un excellent exemple de cette technique est décrit dans l'*Atlas de l'interprétation de photographies aérospatiales multibandes* (Sagdayew *et al.*, 1982).

L'analyse d'image visuelle/manuelle s'appuie fréquemment sur les techniques d'accentuation d'image qui, utilisées convenablement, peuvent améliorer et accélérer l'analyse visuelle en accentuant l'expression des teintes et couleurs prises par différents groupes d'objets, et en les rendant plus identifiables et plus faciles à distinguer et à délimiter. Les techniques d'accentuation sont assurées en général par voie informatique. Après exploitation des algorithmes d'accentuation choisis, les données obtenues sont restituées sur support permanent en noir et blanc, ou en compositions colorées, reproduites à l'échelle souhaitée.

Les techniques numériques d'analyse d'image assistée par ordinateur ont l'avantage d'une manipulation relativement aisée des données, par application d'algorithmes de prétraitement (corrections radiométriques et géométriques), d'accentuation et de classification. L'affichage et la reprographie d'image peuvent être effectuées à différents stades de la manipulation de données. L'analyse d'image numérique requiert des outils adéquats, matériel et logiciels, pour le traitement et l'affichage des données.

Les progrès en informatique ont conduit à développer deux niveaux auxquels l'analyse numérique d'image est conduite: au niveau informatique supérieur, il existe des systèmes d'analyse d'image spécialement conçus et configurés, basés sur l'architecture parallèle. De tels systèmes sont capables de traiter de grands volumes de données et de mettre en oeuvre des algorithmes d'accentuation et de classification (fig. 1.2).

Au niveau informatique inférieur, qui connaît une rapide popularité auprès des usagers, figurent les micro-ordinateurs avec les progiciels disponibles. Ils sont bien adaptés aux sociétés de consultance, aux services d'éducation et aux experts professionnels. La nouvelle génération de micros peut dépasser la capacité mémoire et la puissance de calcul des mini-ordinateurs de naguère. La tendance technologique indique que les micro-ordinateurs équipés en progiciels de télédétection et en accessoires d'impression, de tracé et d'affichage vidéo, vont encore stimuler

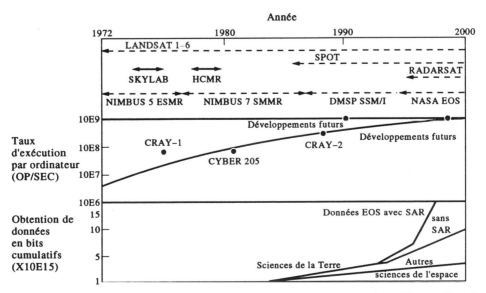

Fig. 1.2 Développement de la puissance informatique, satellites d'observation de la Terre (d'après NASA, 1988)

l'analyse d'image numérique, jusqu'à en faire un outil usuel des opérations cartographiques faisant appel aux données de télédétection.

Les procédés pour extraire l'information des données satellitaires sont décrits en détail dans le chapitre 2 de ce manuel.

1.1.1 Résolution spectrale, spatiale et temporelle des images satellites

Les applications cartographiques des images satellites dépendent de l'aptitude des images, générées par un capteur particulier, à donner une image fidèle des objets étudiés. Ceci se traduit souvent, avec une précision limitée, par les termes "résolution d'image", ou "résolution spatiale", exprimée par la taille minimale des objets ou distances, et décrite comme résolution au sol.

Le terme "résolution d'image" est complexe. Tout d'abord, il s'applique aux propriétés de l'image, qui résultent de l'interaction des caractéristiques du capteur, des procédures de traitement, des caractéristiques de restitution et des conditions d'analyse d'image. Ensuite, ce terme peut aussi renvoyer aux résolutions spectrale et temporelle. La résolution spectrale fait référence à l'amplitude et au nombre des intervalles des longueurs d'onde spécifiques dans le spectre électromagnétique utilisé par un capteur particulier. La résolution temporelle fait référence à la fréquence d'enregistrement d'une zone donnée par un système imageur donné. Les résolutions spectrale et temporelle des capteurs à distance sont essentielles pour les caractéristiques dimensionnelles des données enregistrées, et pour le pouvoir discriminant de celles-ci. Elles sont également importantes pour l'évaluation de la capacité du capteur à détecter les changements. La première application du chapitre 5 illustre les différences de résolution spatiale de la scène des Iles Toronto, enregistrée par quatre systèmes spatiaux: Landsat MSS, TM, et SPOT panchromatique et multibande.

1.1.2 Classification des capteurs satellitaires

Les classifications des capteurs satellitaires sont basées sur différents critères. Les plus communément utilisés sont:

— les caractéristiques des capteurs,
— les caractéristiques des plates-formes et orbites des satellites,
— les domaines d'applications.

Une grande variété de capteurs satellitaires a été lancée depuis 1960, et beaucoup d'entre eux ont été abandonnés, améliorés ou modifiés. Le tableau no. 1.2 liste les principaux systèmes de télédétection par ordre de résolution spatiale décroissante. La résolution temporelle, la largeur de fauchée et les principales applications des images sont listées en fonction de considérations cartographiques (tableau 1.2).

1.2 Satellites météorologiques géostationnaires et à orbite polaire

Les satellites météorologiques opèrent soit à partir d'orbites géostationnaires (GO), à altitude élevée, située à 35 900 km au-dessus de l'Equateur, soit à partir d'orbites basses, proches du pôle (abrév. LEO), à des altitudes entre 200 et 1000 km au-dessus de la Terre (fig. 1.3, tableau 1.3).

Les orbites géostationnaires permettent la synchronisation du déplacement du satellite avec la rotation de la Terre. Il en résulte que le satellite géostationnaire occupe une position apparemment fixe au dessus d'un même point de la surface de la Terre. Un capteur évoluant sur une orbite géostationnaire a un champ de vision qui lui permet de couvrir la totalité d'un disque terrestre de 80 degrés de rayon (environ 6200 km). C'est ainsi que les cinq satellites judicieusement répartis le long de l'équateur à différentes longitudes peuvent couvrir entièrement le globe terrestre entre 80 degrés N et 80 degrés S de latitude. Toutefois, par suite de la courbure terrestre, les régions polaires sont représentées avec une très forte distorsion.

Les satellites géostationnaires assurent les missions suivantes:

— *Missions d'imagerie.* Les images sont produites à partir des données générées par les radiomètres à balayage/défilement opérant dans les parties visibles et infrarouges du spectre. La résolution de l'image au nadir du satellite varie de 0,9 à 2,5 km (selon le type d'engin spatial) dans le spectre visible, et de 5 à 7 km dans l'infrarouge.
— *Missions de diffusion de données.* Les satellites comportent un ou deux transmetteurs en bande S pour relayer les données d'image prétraitées, analogiques ou numériques, vers les stations de réception-utilisateur en onde radio. La transmission

Tableau 1.2 Principaux systèmes de TD spatiale

Satellite System(s) Country	Operational period	Image repeat possibility	Wavelengths frequency	No. of bands	Spatial resolution	Swath width	Applications
GOES Visible and IR Spin Scan Radiometer (VISSR) USA	From October 1975	19 min	Visible Thermal IR	2	0.8 km 9.9 km	Full Earth disc or quarter disc	Meteorology
METEOSAT Radiometer ESA	From November 1977	30 min	Visible Middle IR Thermal IR	3	2.4 km 5 km 5 km	Full Earth disc or sectors	Meteorology Environmental Studies
GMS Visible and IR Spin Scan Radiometer Japan	From July 1977	30 min	Visible Thermal IR	2	1.25/4 km 5/7 km	Full Earth disc or quarter disc	Meteorology
INSAT I Very High Resolution Radiometer India	From April 1982	30 min	Visible Thermal IR	2	2.75 km 11 km	Full Earth disc	Meteorology
TIROS-N/NOAA Advanced Very High Resolution Radiometer (AVHRR) USA	From October 1978	12 hours	Visible Near IR Middle IR Two thermal bands	5	1.1 km	2400 km	Meteorology Oceanography Hydrology Vegetation
DMSP Defence Meteorological Satellite Programme Operational Linescan System (OLS) USA	Block 5D from September 1978	12 hours	Visible and near IR Thermal IR	2	0.6 km	620 km	Meteorology
METEOR 1-PRIRODA Multispectral Scanners USSR	From 1974 to 1980						
MSU-M			Visible Near IR	4	1.7 km	1930 km	Environmental Studies
MSU-S			Visible	2	0.24 km	1380 km	Environmental Studies

Tableau 1.2—(*continued*)

Satellite System(s) Country	Operational period	Image repeat possibility	Wavelengths frequency	No. of bands	Spatial resolution	Swath width	Applications
METEOR-2 PRIRODA	From 1980						
MSU-SK			Visible Near IR	4	0.24 km	600 km	Environmental Studies
MSU-E			Visible Near IR	3	28 m	28 km	Environmental Studies
Fragment Resource—O	Operating		Visible Near IR Middle IR Thermal IR	6	0.8 km	85 km	Environmental Studies
LANDSAT USA Multispectral Scanner (MSS)	From 1972						
Landsat 1–3		18 days	Four bands in visible and near IR Thermal IR (Landsat 3 only)	4	80 m 237 m 80 m	185 km	Land use Vegetation geology Geomorphology Hydrology
Landsat 4–5	From 1982	16 days	(As above)	4			
Thematic Mapper (TM) Landsat 4–5	From 1982	16 days	Visible Near Middle IR Thermal IR	6	30 m 120 m	185 km	Land use Vegetation Geology Geomorphology Cartography
SEASAT Synthetic Aperture Radar (SAR) USA	In 1978	Limited cover	23.5 cm L-bank	1	25 m	100 km	Oceanography
SPOT High Resolution Visible (HRV) France	From 1986	2.5 days off nadir	Visible Near IR Panchromatic	3 1	20 m 10 m	60 km	Land use Agriculture Cartography

System	Date	Coverage	Spectral band	Number	Resolution	Swath	Applications
MOS-1 Japan	From 1987						
MESSR		17 days	Visible Near IR	4	50 m	100 km	Oceanography
VTIR			Visible	1	0.9 km	1500 km	Oceanography
MSR			23.8 and 31.4 GHz		32 km / 23 km	320 km	Oceanography
IRS-IA LISS India	From 1988	22 days	Visible Near IR	4	36.5 m	148 km	Environmental Studies
SPACE SHUTTLE	From 1981						
SIR-A	1981	Limited Cover	23.5 cm L-band		40 m	50 km	Geology Geomorphology Soils
SIR-B USA	1984	Limited Cover	23.5 cm L-band		30 m	20–50 km	Land Use Oceanography
ERS-1	From 1991	Global Coverage	C-band 5.3 GHz		20–30 m	80.4 km (full performance)	Cartography Oceanography Ice Studies
Metric Camera ESA	1983	Limited Cover	Panchromatic		20 m		Cartography
Large Format Camera (LFC) USA	1984	Limited Cover	Panchromatic Colour IR	2			
MOMS West Germany	1984	Limited Cover	Visible Near IR Thermal IR		20 m / 10–20 m	140 km	Cartography Geology Soils Vegetation
COSMOS Space Photography USSR							
Resource F-1 KATE-200	Operating	Limited Cover	Multispectral (4-lenses)		15–30 m		Multidisciplinary Applications
KFA-1000			Panchromatic		5–7 m		Cartography
Resource F-2 MK-4			Multispectral (4 lenses)				
Soyuz-22, Salyut 6 MIR-Mission MKF-6		Limited Cover	Multispectral (6 lenses)		9–13 m		

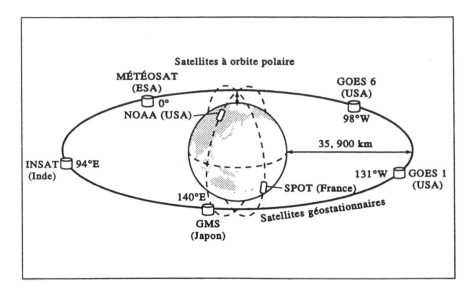

Fig. 1.3 Orbites géostationnaires et polaires pour les satellites météorologiques et d'observation de la Terre.

Tableau 1.3 Altitudes orbitales de certains satellites terrestres

	Miles	Km
Communication satellite (Westar)	22 300	35 680
Space shuttle	115–690	184–1104
Landsat 1, 2 and 3	570	912
Landsat 4 and 5	423	705
Seasat	480	800
Tiros 1 (weather satellite)	480	800
Nimbus (weather satellite)	594	950
GOES (weather satellite)	21 480	35 800

Automatic Picture Transmission (APT), nommée Weather Facsimile Service (WEFAX), permet d'acheminer des données d'image "sectorisée" auprès du vaste groupe des utilisateurs, et d'implémenter celles-ci dans les processus d'analyse et prévision météorologique.

—*Missions d'acquisition de données.* Les satellites sont capables de collecter des données météorologiques, hydrographiques et océanographiques, à partir d'un grand nombre de plates-formes de collecte de données (DCP), stationnaires ou mobiles, et d'acheminer ces données vers la station centrale au sol pour traitement et exploitation.

Les satellites météorologiques à orbite polaire tournent à basse altitude, de 600 à 1000 km. Leur mouvement est héliosynchrone, c'est-à-dire que l'orbite demeure dans un plan fixe par rapport au Soleil, tandis que la Terre défile en-dessous. La couverture du globe s'obtient par orbites successives. A l'altitude de 900 km, le satellite a une période de 103 mn, et deux orbites consécutives sont distantes de 2860 km, et ne fournissent donc qu'une couverture incomplète dans la zone équatoriale.

Les satellites à orbite polaire ont différents capteurs et charges utiles, qui sont généralement aptes à remplir les missions suivantes:

—*Missions d'imagerie.* Les données d'image sont générées par des radiomètres dans les parties visibles et infrarouges du spectre. La résolution au sol varie de 1 à 4 km dans le visible, et de 1 à 7 km dans l'infrarouge le long de la trace du satellite.

—*Missions de sondage.* Les sondages de l'atmosphère sont assurés par des radio-mètres verticaux, qui travaillent dans le visible et l'infrarouge, et fournissent des informations sur la température et l'humidité à différentes altitudes.

—*Missions de transmission.* Les satellites fournissent un service de transmission directe qui assure la réception des données en temps réel auprès de stations au sol situées à une distance radio de réception. Les données images à moyenne résolution sont transmises par APT (Automatic Picture Transmission), tandis que les données images à haute résolution sont transmises par HRPT (High Resolution Picture Transmission).

—*Missions de collecte de données.* Les satellites météorologiques à orbite polaire sont équipés pour acquérir des données à partir des "Plates-formes de Collecte de Données" (PCD).

1.2.1 Satellites météorologiques géostationnaires

1.2.1.1 Satellite géostationnaire opérationnel environnemental (GOES)

Les satellites météorologiques américains (GOES) ont démarré en 1975. Ils appartiennent à l'US National Oceanic and Atmospheric Administration (NOAA) et sont gérés par elle. GOES-7 a été lancé en 1987 et le service des satellites suivants est prévu jusqu'à 2000. Les instruments à bord du satellite comprennent le "Visible and Infrared Spin/Scan Radiometer" (VISSR) afin d'enregistrer des images du disque terrestre dans le visible (0,66–0,7 μm) et l'infrarouge thermique (10,5–12,6 μm), avec une résolution spatiale de 0,8 km pour le visible et 6,9 km pour le thermique. La répétitivité possible est de 19 mn, et la couverture est celle du disque terrestre complet, ou d'un quart. Depuis GOES-3, les satellites sont aussi équipés du sondeur atmosphérique pour collecter l'information sur des profils verticaux de l'atmosphère.

Un système opérationnel GOES consiste en trois satellites: GOES-E (Atlantique-Ouest), GOES-W (Pacifique-Est), et GOES-IO (Océan Indien), qui a connu quelques difficultés techniques. Les données des satellites GOES sont utilisées conjointement avec celles d'autres satellites géostationnaires opérationnels mis en oeuvre par l'ESA, le Japon, l'Inde, l'ex-URSS, pour collecter l'information météorologique à l'échelle du globe.

1.2.1.2 Programme opérationnel METEOSAT

Le programme METEOSAT a été initialisé par les autorités spatiales et météorologiques françaises dans les années 1970. En 1972 le programme a été transféré à l'Agence Spatiale Européenne (ESA), qui a approuvé en 1973 le développement de deux satellites Meteosat. Meteosat-1 a disparu en 1979, mais a été remplacé en 1981 par Meteosat-2 qui est toujours en orbite actuellement. En juin 1988, Meteosat-P2, maintenant dénommé Meteosat-3, est devenu progressivement opérationnel, initialisant le Meteosat Operational Program (MOP), exécuté par l'ESA pour le compte d'EUMETSAT, l'organisation spatiale euro-péenne météorologique. EUMETSAT prévoit d'autres lancements Meteosat en 1992/1993.

Meteosat comporte un radiomètre opérant dans trois bandes: 0,4–1,1 μm (visible/IR), 5,7–7,1 μms (vapeur d'eau) et 10,5–12,5 μms (infrarouge thermique). L'imagerie obtenue couvre une zone comprise entre 55 degrés Nord et Sud. La résolution au niveau de l'équateur et à 0 degré de longitude est de 2,5 km pour le visible/IR et 5 km pour les deux autres bandes. La répétitivité est de 30 mn.

Pour la mission Meteosat vers le milieu des années 1990, dénommée Meteosat Seconde Génération (MSG), Eumetsat prévoit d'ajouter de nouveaux instruments:

—un radiomètre à très haute résolution (500 m) dans le visible;

—un nouveau radiomètre opérant dans le visible et l'infrarouge avec des éléments de sondage;

—un radiomètre à bande large capable de mesurer les rayonnements solaire et terrestre avec une résolution spatiale de 200 km;

—un sondeur infrarouge à très haute résolution spectrale.

1.2.1.3 GMS, satellites géostationnaires du Japon

Le premier satellite géostationnaire du Japon, GMS-1 appelé aussi Hamavari 1, a été lancé en juin 1977, le second GMS-2 en août 1981, lequel a été remplacé, après être tombé en panne en janvier 1984, par GMS-3 lancé en août 1984 (fig. 1.4).

GMS-4 était prévu pour 1989, GMS-5 pour 1993 et GMS-6 pour 1994. Tous les satellites GMS sont positionnés au-dessus de l'équateur et du Pacifique Ouest à 140 degrés de longitude Est (Matthews, 1988).

1.2.1.4 INSAT, satellites géostationnaires de l'Inde

Le premier satellite géostationnaire indien INSAT (IA) a été lancé en avril 1982. INSAT-IA emporte un radiomètre à très haute résolution opérant dans le visible et l'infrarouge. La résolution spatiale était de 2,75 km dans le visible et de 11 km dans l'infrarouge. Le premier INSAT était placé au-dessus de l'équateur à la longitude de 74 degrés Est. Les opérations ont duré jusqu'en septembre 1982. En août 1982, le remplacement fut assuré par INSAT IB à l'aide de la navette spatiale de la

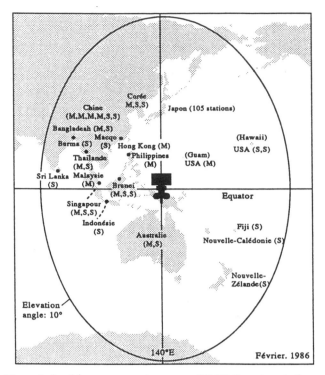

Fig. 1.4 Aire couverte par GMS-3 et stations d'exploitation des données. M, station d'utilisation des données moyenne échelle. S, station d'utilisation des données petite échelle

NASA. Sa position était à la longitude de 94 degrés Est au-dessus de l'équateur.

1.2.2 Les satellites météorologiques à orbite polaire des années 1980

Aux Etats-Unis la 3ème génération des satellites à orbite polaire est représentée par les satellites TIROS-N.

La série des TIROS-N a commencé en 1978. Ils étaient alors dénommés NOAA 6, 7 suivis par les satellites "Advanced TIROS-N" numérotés NOAA 8, 9 et 10. Les satellites NOAA possèdent les "Advanced Very High Resolution Radiometers" (AVHRR) fonctionnant en mode balayage. Ce sont des instruments à 4 ou 5 canaux (tableau 1.4), opérant dans le visible et dans l'infrarouge proche, moyen et lointain, avec une résolution spatiale de 1,1 km.

Les satellites TIROS-N sont placés sur des orbites héliosynchrones, et fournissent des données chaque jour à la même heure locale. Les orbites sont à 850 km d'altitude environ. Le système des satellites TIROS-N consiste en deux satellites opérationnels, l'un traversant l'équateur

Tableau 1.4 Caractéristiques des missions NOAA 6 à NOAA 10

Parameter	NOAA-6, 8 and 10	NOAA 7 and 9
Launch	6/27/79, 3/28/83, 9/17/86	6/23/81, 21/12/84
Altitude (km)	833	833
Period of orbit (min)	102	102
Orbit inclination	98.9 degrees	98.9 degrees
Orbits per day	14.1	14.1
Distance between orbits	25.5 degrees	25.5 degrees
Day-to-day orbital shift	5.5 degrees East	3.0 degrees East
Orbit repeat period (days)	4–5	8–9
Scan angle from nadir	55.4 degrees	55.4 degrees
Optical field of view (mr)	1.3	1.3
IFOV, at nadir (km)	1.1	1.1
IFVO, off-nadir, max. (km)		
along track	2.4	2.4
across track	6.9	6.9
Swath width	2400 km	2400 km
Coverage	Every 12 hr	Every 12 hr
Northbound equatorial crossing (pm)	7:30	2:30
Southbound equatorial crossing (am)	7:30	7:30
AVHRR spectral channels (μm)		
1	0.58–0.68	0.68–0.68
2	0.72–1.10	0.72–1.10
3	3.55–3.93	3.55–3.93
4	10.50–11.50	10.30–11.30
5	channel 4 repeat	11.50–12.50

vers le Sud à 7 h 30, l'autre vers le Nord à 15 h 30 heure locale.

Une importante contribution à la connaissance océanographique a été obtenue grâce aux données collectées par le "Coastal Zone Colour Scanner" (CZCS), qui a fonctionné de 1978 à 1986 à bord du satellite NIMBUS-7. CZCS a acquis plus de 66 000 images, chacune couvrant environ 2 millions de km carrés de surface océanographique. Les spécifications de CZCS sont données au tableau 1.5.

Tableau 1.5 Caractéristiques de NIMBUS-7 et du Coastal Zone Colour Scanner

Orbital altitude	955 km
Equatorial crossing	12:00
Nadir ground resolution	825 m
Swath width	1566 km
Field of view	± 39 degrees
Spectral bands (μm)	
(1) 0.43–0.45	Blue (chlorophyll absorption)
(2) 0.51–0.53	Blue/green (chlorophyll absorption)
(3) 0.54–0.56	Green (yellow substance)
(4) 0.66–0.69	Red (chlorophyll concert.)
(5) 0.70–0.80	Near IR (surface vegetation)
(6) 10.50–12.50	Thermal IR (surface temperature)

En URSS, après 25 missions de la série METEOR-1 de satellites météorologiques (1969–76), les nouveaux satellites METEOR-2 sont entrés en service en 1977. La résolution du scanneur dans le visible est de 2 km. Le stade opérationnel et les applications concrètes ont commencé dans les années 1980.

Les informations acquises grâce aux systèmes satellitaires géostationnaires et à orbite polaire apportent une contribution majeure à la connaissance et à une meilleure compréhension de l'environnement planétaire, continental ou régional. Ces informations servent aux études quantitatives et qualitatives en météorologie, hydrologie et océanographie.

Les images satellites se prêtent aux processus d'analyse et d'interprétation qui génèrent de nombreux produits cartographiques: cartes thématiques représentant différents éléments et caractéristiques de l'atmosphère, de l'hydrosphère et de la biosphère. Les cartes expérimentales produites grâce aux données satellitaires représentent les catégories d'informations suivantes:

Atmosphère
　Températures moyennes des couches isobares
　Teneur totale en vapeur d'eau et sa distribution par couches
　Teneur totale en ozone et sa distribution par couches
　Vitesse et direction du vent dans la troposphère.
Nuages
　Distribution spatiale et structure des nuages
　Hauteur et température des crêtes de nuages
　Teneur totale en eau des nuages
　Localisation et intensité des précipitations
Surface des océans
　Température de la surface océanique
　Localisation des principaux courants océanographiques de surface
　Rugosité de la surface océanique
　Etat de la glace
　Localisation des zones polluées de la surface océanique.
Surface des continents
　Température de la surface continentale
　Degré d'humidité du sol
　Répartition du manteau neigeux
　Localisation des zones mixtes glace-neige
　Distribution des sols et couvert végétal

1.3 Systèmes satellitaires d'observation de la Terre

Les systèmes spatiaux d'observation de la Terre, dénommés aussi Satellites de Ressources Terrestres se sont développés comme des précurseurs des satellites météorologiques, et comme des prolongements de systèmes initiaux tels que Mercury, Gemini, Apollo et Skylab. Ces premières missions impliquaient l'expérimentation de capteurs photographiques et électroniques qui ont fourni la première génération des images spatiales de la Terre, stimulant ainsi recherches et investigations multidisciplinaires.

Les applications cartographiques ont été développées, et des cartes thématiques représentant des synthèses régionales morphostructurales et physiographiques de différentes parties du globe, ont été révisées et enrichies.

La photographie spatiale multibande, précurseur de l'imagerie multibande, a contribué non seulement à une meilleure détection de la configuration spatiale des objets terrestres, mais aussi au développement de techniques d'accentuation et d'analyse d'image, qui sont jusqu'ici essentielles à l'analyse visuelle/manuelle d'images satellites.

L'ère des satellites d'observation de la Terre a commencé au stade préparatoire en 1968, et au stade opérationnel en 1972, lors du lancement du premier satellite ERTS-1 par la NASA. Entre 1972 et 1984, cinq satellites Landsat ont été lancés avec une continuité prévue du programme portant sur au moins deux satellites (tableau 1.6).

Tableau 1.6 Missions Landsat 1 à 6

Satellite	Launched	Retired	Sensors
Landsat 1	23.06.72	1.06.78	MSS, RBV
Landsat 2	22.01.75	30.09.83	MSS, RBV
Landsat 3	5.05.82	30.09.83	MSS, RBV
Landsat 4	16.07.82	—	TM, MSS
Landsat 5	1.03.84	—	TM, MSS

En 1978, le satellite SEASAT a fourni d'importantes données grâce au "Synthetic Aperture Radar" (SAR) (radar à synthèse d'ouverture), prouvant ainsi l'utilité de l'imagerie SAR pour les études d'environnement. Durant la période 1978–80, la "Heat Capacity Mapping Mission" (HCMM) a assuré une large couverture en données thermiques. En 1986, la France a lancé le premier satellite SPOT avec un nouveau type de capteur à haute résolution.

En URSS, les satellites de la série COSMOS sont équipés de scanneurs à 4 bandes et par des chambres photographiques multibandes et monobandes. Des capteurs électroniques et photographiques ont été utilisés sur les Soyouz et stations spatiales Mir-Kvant (Gatland, 1989).

Aux Etats-Unis, à partir de 1981 un nouveau vecteur/plate-forme, la navette spatiale (Space Shuttle), a été utilisé pour quelques missions, au cours desquelles de nouveaux capteurs ont été testés.

Une étape très significative a été franchie en juillet 1991, lors du lancement du satellite ERS-1, correspondant au début du programme européen d'observation de la Terre. Le texte suivant décrit brièvement les caractéristiques des systèmes mentionnés ci-dessus.

1.3.1 Le programme des satellites Landsat

Le système Landsat représente un programme significatif et opérationnel de surveillance des ressources du globe, qui atteint 20 ans de continuité. Le programme initialisé par la NASA en 1967 a conduit à prévoir une série de six satellites nommés ERTS (Earth Resources Technology Satellites), basés sur une version modifiée du satellite météo Nimbus. L'objectif principal était la collecte de données sur les ressources terrestres, sur une base systématique, répétitive, multibande et à résolution moyenne, avec un accès sans discrimination aux données pour toute la communauté internationale.

ERTS-1 a été lancé le 23 juillet 1972, premier satellite inhabité conçu spécialement pour l'acquisition de données relatives aux ressources de la Terre. Avant le lancement du second satellite, le 22 janvier 1975, la NASA rebaptisa le programme ERTS, Landsat, y compris pour ERTS-1 devenu Landsat 1.

Le programme Landsat pour la période 1972–91 fait référence à toutes les opérations liées à 5 satellites lancés, et aux futurs satellites proposés Landsat 6 et 7. Par suite des différences existant entre orbites des satellites, configuration des capteurs, et spécifications des données acquises, les satellites Landsat se regroupent en 2 générations: d'abord Landsat 1, 2 et 3, puis Landsat 4 et 5.

Première génération des satellites Landsat 1, 2 et 3. Les 3 premiers satellites Landsat possèdent la même configuration (ou très semblable) de 2 capteurs indépendants: le scanneur multibande (MSS) et la chambre "Return Beam Vidicon" (RBV). La réponse des capteurs à l'énergie électromagnétique réfléchie est enregistrée sous forme de données multibandes, à 4 bandes pour MSS, et à 3 bandes pour RBV, avec une résolution de pixel à 79 m pour le MSS de Landsat 1, 2 et 3 et le RBV de Landsat 1 et 2, et à 30 m pour le RBV de Landsat 3 (tableau 1.7).

Les satellites Landsat occupent une orbite quasi-circulaire, quasi-polaire, couvrant ainsi le globe jusqu'aux latitudes 80 degrés N et S, avec une période de résolution de 103 mn, et coupant l'équateur à 9 h 30 heure locale. Le cycle de couverture du globe pour cette première génération de satellites Landsat requiert 18 jours, ce qui autorise 20 cycles par an.

La source primaire de données des Landsat 1, 2 et 3 est fournie par le scanneur optico-mécanique (MSS). L'échantillonnage du signal analogique par les convertisseurs analogiques/

Tableau 1.7 Spécifications des capteurs de Landsat 1 à 3

Sensor	Band	Spectral sensitivity (μm)
Landsat 1 and 2		
RBV	1	0.475–0.575 (green)
RBV	2	0.58–0.68 (red)
RBV	3	0.69–0.83 (near IR)
MSS	4	0.5–0.6 (green)
MSS	5	0.6–0.7 (red)
MSS	6	0.7–0.8 (near IR)
MSS	7	0.8–1.1 (near IR)
Landsat 3		
RBV		0.5–0.75 (panchromatic)
MSS	4	0.5–0.6 (green)
MSS	5	0.6–0.7 (red)
MSS	6	0.7–0.8 (near IR)
MSS	7	0.8–1.1 (near IR)
MSS	8	10.40–12.60 (thermal IR)

numériques (A/N) génère un intervalle nominal au sol de 56 m, conduisant ainsi à une matrice de cellules de 56×79 m. Toutefois, la valeur de brillance pour chaque pixel est dérivée de la matrice à pleine résolution 79×79 m au sol résultant du champ instantané de prise de vues (Instantaneous Field of View, IFOV). Les données continues MSS pour une fauchée de 185 km sont découpées en scènes couvrant approximativement 185×185 km. Chaque scène comprend 2340 lignes de balayage avec environ 3240 pixels par ligne, ce qui fait au total quelque 7 581 600 pixels par canal. La représentation d'une seule scène en mode multibande (4 bandes) contient ainsi plus de 30 millions de pixels sous forme numérique. Les données MSS obtenues en sortie du convertisseur A/N à bord du satellite sont codées sur 6 bits, soit des valeurs de 0 à 63, mais sont ensuite ré-étalées de 0–127 pour les bandes visibles, lors du traitement au sol.

Les données secondaires RBV, obtenues grâce aux trois chambres multibandes (Landsat 1 et 2) et des deux panchromatiques et proche IR (Landsat 3), sont moins significatives malgré le gain en résolution (30 m) des chambres RBV de Landsat 3.

Au début de Landsat-1, il y avait seulement 4 stations de réception: 3 aux USA et 1 au Canada. Les données des zones hors d'atteinte de ces stations de réception ont été enregistrées à bord du satellite et transmises lorsque celuici passait en vue de l'une des stations US.

Les limitations du système ont été lentement surmontées grâce à la mise en service de stations de réception dans différents pays. Ces stations assurent également le traitement des données, ainsi que la diffusion des images et des données.

Aux Etats-Unis, les produits de Landsat 1, 2 et 3 sous forme d'images multibandes sur film standard noir et blanc, de compositions colorées, et de bandes compatibles ordinateurs (CCT) ont été diffusées auprès des communautés d'utilisateurs par le "Earth Resources Observation System" (EROS) Data Center à Sioux Falls, South Dakota. En 1984, le Congrès américain a décrété l'"Acte de commercialisation de la télédétection terrestre" visant à transférer Landsat de l'administration publique (NOAA) au secteur privé. En 1985, EOSAT, nouvelle filiale de Hughes et de RCA, a été chargée de diffuser toutes les données Landsat, ainsi que le contrôle opérationnel de Landsat 4 et 5 et la préparation des futurs Landsat 6 et 7.

Seconde génération des satellites Landsat: 4 et 5. La seconde génération des satellites Landsat a commencé en 1982 lors du lancement du 4ème satellite. Celui-ci fut suivi par Landsat 5, lancé en mars 1984. Landsat 4 et 5 sont toujours en service (août 1991).

Landsat 4 et 5 sont différents à beaucoup d'égards. La conception de la plate-forme est différente de celle du type de Nimbus, utilisée pour Landsat 1–3. L'orbite des satellites est plus basse, environ à 705 km, ce qui conduit à une répétitivité plus grande, à 16 jours. L'orbite a un angle d'inclinaison de 98,2 degrés par rapport à l'équateur. La traversée de l'équateur le long du trajet Nord-Sud se fait à 9 h 45 heure locale. Chaque révolution dure 99 mn, soit 14,5 révolutions par jour. Les orbites de Landsat 4 et 5 ont été conçues de façon à permettre une répétitivité de 8 jours, grâce à une couverture alternative par chaque satellite (fig. 1.5).

L'innovation la plus significative concerne les capteurs. Landsat 4 et 5 comportant toujours l'ancien MSS, avec un pixel au sol de 82×82 m, mais en plus de MSS, il y a un nouveau capteur dénommé Thematic Mapper (TM).

Le Thematic Mapper est un radiomètre à balayage à 7 bandes et à haute résolution. Les spécifications des bandes sont données en tableau 1.8.

Les données du Thematic Mapper ont une résolution de 30 m pour les bandes 1 à 5, et 7, et de 120 m pour la bande 6. L'intervalle de

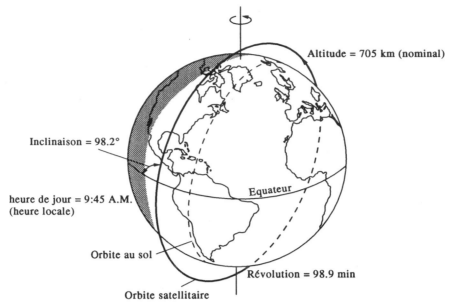

Fig. 1.5 Orbites de Landsat 4 et 5.

Tableau 1.8 Spécifications des capteurs de Landsat 4 et 5

Sensor	Band	Spectral sensitivity (μm)
TM	1	0.45–0.52 (blue–green)
TM	2	0.52–0.60 (green)
TM	3	0.63–0.69 (red)
Tm	4	0.76–0.90 (near IR)
TM	5	1.55–1.75 (mid IR)
TM	6	10.40–12.50 (far IR)
TM	7	2.08–2.35 (mid IR)
MSS	1	0.5–0.6 (green)
MSS	2	0.6–0.7 (red)
MSS	3	0.7–0.8 (near IR)
MSS	4	0.8–1.1 (near IR)

quantification des données TM comporte 256 niveaux (8 bits) contre 64 pour MSS. Les bandes spectrales de TM sont adaptées à des applications et à des cartographies multi-disciplinaires (tableau 1.9).

Interprétation d'image et cartographie thématique basée sur les données Landsat. Depuis le début de l'ère Landsat en 1972 la communauté scientifique tout entière à travers le monde a tiré profit de l'accès sans restriction à des données nouvelles et significatives pour l'environnement. De nombreuses investigations ont exploité les caractéristiques spectrales, spatiales et

Tableau 1.9 Bandes spectrales du Thematic Mapper et leurs applications

Band	Wavelength (μm)	Nominal spectral location	Main applications
1	0.45–0.52	Blue	For water body penetration, coastal mapping, soil and vegetation discrimination, forest type mapping
2	0.52–0.60	Green	For vegetation discrimination and vigour assessment
3	0.63–0.69	Red	Plant species studies cultural feature typology
4	0.76–0.90	Near IR	Vegetation studies, vigour assessment, delineation of water bodies
5	1.55–1.75	Mid IR	Vegetation and soil moisture studies. Snow/clouds discrimination
6	10.40–12.50	Thermal IR	Vegetation stress, analysis, soil moisture studies, thermal mapping
7	2.08–2.35	Mid IR	Discrimination of minerals and rock types. Studies of vegetation moisture content

temporelles des données Landsat. Les données numériques ont orienté les investissements, et la recherche s'est trouvée liée aux techniques d'accentuation numérique d'image et de classification. Les données Landsat ont été acceptées comme un composant obligé de la recherche géoscientifique, et adoptées par de nombreuses disciplines traitant des différentes parties de la géosphère. Les applications des données Landsat vont des études de structures profondes et de géologie de surface à celles de géomorphologie, d'hydrographie, d'utilisation/ occupation du sol, aux évaluations de récoltes, aux classifications de forêts et aux détections de changements urbains. Les résultats des recherches figurent dans nombre de publications listées dans *Geo-Abstracts* ou d'autres sources telles que *Remote Sensing Yearbooks* (1987, 1988/89) et les publications de Carter (1986) et Hyatt (1988).

Les iconocartes satellitaires accompagnées de textes descriptifs ont été publiées dans beaucoup de pays. Quelques exemples sont présentés dans la publication ACI rédigée par J. Denègre (1988). Des collections de cartes élaborées à partir de données Landsat ont été aussi publiées sous forme d'atlas. Le contenu et les méthodes d'élaboration sont variables, depuis la rédaction classique de cartes au trait basées sur l'interprétation visuelle/manuelle d'images Landsat (exemple: Cheng-Shu Peng, 1986), jusqu'à l'établissement de jeux de cartes numériques thématiques, basées sur l'analyse et la classification assistée par ordinateur (exemple: Adeniyi et Bullock, 1988). Les expériences européennes avec le Thematic Mapper ont été résumées par Guyenne et Calabresi (ESA SP-1102).

Le *Manuel de l'utilisateur des données Landsat* (US Geological Survey, 1979) et le *Guide pédagogique Landsat* (Short, 1982) fournissent une bonne information de départ sur le système Landsat. Les textes consacrés à la télédétection par Lillesan et Kiefer (1987), Sabins (1987) et Campbell (1987) contiennent des développements traitant du système Landsat. Une initiation à l'analyse numérique d'image est présentée par Jensen (1986) et Richards (1986).

1.3.2 *Heat Capacity Mapping Mission (HCMM)*

La "mission de cartographie thermique" (HCMM) s'est déroulée du 26 avril 1978 au 31

août 1980. Le satellite HCMM a été le premier à évaluer les propriétés thermiques de la surface de la Terre. Il était placé sur une orbite circulaire, héliosynchrome, inclinée de 97,6 degrés, et à l'altitude de 620 km. Le capteur de HCMM, le "Radiomètre thermographique" (HCMR), opérait dans les régions du visible et de l'infrarouge proche et thermique (tableau 1.10), avec des résolutions respectives de 500 et 600 m.

HCMM était une mission expérimentale, qui n'assurait pas une couverture totale du globe. L'absence d'un enregistreur de bord a restreint cette couverture à l'Amérique du Nord, l'Europe et l'Australie, qui se trouvaient dans le champ des stations de réception (NASA, 1980). Pendant les 28 mois de la vie du satellite, plus de 37 600 images standards furent obtenues. L'interprétation, l'analyse et la cartographie expérimentale ont été faites en relation avec les structures géologiques, la lithologie et l'inertie thermique. L'information de base sur la mission HCMM et les caractéristiques des résultats ont été présentés et publiés par Short et Stuart (1982).

Tableau 1.10 Spécifications du radiomètre de cartographie des températures (HCMR)

Orbital altitude	620 km
Angular resolution	0.83 mrad
Instantaneous field of view	0.6 × 0.6 km at nadir (IR)
	0.5 × 0.5 at nadir (visible)
Scan angle	60 degrees (full angle)
Scan rate	1.19 samples per resolution element at nadir
Sampling interval	9.2 μsec
Swath width	716 km
Information bandwidth	53 kHz per channel
Thermal channel	10.5 to 12.5 μm
Usable range	260 to 340 degrees K
Visible channel	0.55 to 1.1 μm
Dynamic range	0 to 100 percent albedo
Scan mirror	45 degrees elliptical, flat
Telescope diameter	20 cm
Calibration	
Infrared	View if space, seven-step staircase. Electronic calibration and blackbody calibration once each scan
Visible	Preflight calibration

1.3.3 *La mission SEASAT*

Lancé le 26 juin 1978, le satellite Seasat a été le premier à emporter un radar à synthèse d'ouverture (SAR) conçu pour des applications civiles. Seasat, qui a fonctionné seulement 106 jours, jusqu'au 10 octobre 1978, occupait une orbite polaire avec une inclinaison de 108 degrés et une altitude de 790 km. La durée de révolution était de 100 mn, soit 14,3 révolutions par jour. Le cycle complet nécessitait 152 jours.

Le radar Seasat utilise la longueur d'onde de 23,5 cm et la polarisation HH. Les données obtenues ont une résolution au sol d'environ 25 m.

Les données radar Seasat traitées par voie optique ont été utilisées pour produire diverses mosaïques non contrôlées sur la Californie, la Floride, la Jamaïque, le Royaume-Uni et l'Islande. Plus de 300 scènes corrigées numériquement (100×100 km) et 400 corrigées optiquement (100×400 km) sont disponibles à la NOAA-Centre National de Données Scientifiques de l'Espace (NSSDC).

Seasat utilise aussi quatre autres capteurs:

—un radar altimètre pour déterminer les conditions de surface des océans;
—un radar-diffusiomètre pour mesurer la vitesse et la direction du vent;
—un radiomètre à micro-ondes pour mesurer la température de la surface des océans, la pluviométrie et le degré hygrométrique;
—un radiomètre dans le visible et l'IR pour mesurer la température de la surface des océans, et enregistrer cette surface et celle du littoral sous forme d'images.

Le radar altimètre fournit des données sur la topographie de la surface des océans avec une précision altimétrique relative de 10 cm. Le tableau 1.11 indique les caractéristiques principales de Seasat.

Des informations complémentaires sur Seasat peuvent être obtenues dans le manuel des utilisateurs des données du radar à ouverture synthétique Seasat ainsi que dans les publications de Ford *et al.* (1980) et Fu et Hold (1982).

1.3.4 Le système SPOT

Le système SPOT est la contribution française au programme satellitaire d'observation de la Terre consacré aux ressources terrestres et aux applications cartographiques. Le 22 février 1986, la France lança avec une fusée Ariane le premier des quatre "satellites probatoires d'observation de la Terre" (SPOT). Celui-ci est devenu opérationnel en mai 1986. Le programme SPOT est géré par le Centre National d'Etudes Spatiales (CNES), qui est responsable de son développement et de sa mise en oeuvre. SPOT-2 a été lancé en 1990 comme réplique exacte de SPOT-1.

La première génération de satellites SPOT fonctionne à partir d'une orbite polaire, héliosynchrone, inclinée de 98,7 degrés et à une altitude de 825 km. Le passage à l'équateur a lieu à 10 h 30, heure locale. La répétitivité du cycle SPOT est de 26 jours. Toutefois SPOT possède une capacité accrue de répétition grâce au dépointage de l'optique du système (tableau 1.12).

La visée oblique sur une zone intéressante est rendue possible en déplaçant la fauchée verticale par dépointage de miroirs télé-commandés (vers l'Est ou vers l'Ouest), par intervalles de 1 à 27 degrés, permettant ainsi de cibler toute zone à l'intérieur d'une bande large de 950 km centrée sur la trace du satellite.

Tableau 1.11 Cartactéristiques principales de Seasat

Sensor	Observables	Demonstrated accuracy	Demonstrated range of observables
Altimeter	Altitude	8 cm (precision)	< 5 m
	Wave height	10% or 0.5 m	0 to 10 m
	Wind speed	2 m/sec	0 to 10 m/sec
Scatterometer	Wind speed	1.3 m/sec	4 to 26 m/sec
	Wind direction	16 degrees	0 to 360 degrees
Scanning microwave radiometer	Sea surface temperature	1.0 degree C	10 to 30 degrees
	Wind speed	2 m/sec	0 to 25 m/sec
	Atmospheric water vapour	10% or 0.2 g/cm^2	0 to 6 g/cm^2
SAR	Wave length	12%	Wavelength < 100 m
	Wave direction	15 degrees	0–360 degrees

Tableau 1.12 Caractéristiques de base de SPOT

Orbit	Circular at 832 km Inclination: 98.7 degrees Descending mode at 10:30 a.p. Orbital cycle: 26 days

Sensor:
  High resolution
	Two identical instruments
	Pointing capability: 27 degrees
	  East or West of the orbital
	  plane

Visible (HRV)	Ground swath: 60 km each at
	  vertical incidence
	Pixel size:
	  10 m in panchromatic mode
	  20 m in multispectral mode
	Spectral channels (μm)
	  Panchromatic: 0.51–0.73
	  Multispectral: 0.50–0.59
	    0.61–0.68
	    0.79–0.89

Image transmission	Two onboard recorders with
	  24 min capacity, each direct
	  broadcast at 8 Gz
	  (50 Mbits/sec)

Weight	1759 kg
Size	2 × 2 × 3.5 m plus solar panel (9 m)

Cette technique rend possible l'enregistrement d'une même scène à partir d'orbites différentes. Ainsi, à l'équateur, une même zone peut être enregistrée 7 fois au cours du cycle de 26 jours, soit 98 fois par an, soit encore une accessibilité moyenne de 3,7 jours. A la latitude de 45 degrés, une même zone peut être enregistrée 11 fois par cycle, soit 154 fois par an, soit encore une accessibilité moyenne de 2.4 jours, avec un intervalle maximal de 4 jours, et minimal de 1 jour.

La stéréoscopie peut être obtenue par couplage de deux images de la même zone, enregistrées à partir d'orbites différentes et sous des angles de visée différents (tableau 1.13).

Les capteurs de SPOT consistent en deux systèmes identiques de capteurs à barrettes à Haute Résolution Visible (HRV), chacun couvrant une bande de terrain large de 60 km avec un recouvrement de 3 km, lorsque les capteurs sont en visée verticale. Chaque capteur HRV peut enregistrer selon deux modes. En mode panchromatique (P), le HRV fournit 6000 pixels correspondant à une résolution au sol de 10 m, dans la bande de 0,51 à 0,73 μm, et codés sur 6 bits. En mode multibande (XS), le HRV fournit 3000 pixels avec une résolution au sol de 20 m, enregistrés dans les bandes de 0,50–0,59, 0,61–0,78 et 0,79–0,89 μm, et codés sur 8 bits.

Les données SPOT sont transmises aux stations de réception de Toulouse (France), Kiruna (Suède), Prince-Albert et Gatineau (Canada), Hyderabad (Inde), Maspalomas (Espagne). D'autres stations sont opérationnelles ou prévues en Chine, au Bangladesh, au Brésil, en Argentine et en Australie (fig. 1.6).

Pour les zones en dehors de portée des stations de réception, les données SPOT peuvent être enregistrées, sur demande, à bord du satellite, avec une capacité de durée de 20 mn. Ces données sont ensuite retransmises à la station de Toulouse. Les données SPOT sont distribuées aux usagers par SPOT-IMAGE en standard sur bandes magnétiques compatibles ordinateur (CCT) en densité 6250 ou 1600 bpi, ou sur films 241 × 241 mm, présentant ainsi la scène entière à l'échelle de 1:400 000 pour le niveau 1. Le format de base pour les produits de niveau 2 est de 350 × 350 mm pour les échelles de 1:400 000 et 1:200 000, et de 700 × 700 mm pour échelles de 1:200 000 et de 1:100 000. Les films peuvent être commandés en noir et blanc (en mode P ou pour chacune des trois bandes du mode XS), ou en composition colorée pour le mode XS.

Les quatre niveaux de base en matière de correction et de prétraitement sont utilisés pour

Tableau 1.13 Caractéristiques de scènes de SPOT

	XS mode	P mode
Scene dimensions (nadir viewing)	60 × 60 km	60 × 60 km
Pixel size	20 × 20 m	10 × 10 m
Number of spectral bands	3	1
Dimensions of preprocessed scenes:		
Number of pixels per line (raw scene to level 2)	3 × (3000–5200)	6000–10 400
Number of lines per scene (raw scene to level 2)	3 × (3000–4900)	6000–9800
Volume (8-bit bytes)	27–76.5 Mb	36–100 Mb

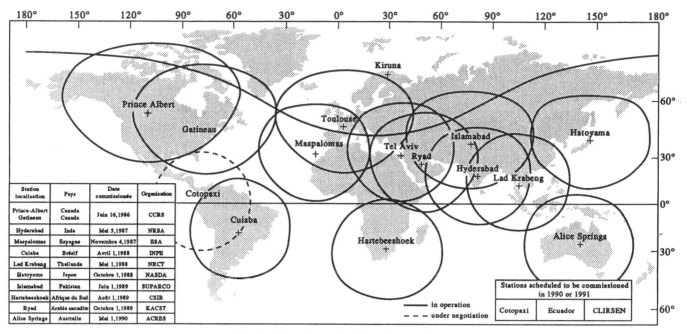

Fig. 1.6 Stations de réception SPOT.

la scène SPOT avec une couverture nominale de 60×60 km, en mode panchromatique (P) ou multibande (XS).

Le catalogue principal de toutes les images réceptionnées depuis 1986 contient les références de plus de 2 000 000 scènes dont 25% ont moins de 10% de couverture nuageuse et 30% ont une couverture inférieure à 25%. La publication "les nouvelles de SPOT", par SPOT Image, fournit régulièrement des informations sur l'état des missions SPOT.

Les données SPOT ont été testées de façon intensive et exploitées pour des applications cartographiques. Le principal avantage des données SPOT, comparées aux données MSS ou TM, est lié aux caractéristiques suivantes:

—large éventail de conditions de prise de vues et de mode spectral,
—capacité stéréoscopique,
—résolution spatiale supérieure, nécessaire en cartographie thématique, notamment dans les zones à fort morcellement du terrain et à forte diversification des conditions de couverture/occupation du sol.

Les résultats d'applications cartographiques des données SPOT ont été présentés dans plusieurs articles: Welch (1985), Gugan (1987), Rochon et Toutin (1987). L'introduction de SPOT dans les SIG a été décrite par Denègre (1987). Un panorama exhaustif des résultats de SPOT après 2 années de fonctionnement a été établi par Rivereau et Pousse (1988).

Le système SPOT est programmé jusqu'en 2005 au moins, avec des spécifications améliorées:

1. Résolution spectrale supérieure avec un capteur à moyen infrarouge (1,58–1,75 μm) pour SPOT-4 (1996).
2. Instrument végétation à large fauchée (2200 km) et à résolution de type statistique (1 km).
3. Résolution géométrique au sol supérieure (5 m) pour SPOT-5 (2001).
4. Stéréoscopie avant-arrière pour SPOT-5.

1.3.5 MOS-1, satellite d'observation de la marine japonaise

En février 1987, le Japon a lancé son premier satellite MOS-1, dédié à la collecte de données océanographiques. Le satellite a été mis sur une orbite quasi-polaire, à une altitude de 909 km avec une inclinaison de 70 degrés (Mattews, 1988).

MOS-1 est équipé des trois capteurs suivants:

—Un radiomètre imageur comportant quatre canaux dans le domaine du visible et du proche infrarouge. Le champ d'observation est de 78 km avec une résolution au sol de 45 m.
—Un radiomètre à quatre canaux dont une bande se situe dans le visible, de 0,5 à 0,7 μm, avec une résolution de 870 m et les

trois autres bandes dans le domaine de 6,0 à 12,5 μm avec une résolution de 2600 m.

—Un radiomètre comportant deux canaux dans le domaine des micro-ondes, opérant à des fréquences de 23 GHz et de 31 GHz. Il est utilisé pour surveiller la température de la surface de la mer.

Le nouveau scanneur MESSR de MOS-1 opère dans quatre bandes avec une résolution de 50 m (Tsuchiya *et al.*, 1987) (tableau 1.14).

1.3.6 La mission du satellite indien, IRS

Le 17 mars 1988, l'Inde a lancé son satellite IRS-1 depuis la base de lancement de Baikonour en URSS Le lanceur était la fusée de trois étages, Vostok. Le satellite est situé sur une orbite héliosynchrone avec une inclinaison de 99,02 degrés et à une altitude de 904 km. Il fournit une couverture globale complète pour les latitudes comprises entre 81 degrés nord et 81 degrés sud. Le passage à l'équateur a lieu à 10 h 25, au noeud descendant. Le cycle du satellite comprend 307 orbites et est effectué en 22 jours (fig. 1.7).

La charge utile du capteur est constituée de deux chambres push-broom: un capteur à barrette à balayage automatique (LISS-II) de 36,5 m de résolution et une chambre (LISS-I) de 72,5 m de résolution au sol, utilisant une

Tableau 1.14 Spécifications de MOS-1

Item	Sensor		
	MESSR	VTIR	MSR
Measurement objectives	Sea surface colour	Sea surface temperature	Water content of atmosphere
Wavelength (μm)	0.51–0.59	0.5–0.7	231 K 311 K
	0.61–0.68	6.0–7.0	
	0.72–0.80	10.5–11.5	
	0.80–1.10	11.5–12.5	
Frequency (Gz)	—	—	23.8–31.4
Geometric resolution (IFOV in km)	0.5	0.9	32–33
Radiometric resolution	39 dB	55 dB 0.5 K	1 K 1 K
Swath width (km)	100	1500	320
Scanning method	Electronic	Mechanical	Mechanical

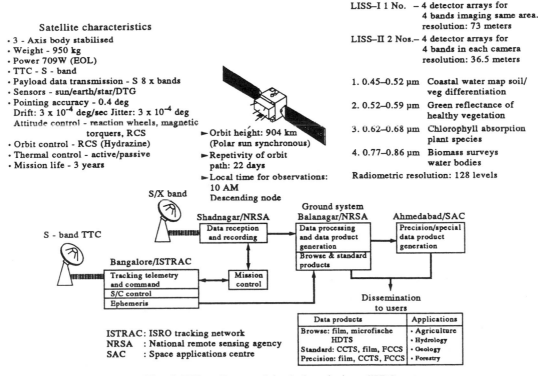

Fig. 1.7 Vue d'ensemble de la mission d'IRS.

barrette CCD de 2048 détecteurs. Chacune des chambres fournit des images dans quatre bandes spectrales situées dans le visible et le proche infrarouge (0,45–0,86 μm). Pour les images obtenues par la chambre LISS-I, la fauchée est de 148,48 km tandis que pour la chambre LISS-II, elle est de 74,24 km pour chacune des images adjacentes avec un recouvrement de 1,5 km au niveau de la trace du satellite. L'énergie reçue est répartie en 128 niveaux. Les données produites par LISS-I et LISS-II comprennent des films panchromatiques négatifs ou positifs de 70 mm et 240 mm, des compositions colorées et des bandes magnétiques à 9 pistes de 1600 bpi contenant les données numériques.

Les données sont disponibles dans trois centres:

—Système de réception de données (DRS) de Shadnagar, Hyderabad.
—Système de production de données de Balanagar et de Ahmedabad (SAC).

La distribution des données collectées par IRS-I a commencé en mai 1988 (Equipe Projet IRS, 1988).

1.4 Les satellites russes pour l'environnement et leurs capteurs

Les satellites russes pour l'environnement appartiennent à la catégorie des Meteor-Priroda et Kosmos; ils sont non habités. Les satellites Meteor-Priroda sont équipés de scanneurs optico-mécaniques alors que les satellites Cosmos acquièrent leurs informations sur l'environnement par des photographies spatiales. Les photographies spatiales sont aussi régulièrement obtenues lors des missions habitées de Soyouz-Salyout ainsi que depuis la station orbitale MIR.

Les recherches et le contrôle des ressources terrestres effectués par le système Meteor-Priroda ont débuté en 1974. Des satellites météorologiques de la série des Meteor ont été modifiés et rebaptisés Meteor-Priroda. Ce système a été opérationnel jusqu'en 1980 depuis des orbites inclinées à 82 et 98 degrés et d'altitude respectivement 900 km et 650 km.

La première génération de satellites de la série des Meteor-Priroda était équipée de deux scanneurs de résolution différente. Le scanneur MSU-M fournissait des données dans quatre bandes spectrales, du visible jusqu'au proche infrarouge (0,5–1.0 μm). La résolution au nadir était de 1700 m, à l'altitude nominale de 650 km. Pour une trajectoire, le balayage obtenu était de 1930 km de large.

Le second scanneur, de nom de code MSU-S, était un scanneur à deux bandes opérant dans le domaine de 0,5 à 0,7 et de 0,7 à 1,0 μm. La résolution au sol, au niveau de la ligne des nadirs, était de 142 m, avec un balayage sur la trajectoire de 1380 km de large. Les données obtenues par les satellites Meteor-Priroda étaient reçues et traitées par des stations au sol situées à Moscou, Novosibirsk et Chabarovsk. Les données fourniers par MSU-S étaient aussi enregistrées au niveau de stations de réception inhabitées, localisées en différents endroits de l'URSS ainsi qu'à bord des stations orbitales soviétiques.

EN 1980, l'URSS a lancé le premier satellite de seconde génération, Meteor 2-Priroda, qui a été équipé des trois senseurs supplémentaires suivants:

—Un scanneur optico-mécanique: MSU-SK. Il était conçu comme un instrument expérimental, multispectral, de quatre bandes, opérant dans le visible et le proche infrarouge (0,5–0,6, 0,6–0,7, 0,7–0,8 et 0,8–1,0 μm). La résolution au sol est de 243 m pour la ligne du nadir, à 650 km d'altitude. La largeur du champ observé était de 600 km.

—Un scanneur de haute résolution à trois bandes spectrales: MSU-E. Il fonctionne en utilisant une barrette de détecteurs fixe. Les bandes spectrales s'étendaient entre: 0,6–0,7, 0,7–0,8 et 0,8–1.0 μm. L'IFOV du scanneur était seulement de 2,5 degrés ce qui induisait un champ d'observation réduit à 28 km. La résolution au sol était de 28 m.

—Un troisième scanneur expérimental à huit bandes: FRAGMENT. Il fonctionne sur les bandes: 0,5–0,7, 0,5–0,6, 0,6–0,7, 0,7–0,8, 0,8–1,1, 1,5–1,8 et 2,2–2,4 μm. Depuis une altitude de 650 km, FRAGMENT offre un champ d'observation de 85 km de large avec une résolution au sol de 80 m. FRAGMENT a été construit avec l'assistance technique de Zeiss Jena en RDA.

Les produits obtenus par les capteurs FRAGMENT étaient distribués sous forme de bandes magnétiques ou d'images noir et blanc à l'échelle du 1 : 1600 000.

Les satellites issus de la famille des Cosmos sont conçus pour la photographie spatiale. Ils

0   1   2   3 km

N

Fig. 1.8 Photographie spatiale à l'aide de la chambre KFA-1000. Varsovie. Pologne.

sont rangés en deux catégories nommées: RESSOURCE O et RESSOURCE F.

Les satellites de la catégorie des RESSOURCE O sont situés sur des orbites circulaires inclinées à 82 degrés. Les altitudes des orbites vont de 230 km à 240 km. Les photographies spatiales sont prises par deux types de chambres: KATE-200 et KFA-1000.

La chambre Kate-200 est une chambre multispectrale comprenant quatre objectifs ayant une focale de 200 mm. Trois chambres sont utilisées avec des films noir et blanc. Un filtrage permet d'obtenir des enregistrements dans le domaine de 0,5–0,6, 0,6–0,7 et 0,7–0,8 μm. La quatrième chambre est utilisée avec un film spectro-zonal. La dimension des photos obtenues est de 180×180 mm. L'échelle approximative des négatifs est le $1:300\,000$ et la résolution au sol est de 15 à 30 m. La qualité des négatifs originaux permet de faire des agrandissements.

La chambre KFA-1000 est utilisée seule. Elle est équipée avec un objectif de 1000 mm de focale. Les photographies ont une dimension de 300×300 mm. A l'origine, la photographie spatiale pratiquée avec la chambre KFA-1000 utilisait des films panchromatiques à l'échelle du $1:240\,000$ et une résolution au sol de 5 à 7 m. (fig. 1.8). Les négatifs de très bonne qualité permettaient des agrandissements jusqu'à l'échelle du $1:24\,000$. Les chambres KFA-1000 ont été utilisées dans une configuration Trimetrogon, avec une chambre orientée verticalement et deux chambres en position oblique. Dans cette configuration, il était possible d'obtenir un champ d'observation de 220 km de large.

Avec l'amélioration de la qualité des films couleurs spectro-zonaux, la chambre KFA-1000 est maintenant utilisée plus souvent pour la photographie spatiale spectro-zonale. Ce type de photographie fournit des informations enregistrées sur deux couches d'émulsions sensibles dans les domaines de 0,57–0,68 et 0,68–0,81 μm. La résolution au sol de la chambre KFA-1000, en utilisant un film spectro-zonal est approximativement de 10 m. La chambre KFA-1000 est aussi utilisée en couple avec une inclinaison de 19 degrés pour chacune des chambres et un recouvrement partiel. Cette configuration permet d'avoir accès à un champ d'observation de 150 km de large. On a pu constater l'obtention de couples stéréoscopiques ayant un recouvrement de 80%.

Les satellites de la catégorie des RESSOURCE F sont en orbite à une altitude de 180 à 450 km. Les orbites sont inclinées à 82 degrés. Ce sont des chambres spatiales MK-4 qui sont embarquées à bord de ces satellites. MK-4 est une chambre multispectrale équipée d'objectifs de 300 mm de distance focale. Les photographies obtenues ont un format 180×180 mm et sont à l'échelle approximative du $1:600\,000$. Elles sont réalisées sur film noir et blanc avec différents filtres. La reproduction en couleur est réalisée par un système supplémentaire, par exemple par le Rectimat-CM, un agrandisseur de couleur et un correcteur précis pour l'imagerie multi-spectrale. Des impressions spectro-zonales sont simulées à des fins d'interprétation.

C'est lors de plusieurs missions Intercosmos russes et internationales que des photographies ont été obtenues en utilisant la chambre multispectrale MKF-6 construite par Zeiss Jena en RDA. La chambre MKF-6 est composée de six objectifs d'une distance focale de 125 mm. La dimension de l'image obtenue est de 55×81 mm. C'est une chambre à compensation de filé. Un travail expérimental effectué en relation avec la mission Soyouz 22 a montré que, pour une hauteur de vol de 250 à 260 km, l'échelle des photographies était approximativement de $1:2\,000\,000$ et la résolution au sol était de 13 m. Des agrandissements réalisés jusqu'à l'échelle du $1:100\,000$ ont été satisfaisants pour des besoins cartographiques.

1.5 Missions utilisant la navette spatiale comme lanceur

Les développements de la navette spatiale de la NASA, visant à en faire une plate-forme réutilisable, ont fourni une solution pour réaliser des tests expérimentaux de nouveaux détecteurs électroniques et photographiques destinés à de futures missions. De 1981 jusqu'à la tragique explosion de la navette spatiale Challenger, le 28 janvier 1986, les expériences effectuées ont utilisé les radars imageurs de la navette, SIR-A et SIR-B, Le scanneur multispectral opto-électronique et modulable (MOMS), ainsi que les deux chambres spatiales: la chambre métrique et la Chambre à Large Format (LFC).

1.5.1 Missions des radars imageurs de la navette spatiale

La première expérience effectuée avec le premier radar imageur de la navette, SIR-A, a eu lieu en novembre 1981 à l'altitude de 260 km. SIR-A utilisait la longueur d'onde L (23,5 cm) avec la polarisation HH. Les images acquises couvrent un champ d'observation de 50 km de large avec une résolution au sol de 40×40 m.

C'est approximativement 10 millions de kilomètres carrés qui ont été couverts et utilisés essentiellement à des fins de cartographie géologique. Les données de SIR-A sont disponibles au National Space Science Data Centre de la NASA (NSSDC). Des informations concernant le satellite SIR-A peuvent être trouvées dans les publications de Chimino et Elachi (1982), Holmes (1983) et Sabins (1983).

L'expérience effectuée avec SIR-B a eu lieu en octobre 1984. On a aussi travaillé dans la longueur d'onde L et une polarisation HH. La principale différence entre SIR-A et SIR-B réside dans la configuration de l'antenne. Pour SIR-B, cette antenne était mobile, pouvait être inclinée et pouvait envoyer des signaux radar en direction de la terre selon des angles d'observation variables (entre 15 et 60 degrés). Cela permit l'acquisition d'images radars stéréoscopiques. La résolution de SIR-B était de 25 m (azimuth) et la résolution variait de 14 m pour un angle d'observation de 60 degrés à 46 m pour un angle d'observation de 15 degrés. Les données obtenues par SIR-B sont aussi disponibles auprès du NSSDC.

1.5.2 Le scanneur multispectral opto-électronique modulable (MOMS)

Le scanneur multispectral optoélectronique modulable (MOMS) a été développé en RFA dans le but d'être utilisé aussi bien à bord d'un avion que lors de missions spatiales. MOMS a été utilisé à l'occasion de deux missions de la navette spatiale en juin 1983 et février 1984.

MOMS est constitué d'une barrette de détecteurs à optique multilinéaire opérant selon le principe push-broom. MOMS a deux canaux: l'un dans le visible (0,575–0,625 μm) et l'autre dans le proche infrarouge (0,825–0,975 μm). Chaque ligne balayée est composée de 6912 pixels fournissant une résolution spatiale de 20 m. Pendant la mission expérimentale, le temps d'enregistrement était limité à une vingtaine de minutes. Les enregistrements obtenus sont localisés entre les latitudes 28 degrés nord et 28 degrés sud et couvrent approximativement 1 800 000 km^2 par bandes de 140 km.

Le traitement et le stockage des données acquises par MOMS est réalisé par le centre allemand des données de télédétection d'Oberpfaffenhofen. Les produits disponibles sont:

—images Quick look,
—images sur bandes magnétiques au format 1600 bpi(données brutes et corrigées),
—images noir et blanc (tirage sur film ou papier).

Les utilisations expérimentales des données MOMS dans un but cartographique ont montré que les informations fournies par MOMS sont convenables pour la réalisation d'une cartographie thématique dans les différentes disciplines géoscientifiques jusqu'aux échelles du 1 : 50 000 (Bodechtel, 1986).

1.5.3 La chambre métrique de Spacelab

La chambre métrique est une modification de la chambre de télédétection aérienne Zeiss RMK A 30/23 pour le système Spacelab. Les spécifications de cette chambre sont données dans le tableau 1.15.

La chambre métrique a été utilisée pour la première fois à l'occasion d'une mission Spacelab lancée lors d'un vol de la navette spatiale de la NASA en novembre 1983. La navette transportait le Spacelab, de conception européenne, qui permet à 5 astronautes de travailler. La photographie spatiale est l'une des plus importantes des 37 expériences réalisées à bord du Spacelab. Pendant ce vol, la chambre métrique a été utilisée pendant une durée effective de 3 heures et à une altitude de 250 km. Deux types de films ont été utilisés: le film Kodak 2443, infrarouge en fausses couleurs et le film Kodak XX, noir et blanc aérographique. Au cours de la mission, ce sont 550 photographies couleurs et 480 photographies noir et blanc qui ont été prises. Les images sont à l'échelle du 1 : 820 000 et couvrent environ 11 millions de kilomètres carrés.

Les recherches concernant ces photographies ont concerné essentiellement leur utilisation pour la cartographie topographique, la révision cartographique et la confection de cartes

Tableau 1.15 Spécifications de la chambre métrique

Type	Modified Zeiss RMK A 30/23
Lens	Topar A 1 with 7 lens elements
Calib. focal length	305–128 mm
Max. distortion	6 μm (measured)
Resolution	391 p/mm AWAR on Aviphot Pan 30 film
Film flattening	By blower motor incorporated in the camera body
Shutter	Aerotop rotating disc shutter (between the lens shutter)
Shutter speed	1/250–1/1000 s in 31 steps
F/STOPS	5.6–11.0 in 31 steps
Exposure frequency	4–6 and 8–12 s
Image format	23 × 23 cm
Film width	24 cm
Film length	150 m = 550 images frames
Dimensions	46 × 40 × 52 cm
Dim. camera magazine	32 × 23 × 47 cm
Mass	54.0 kg
Mass camera magazine	24.5 kg (with film)

thématiques. Des évaluations photogrammétriques ont montré qu'une précision planimétrique de moins de 20 m pouvait être atteinte. On a pu obtenir, grâce à ces clichés, des courbes de niveau tous les 100 m en terrain pentu et tous les 50 m en terrain plus plat. Cela convient pour une cartographie jusqu'à l'échelle du 1:100 000. (Schroeder, 1986; Konecny, 1986). La cartographie thématique a été gênée car ces photos ont été prises tard dans la saison (novembre-décembre).

1.5.4 La chambre à large format (LFC)

La Chambre à Large Format (LFC) a été construite à des fins cartographiques, pour la NASA, pour les missions de la navette.La LFC est une chambre de précision avec compensation de filé, de distance focale de 305 mm dont les images sont de taille 230 × 460 mm avec la plus grande dimension orientée dans la direction de vol (Doyle, 1978).

La LFC a été utilisée pour la première fois lors de la mission de la navette spatiale en octobre 1984. Elle a été utilisée à une altitude de 235 km et 375 km, fournissant respectivement une couverture au sol de 180 × 360 km et 285 × 570 km. Des couvertures stéréoscopiques avec des recouvrements de 20, 40, 60 et 80% ont été également obtenues.

Les photographies LFC sont distribuées par le Chicago Aerial Survey Inc., 2140 Wolf Road, Des Plains, Ill. 60018,USA.

Des informations complémentaires concernant les chambres spatiales contemporaines et leurs performances sont présentées dans les actes de deux conférences:

—Symposium de la commission II ISPRS, "Photogrammetric and Remote Sensing Systems for Data Processing", Baltimore, USA, 26–30 mai 1986.
—Symposium de la commission I ISPRS, "Progress in Imaging Sensors", Stuttgart, FRD, 1–5 septembre 1986.

Des données comparatives ont aussi été présentées par Szangolies (1987) et Kromer (1987) (tableau 1.16).

1.6 Synthèse

Les systèmes opérationnels courants des satellites fournissent les données de télédétection standards suivantes:

I. *Données satellitaires et images utiles à la cartographie thématique*
   1. NOAA: deux fois par jour, résolution d'1 km.
   2. Meteosat: toutes les trente minutes, résolution d'1 km.
   3. Landsat MSS: tous les 16 à 18 jours depuis 1972, résolution de 80 m.
   4. Landsat TM: tous les 16 jours, depuis 1982, résolution de 30 m.
   5. IRS-1: tous les 22 jours depuis 1988, résolution de 36 à 72 m.
   6. Photographie Spatiale: chambre MKF-6 depuis 1976, résolution de 15 à 30 m.
II. *Les données satellitaires et images utiles à la cartographie thématique et topographique*
   1. Données SPOT: tous les 26 jours depuis 1986, résolution de 10 à 20 m.
   2. Photographie spatiale: chambre métrique, chambre à large format, KATE-200, KFA-1000.

En référence aux besoins cartographiques fondamentaux pour la cartographie à différentes échelles à savoir:

—la précision planimétrique,
—la précision altimétrique,
—la détectabilité des objets,

Tableau 1.16 Comparaison des caractéristiques techniques des chambres spatiales

Cameras–Missions	(1)	(2)	(3)	(4)	(5)
Operation time	1976–82	1982–86	from 1987	1984	1983
Calibrated focal length (mm)	125	125	125	305	305
Image size (mm × mm)	55 × 81	55 × 81	55 × 81	230 × 460	230 × 230
Forward motion compensation	yes	yes	yes	yes	no
Spectral bands (μm)	0.48*	0.48*	0.48*	0.4–0.9	0.53–0.7
	0.54*	0.54*	0.54*		(for PLA)
	0.60*	0.60*	0.60*		0.53–0.9
	0.66*	0.66*	0.66*		(for MLA)
	0.72*	0.72*	0.72*		
	0.84†	0.84†	0.84†		
Resolving power (l/min)	150–220	150–220	150–220	80	30
Flight height (km)	250–260	350	240–300	220–370	250
Image scale (in Mio)	1/2.0	1/2.8	1/1.9	1/0.7	1/0.8
			1/2.4	1/1.2	
Area per photopair (in km)‡	70 × 160	100 × 225	70 × 150	220 × 165	125 × 185
			85 × 190	365 × 275	
Ground resolution (m)	13	9–13	9	10	20
Theoretical accuracy of height measurement (m)‡	42	58	40–50	7–12	16
Theoretical accuracy of coordinate measurement (m)	6	8	6–7	4–6	4
Enlargement for map scale to 1/100 000	20	28	19–24	7–12	8

(1) MKF-6 from Jena Soyuz 22-Mission, USSR. (2) Salyut 6-Mission, USSR. (3) MIR-Mission, USSR. (4) LFC–NASA Challenger Mission, USA. (5) Metric camera Option feintechnik, GmbH/FRG. Spacelab-mission, ESA.

*Amplitude: ± 0.02.

†Amplitude: ± 0.05.

‡66% overlap.

les commentaires suivants, adaptés de la présentation de Konecny (1989) semblent être appropriés:

> Si la précision planimétrique ne semble pas représenter un réel problème, la plupart des préoccupations concernent la précision altimétrique et la détectabilité des objets. La meilleure précision altimétrique est obtenue à l'aide de données SPOT (approximativement 5 m). La meilleure détectabilité des objets est constatée sur les photographies spatiales qui en regard de la très bonne résolution et du terrain visible est supérieure aux détecteurs numériques.

1.7 Les futures missions des satellites et leur intérêt cartographique

Les développements futurs dans la télédétection satellitaire dépendent de l'interaction entre de nombreux facteurs techniques, économiques et politiques qui changent aux échelles globales et régionales à un taux difficile à prévoir. Il ne fait pas de doute que l'amélioration du climat politique général, la réduction des budgets militaires ainsi qu'une prise de conscience sociale croissante sur les conditions de l'environnement de la biosphère, sont favorables à l'intensification des missions satellitaires nationales et internationales.

Les développements vont sans doute concerner:

—la construction des détecteurs,
—la conception des plates-formes,
—la réception des données, leur traitement et leur analyse,
—les structures socio-économiques: commercialisation, coopération internationale, missions en partenariat.

Plusieurs futures missions satellitaires ont été prévues aux Etats-Unis, en Europe, au Japon, au Canada, au Brésil, en Inde et en Chine. Les informations techniques préliminaires et les cahiers des charges sont fréquemment revus.

La prochaine décennie sera caractérisée par la poursuite des missions actuelles en modifiant ou ajoutant de nouveaux détecteurs, en augmentant le nombre de données acquises, le traitement et la distribution des données.

En ce qui concerne les missions d'après l'an

2000, certaines idées et concepts ont été formulés, même si l'avenir reste plus incertain.

Dans la partie suivante, nous passerons en revue les futurs développements concernant les trois types de satellites:

—les satellites météorologiques,
—les satellites pour une étude générale de l'environnement,
—les satellites à vocation cartographique.

1.7.1 Les futurs satellites météorologiques

Des satellites géostationnaires sont prévus par les Etats-Unis, l'Agence Spatiale Européenne, le Japon et la Russie. Aux Etats-Unis, la NOAA prévoit une nouvelle génération de satellites GOES-NEXT qui sont actuellement en phase de développement.

En Europe, l'ESA prévoit, dans le cadre de son Programme Opérationnel Météosat (MOP), deux lancements supplémentaires de satellites Météosat. Ces satellites codés MOP-2 et MOP-3 seront lancés respectivement en 1990 et 1991 fournissant un éventail de services à leurs utilisateurs jusqu'au moins en 1995.

Les Japonais ont prévu de lancer GMS-4 en 1990 et GMS-5 en 1994. Ces satellites vont fournir des informations grâce à des radiomètres opérant dans le visible et l'infrarouge. L'ex-URSS continuera ses séries de Météor-2.

Les satellites météorologiques polaires vont aussi continuer leur travaux. Aux USA, NOAA va produire des séries TIROS-N (NOAA K, L, M, N, O, P). En plus, le Defense Meteorological Satellite Program (DMSP) continuera à fournir des données.

1.7.2 Les futurs satellites pour l'étude générale de l'environnement

Cette catégorie représente le groupe le plus important et le plus diversifié. Il comprend les satellites avec leurs différents capteurs (actifs et passifs), et les outils destinés aux applications générales multidisciplinaires ou à objectifs plus spécifiques comme la surveillance de la végétation.

Les satellites proposés seront sur des orbites quasi-polaires aussi bien que sur des orbites héliosynchrones. Cela permettra une couverture répétitive et des activités internationales partagées, comme celles proposées pour l'Earth Observing System (EOS), ou des orbites basses convenables pour les missions habitées utilisant des plates-formes spatiales équipées de détecteurs électroniques et photographiques.

Les projets des Etats-Unis incluent la continuation de la série des Landsat, le développement de nouveaux détecteurs constituant la charge utile des plates-formes en orbite polaire constituant les éléments des stations orbitales, et les futures missions de la navette spatiale.

Le programme Landsat mis au point par EOSAT a entériné les prévisions pour Landsat 6 et 7, mais des détails concernant les spécifications de la plate-forme et des détecteurs restent à préciser. L'une des propositions pour le détecteur de Landsat 6 fait mention du Enhanced Thematic Mapper (ETM), similaire au détecteur TM mais avec une bande panchromatique supplémentaire (0,5–0,85 μm) et une résolution au sol de 15 m. Landsat 7 pourrait être équipé d'ETM ayant les possibilités de détection d'un capteur multibande dans l'infrarouge thermique. EOSAT recherche et détermine aussi les performances de la barrette de détecteurs multispectrale (MLS) pour des applications futures. Plusieurs organisations internationales et EOSAT sont d'accord sur le fait que Landsat 7 devra être doté des possibilités de prises de vues stéréoscopiques. Après Landsat 7, nous savons seulement qu'il existe une volonté générale de continuer la collecte de données de type Landsat.

Un nombre important de détecteurs ont été conçus pour les nouvelles missions de la navette spatiale. Des développements complètement nouveaux comprennent:

MAPS: mesure de la pollution de l'air depuis l'espace.
FILE: identification de caractéristiques et expérience de localisation.
SIR-C: radar imageur de la navette spatiale fonctionnant dans la bande C.
SISEX: Spectromètre imageur de la navette spatiale.
LIDAR: Mesure et détection d'intensité lumineuse.

Topex-Poséidon, programme mis au point par les USA et la France s'intéresse à l'étude de la topographie de la surface de l'océan et à la circulation des eaux. Il est prévu de lancer le programme en 1992 pour une durée de 3 ans. Des radars altimètres et des systèmes de localisation sont prévus pour cette mission.

Le projet d'une station orbitale a débuté en 1984 aux USA en collaboration avec l'Europe, le Canada et le Japon. La composante européenne de ce programme est dénommée Colombus. Deux des plates-formes en orbite polaire qui seront des composantes de la station orbitale, seront construites à l'aide de la navette. L'une d'entre elles sera préparée par l'ESA et sera opérationnelle sur une orbite permettant un passage à l'équateur tous les matins à 10 h 00. La seconde plate-forme polaire sera mise en oeuvre par la NASA sur une orbite similaire mais avec un passage à l'équateur à 2:00 de l'après-midi. Les plates-formes polaires fonctionneront à 850 km d'altitude. La longue liste de l'instrumentation proposée montre l'intérêt suscité pour les observations météorologiques, atmosphériques, océaniques et terrestres.

Les futures activités européennes (ESA) dans ce domaine des satellites pour l'étude de l'environnement, culmine avec le programme ERS: Satellite Européen Pour l'Environnement. La mission ERS utilise 2 satellites, ERS-1 lancé en juillet 1991 puis ERS-2 en 1994 avec des opérations assurées en continu jusqu'en 1996. Les missions ERS sont principalement orientées vers la surveillance des glaces et eaux des océans polaires. ERS-1 est mis sur une orbite héliosynchrone basse, à une altitude de 800 km. L'instrumentation d'ERS-1 inclut les instruments suivants :

—un radar altimètre,
—un Instrument Actif dans le domaine des micro-ondes,comprenant un diffusiomètre à vent sur la bande C avec un Radar à Synthèse d'Ouverture (SAR) (tableau 1.17),
—un radiomètre à balayage suivant la trajectoire.

Le Japon, non concerné jusque là par la commercialisation, prévoit une contribution internationale substantielle pour la future surveillance globale de l'environnement. Pour la seconde moitié des années 1990, l'agence pour la technologie et l'espace (STA), recommande le suivi des trois projets suivants:

—ADEOS (Satellite Avancé d'Observation de la Terre). ADEOS est prévu pour 1995 avec deux détecteurs: AVNIR (Advanced

Tableau 1.17 Paramètres du SAR de ERS-1

Frequency	L-band
Polarization	HH linear
Orbit	570 km, Sun-synchronous
Off nadir angle	35 degree
Swath width	75 km
Resolution	18 m × 18 m (swath centre)
Multilook number	3
Signal-to-noise ratio	7 dB
Signal-to-ambiguity ratio	14 dB
Output data rate	60 Mps

Transmitter and receiver
Frequency	1275 MHz
Band width	15 MHz
Pulse width	35 μs
PRF	1505–1605 Hz
RF peak power	1050 W (minimum)
Noise figure	4–6 dB (maximum
Gain control	70–92 dB

Antenna
Solar paddle deployment mechanism
Microstrip array for radiating elements

Visible and Near IR Scanner), scanneur de quatre bandes spectrales avec un IFOV de 8 à 16 m, la possibilité de dépointage de plus ou moins 40 degrés et avec deux AO (demande d'opportunité), des détecteurs dans douze bandes spectrales dans le visible et l'infrarouge thermique, avec un IFOV de 700 m et une fauchée de 1400 km.
—NPOP-1: deux détecteurs ITIR (Intermediate Infrared Radiometer), ASMR (Advanced Visible and Near IR Scanner). Les deux types de détecteurs sont proposés pour la plate-forme en orbite polaire de la NASA (NPOP-1) prévue pour 1996.
—TRMM (Mission pour la Mesure des pluies tropicales en région Tropicale). Cette mission est encore au stade de l'étude en tant que projet mené en liaison avec les Etats-Unis. Le lancement est approximativement prévu pour 1996. Les objectifs sont: la mesure des précipitations, la circulation de l'eau, l'interaction océan-atmosphère en zone tropicale.

Plus réalisable dans un futur proche: le projet japonais J-ERS-1 prévu pour 1990. C'est un projet complémentaire des missions MOS 1 et 2 avec l'instrumentation suivante: L-band SAR combiné avec des mesures optiques par radiomètre infrarouge à courte longueur d'onde

(SWIR), et par des radiomètres opérant dans le visible et le proche infrarouge (VNIR).

Le Canada projette d'envoyer RADARSAT en 1994, premier satellite canadien d'observation de la Terre. Radarsat sera lancé par une fusée Delta sur une orbite polaire circulaire avec une inclinaison de 99 degrés. L'altitude d'une orbite sera de 800 km, le cycle de l'opération de 24 jours au passage à l'équateur, avec 14,4 orbites par jour. La charge utile intégrera un radar à ouverture synthétique opérant à 5,3 GHz (bande C) avec un azimut et un domaine de résolution de 30 m. Radarsat apportera la flexibilité dans la couverture des zones, l'angle d'observation et la capacité à zoomer pour obtenir différents niveaux de détails. L'antenne SLAR sera capable de pointer son faisceau dans un champ d'observation de 500 km et entre 20 et 50 degrés de chaque côté de la trace du satellite. Ce mode d'opération permet d'obtenir une couverture complète du Canada toutes les 72 heures et une couverture quotidienne de l'Arctique.

La participation russe aux futures études d'un satellite pour l'environnement du globe est difficile à prévoir. D'une part, les capacités techniques russes concernant les lanceurs de satellites, les plates-formes et la conception de détecteurs ont été récemment démontrées avec la fusée Energia, la station spatiale MIR, la navette spatiale Buran ainsi que par un nouveau système radar pour satellite. D'autre part, les mises au point économiques et sociales relatives à la perestroika pourraient entraîner des révisions de nombreux programmes spatiaux au niveau de leur priorité et des calendriers. Avec la poursuite du climat politique actuel, on peut espérer une coopération internationale plus productive, l'accès aux données de télédétection collectées par les Russes, et à plus de collaboration pour les programmes spatiaux.

Les projets concernant de futurs satellites d'environnement au Brésil, en Inde et en Chine sont moins probables. Les changements des spécifications des capteurs, des lanceurs, des objectifs et des caractéristiques des missions sont fréquents.

1.7.3 Les futures missions des satellites cartographiques

Les futures opérations des satellites cartographiques peuvent être décrites par deux types de missions:

—les futures missions de SPOT et des satellites de cartographie stéréoscopique,
—la future photographie spatiale depuis des plates-formes orbitales et les stations spatiales (Plate-forme Polaire, navette spatiale, satellites Cosmos, Buran, stations Mir).

Les futures missions SPOT: SPOT 3, 4 et 5 sont prévus respectivement pour 1993, 1996 et 2000. Sont prévues des modifications des plates-formes et des spécifications des capteurs. Le changement le plus important concerne SPOT 4 et 5, tandis que SPOT 3 aura une configuration similaire à celle de SPOT 2.

En ce qui concerne la future génération de satellites SPOT, il est prévu de rajouter une bande spectrale dans le moyen infrarouge du spectre (1,58 à 1,75 μm) avec une résolution de 20 m afin d'améliorer les possibilités d'étude de la végétation. Il est aussi question d'ajouter un nouvel instrument à grand champ d'observation avec une fauchée de 2000 km et une résolution au sol de 1 km pour des études régionales, et de 4 km pour des investigations sur des territoires plus étendus. L'instrument de base pour les applications topographiques sera toujours le HRV de résolution 10 et 20 m et offrant la possibilité d'avoir la vision stéréoscopique avant-arrière.

Les futurs systèmes des satellites pour la cartographie stéréoscopique ont été décrits par Light (1989) et Colvocoresses (1990). Selon ce dernier, les spécifications optimales de ces futurs satellites seraient les suivantes:

Orbite: 918 km d'altitude. Une altitude inférieure est acceptable (jusqu'à 518 km). Circulaire, héliosynchrone permettant une couverture quotidienne de 2 trajectoires contiguës, avec un passage au noeud descendant entre 9:00 et 9:45 du matin, heure locale.

Champ d'observation: fauchée de 180 km que l'on peut diminuer afin d'obtenir une couverture stéréoscopique et multispectrale à haute résolution.

Capteurs: trois barrettes de détecteurs. Des optiques à réfraction, à visée avant verticale et arrière. Trois ou quatre bandes spectrales prises dans le visible et le proche infrarouge de manière à ce qu'un refroidissement cryogénique des détecteurs ne soit pas nécessaire. Un enregistrement de 256 niveaux de gris (8 bits). A bord du satellite, la compression des données ne devra pas être destructive et devra ne pas dépasser plus de 32 ou 64 niveaux de gris (5 ou 6 bits).

Résolution spatiale: pixels de 10 m dont la combinaison peut être opérée à bord afin d'obtenir la résolution plus faible et plus classique de 20 m. L'utilisation de la stéréoscopie permettra d'atteindre une résolution située au-delà des 10 m indiqués par la taille des pixels.

Détermination et contrôle d'attitude: contrôle à 0,1 degré près, essentiellement pour maintenir le satellite à la verticale de sa trace au sol et une possibilité de dépointage de part et d'autre de cette trace de l'ordre de 10^{-6} degrés par seconde. Détermination de l'attitude à 5 secondes d'arcs près par mesure par rapport aux étoiles.

Position du satellite: fournie par le Système de Positionnement Global (GPS) ou son équivalent avec une précision de 3 à 5 m dans les trois coordonnées de base.

Transmission des données: 50 à 100 Mb/s à 32 ou 64 niveaux de gris (5 ou 6 bits) transmis à un réseau de stations au sol sur la bande X, au moyen d'une antenne omni-directionelle. Stockage des données sur des enregistreurs embarqués.

Décrit par Colvocoresses (1990) le système de satellite pour la cartographie stéréoscopique doit fournir de l'information convenable pour des cartes au 1:50 000 avec des courbes de niveau équidistantes de 20 m. La réalisation du projet doit coûter environ 10 ans d'effort et 1 milliard de dollars américains.

Les futures missions de photographie spatiale représentent certainement un domaine d'activités potentiel. Les chambres spatiales existantes: la chambre métrique, la Chambre à Large Format (LFC) et les chambres russes KATE-200 et KFA-1000 pourront être utilisées lors de futures missions sur plates-formes spatiales, lors d'opérations de la navette spatiale, et enfin par les missions de la station spatiale MIR. Excepté les efforts de l'ex-URSS, les autres organisations impliquées dans les applications de la cartographie satellitaire manifestent un intérêt limité pour la photographie spatiale, malgré les évaluations positives de la photographie spatiale expérimentale issue de la LFC et de la chambre métrique.

1.8 Cadre des activités futures

Malgré de nombreux efforts et une coopération internationale, en particulier dans le domaine des études météorologiques, les activités actuelles en matière de satellites de télédétection destinés à la cartographie et à l'environnement rappellent beaucoup le vieux scénario d'un Monde politiquement et économiquement divisé.

Les organisations en présence, en s'occupant à la fois des missions et de nombreuses activités bureaucratiques superficielles, entravent la réelle progression vers une coopération scientifique globale. Cependant, des pas ont été accomplis dans la bonne direction. Des organisations comme l'Organisation Météorologique Mondiale (WMO), l'Organisation pour l'Alimentation et l'Agriculture (FAO), le Programme des Nations Unies de l'Environnement (UNEP), et d'autres, comme l'International Geosphere-Biosphere Program (PIGB), vont dépendre fortement des données-satellites et de la gestion mondiale de l'environnement. Des organisations telles que ESA, NASA, NOAA prévoient davantage de partenariat. En ex-URSS, la nouvelle tendance à la commercialisation a conduit à la vente par Priroda et Sojuzkarta de photographies spatiales des chambres multispectrales MK-4 et KFA-1000. L'intérêt pour l'environnement global, qui nécessite une coordination et des actions internationales, pourrait contribuer à mettre sur pied de meilleures stratégies ainsi que des missions de coopération.

Les objectifs cartographiques des missions satellites retrouvent leurs priorités à cause de la prise de conscience d'une couverture cartographique globale insatisfaisante, d'une forte demande de révision cartographique efficace et d'un besoin de réalisation de nouvelles cartes thématiques montrant la diversité de la géosphère. Les buts cartographiques des missions satellitaires pourraient être atteints plus rapidement si les pays industrialisés dirigeants, qui possèdent les moyens techniques et économiques, maintenaient leurs engagements, consolidaient et coordonnaient mieux leurs contributions et apportaient leur soutien aux projets de satellites de télédétection et de cartographie.

Références

ADENIYI, P. O. and BULLOCK, R. A. (Eds.). 1988. *Seasonal Land Use and Land Cover in Northern Nigeria; An Atlas of the Central Sokoto-Rima Basin.* Department of Geography, University of Waterloo, Occasional Paper No. 8.

ALLAN, T. D. 1983. *Satellite Microwave Remote Sensing.* Chisterts. Ellis Horwood.

BARKER, J. L. (Ed.). 1984. *Landsat-4. Investigation Summary, including Deembo 1983 Workshop Results.* NASA-CP-2326, Vol. 1–2.

BARKER, J. L. (Ed.). 1985. *Landsat-4. Science, Characterization, Early Results.* NASA-CP-2355, Vol. 1–4.

BAUDOIN, A. 1992. Improvements of the SPOT system at the turn of the next century. *Proceedings 17th ISPRS Conference,* Washington.

BODECHTEL, H. 1986. Thematic mapping of natural resources with the modular optoelectronic multispectral scanner (MOMS). In K. H. Szekielda (Ed.), *Satellite Remote Sensing for Resource Management.* Graham & Trotman Ltd, London.

CAMPBELL, J. B. 1987. *Introduction to Remote Sensing.* Guilford Press, New York and London.

CARTER, D. J. 1986. *The Remote Sensing Sourcebook.* McCarta Ltd, London.

CERCO. 1988. The SPOT system and its cartographic applications. *Proceedings of the Conference,* Saint-Mandé, 6–15 June 1988.

CHEN, H. S. 1985. *Space Remote Sensing Systems.* Academic Press, New York.

CHEN-SHU-PENG. 1986. *Atlas of Geo-Science Analysis of Landsat Imagery in China.* Science Press, Beijing.

CIMINO, J. B. and ELACHI, C. (Eds.). 1982. *Shuttle Imaging Radar-A (SIR-A) Experiment.* Jet Propulsion Laboratory Publication 8277. Pasadena, California.

COLVOCORESSES, A. P. 1990. An operational Earth mapping and monitoring satellite system: a proposal for Landsat 7. *Photogrammetric Engineering and Remote Sensing,* Vol. 56, No. 5, pp. 569–571.

DENÈGRE, J. 1987. Apport de SPOT aux systèmes d'information géographique. *Colloque SPOT-1. Image Utilization Assessment Results,* Paris, pp. 1459–1466.

DENÈGRE, J. (Ed.). 1988. *Thematic Mapping from Satellite Imagery: An International Report.* Elsevier Applied Science Publishers Ltd, Amsterdam.

Dierke Weltraumbild-Atlas. 1981. George Westermann Verlag, Braunchweig.

DOYLE, F. J. 1979. The large format camera for shuttle. *Photogrammetric Engineering and Remote Sensing,* Vol. 45, No. 1, pp. 200–203.

DRURY, S. A. 1990. *A Guide to Remote Sensing.* Oxford University Press.

EHLERS, M., EDWARDS, G. and BÉDARD, Y. 1989. Integration of remote sensing with geographical information system; a necessary evolution. *Photogrammetric Engineering and Remote Sensing,* Vol. 55, No. 11, pp. 1619–1627.

FISHER, P. F. and LINDENBERG, RICHARD E. 1989. On distinction among cartography, remote sensing and geographical information system. *Photogrammetric Engineering and Remote Sensing,* Vol. 55, No. 10, pp. 1431–1434

FORD, J. P. *et al.* 1980. *Seasat Views North America, the Caribbean,and Western Europe with Imaging Radar.* Jet Propulsion Laboratory Publication 80–87. Pasadena, California.

FU, L. and HOLT, B. 1982. *Seasat Views: Oceans and Sea Ice with Synthetic Aperture Radar.* Jet Propulsion Laboratory Publication 81-120 Pasadena, California.

GATLAND, KENNETH. 1989. *The Illustrated Encyclopedia of Space Technology,* 2nd ed. Salamander Books Ltd.

GUGAN, D. J. 1987. Practical aspects of topographic mapping from SPOT imagery. *Photogrammetric Record,* Vol. 12 (69), pp. 349–355.

GUYENNE, T. D and CALABRESI, G. 1989. *Monitoring Earth's Environment.* ESA-SP-1102.

HARRIS, R. 1987. *Satellite Remote Sensing.* Routledge & Kegan Paul, London and New York.

HOLMES, A. L. 1983. Shuttle imaging radar-A, information and data availability. *Photogrammetric Engineering and Remote Sensing,* Vol. 49, pp. 65–67.

HYATT, E. 1988. *Keyguide to Information Sources in Remote Sensing.* Mansell Publ. Ltd, London and New York.

IRS PROJECT TEAM. 1988. *The Indian Remote Sensing Satellite System. Remote Sensing Yearbook 1988/89,* pp. 59–72. Taylor & Francis Ltd, London and New York.

JENSEN, J. R. 1986. *Introductory Digital Image Processing.* Prentice-Hall, Englewood Cliffs, New Jersey.

KONECNY, G. 1986. First results of the European spacelab photogrammetric camera mission. In K. H. Szekielda (Ed.), *Satellite Remote Sensing for Resource Development.* Graham & Trotman Ltd, London.

KONECNY, G. 1989. Recent development in remote sensing. Invited paper, 14th World Conference of the International Cartographic Association, Budapest.

KROMER, J. 1987. Suitability of space photographs for map production and map revision. *Jena Journal for Photogrammetrists and Surveyors,* Vol. 1, pp. 12–16.

LIGHT, D. L. 1989. Remote sensing for mapping. American Society for Photogrammetry and Remote Sensing. *Proceedings ASPRS/ACSM Annual Convention*, Vol. 3, pp. 50–74.

LILLESAND, T. M. and KIEFER, R. W. 1987. *Remote Sensing and Image Interpretation*, 2nd ed. John Wiley & Sons, New York.

MATTHEWS, J. 1988. Images of remote sensing in Japan. In *Remote Sensing Yearbook 1988/89*, pp. 29–47. Taylor & Francis Ltd, London and New York.

MOIK, J. G. 1980. *Digital Processing of Remotely Sensed Images*. NASA-SP-431.

Multilingual Dictionary of Remote Sensing and Photogrammetry. 1984. American Society of Photogrammetry.

NASA. 1977. *Skylab Explores the Earth*. NASA-Sp-380.

NASA. 1982. *Meteorological Satellites: Past, Present and Future*. NASA-CP-2227.

NASA. 1986. *Earth System Science Overview. A Program for Global Change*. Washington, D.C. NASA.

NASA. 1988. *Earth System Science: A Closer View*. Report of the Earth System Science Committee, Washington, D.C. NASA.

PALUDAN, T. and CSATI, E. 1978. *Euludus Map; An International Land Resources Map Utilizing Satellite Imagery*. NASA-TP-1371.

RICHARDSON, J. A. 1986. *Remote Sensing Digital Image Analysis*. Springer Verlag, Heidelberg.

RIVEREAU, J. C. and POUSSE, M. 1988. SPOT after two years in operation. *CERCO. Proceedings of the Conference "The SPOT and its Cartographic Applications"*, 6–15 June 1988.

ROCHON, G. and TOUTIN, T. 1986. SPOT, a new cartographic tool. *International Archives of Photogrammetry and Remote Sensing*, Vol. 26 (4), pp. 192–205.

SABINS, F. F. 1983. Geologic interpretation of space shuttle radar images of Indonesia. *Amer. Associatia. of Petroleum Geology Bulletin*, Vol. 67, pp. 2076–2099.

SABINS, F. F. 1987. *Remote Sensing: Principles and Interpretation*, 2nd ed. W.H. Freeman and Co., New York.

SAGDAYEW, R. S., SALISHCHEW, K. A. and KOUTZLEBEN, ?. (Eds.). 1982. *Atlas of Interpretation of Multispectral Aerospace Photographs; methods and results*. Academia Verlag, Berlin; Publishing House "Nauka", Moscow.

SCHROEDER, M. 1986. Spacelab metric camera experiments. In K. H. Szekielda (Ed.), *Satellite Remote Sensing for Resource Development*. Graham & Trotman Ltd, London.

SHEFFIELD, C. 1981. *Earthwatch; A Survey of World from Space*. Macmillan, New York.

SHEFFIELD, C. 1983. *Man on Earth*. MacMillan, New York.

SHORT, N. M. 1982. *The Landsat Tutorial Workbook: Basics of Satellite Remote Sensing*. NASA-RP-1078.

SHORT, N. M. and BLAIR, ROBERT W. (Eds.). 1986. *Geomorphology from Space: A Global Overview of Regional Landforms*. NASA-SP-486.

SHORT, N. M. and STUART, L. A. 1982. *The Heat Capacity Mapping Mission (HCMM) Anthology*. NASA-SP-465.

SOUTHWORTH, C. S. 1985. Characteristics and availability of data from Earth imaging satellites. *US Geological Survey Bulletom*. 1631.

SZANGOLIES, K. 1987. Acquisition and use of space photographic data for mapping. *Jena Journal for Photogrammetrists and Surveyors*, Vol. 1, pp. 2–4.

SZEKIELDA, K. H. (Ed.). 1986. *Satellite Remote Sensing for Resource Development*. Graham & Trotman Ltd, London.

TSUCHIYA, K., ARAI, KOHEY and IGARASHI, TAMOTSU. 1987. Marine observation satellite. *Remote Sensing Reviews*, Vol. 3, pp. 59–101.

US GEOLOGICAL SURVEY. 1979. *Landsat Data User's Handbook*. USGS.

VAUGHAM, W. W. (Ed.). 1982. *The Conception, Growth, Accomplishments and Future of Meteorological Satellites*. NASA-CP-2257.

WELCH, R. 1985. Cartographic potentials of SPOT image data. *Photogrammetric Engineering and Remote Sensing*, Vol. 51, pp. 1085–1091.

Chapitre 2
PROCESSUS D'EXTRACTION D'INFORMATIONS À PARTIR DES DONNÉES SATELLITES

Rédigé par
Donald T. Lauer

US Geological Survey, Sioux Falls, South Dakota 57198, USA

Traduit par
Laurent Breton,* Patrick Dayan,* Jean-Pierre Delmas,* Alain Jacqmin* et
J. Denègre†

*Institut Géographique National, B.P. 68, 94160 Saint-Mandé, France
†CNIG, 136 bis rue de Grenelle, 75700 Paris, France

TABLE DES MATIERES

2.1 Introduction

2.1.1 But du chapitre

Les cartographes, tout comme les spécialistes des ressources, les ingénieurs, les planificateurs et les responsables de l'aménagement du territoire ont développé l'usage des images-satellites pour la cartographie thématique. Les applications concernent des domaines tels que la production cartographique, les recherches géologiques et hydrologiques, l'inventaire et l'aménagement du territoire, ou la surveillance de l'environnement. Le but de ce chapitre du manuel est de rappeler brièvement les différentes techniques visant à extraire des informations thématiques des données satellitaires. Ces processus sont présentés en deux parties. D'abord sont rappelés les principes, méthodes et techniques d'interprétation visuelle d'images, selon lesquels le volume et la qualité de l'information extraite dépendent largement de l'expérience et de l'habileté de l'imagiste. Ensuite seront présentés les principes, les méthodes et les procédures d'analyse d'image assistée par ordinateur. Dans ce cas les performances de l'informatique sont exploitées par l'analyste d'image pour élaborer des cartes thématiques d'après les données-satellites. Le texte présenté dans cette partie n'abordant que brièvement ces techniques de production, le lecteur désirant approfondir le sujet pourra se reporter à d'autres sources, comme le *Manual of Remote Sensing*, deuxième édition, volumes I et II, 1983 de l'"American Society of Photogrammetry and Remote Sensing".

2.1.2 Image, trait et carte thématique

Il est important de noter que l'utilisation des données satellitaires pour la production cartographique aboutit soit à une "iconocarte" (carte sur fond d'image) du paysage, soit à une carte au trait des caractères importants, soit enfin à une représentation thématique complète de la surface du terrain. En effet, le processus de production peut être limité à une correction géométrique numérique et à une accentuation radiométrique de l'image conservant la teinte continue de la surface terrestre, tout en combinant éventuellement l'image corrigée avec des informations cartographiques provenant d'autres sources; mais l'extraction d'une information thématique complète ou partielle provenant d'une iconocarte doit être réalisée par l'utilisateur final en employant des techniques d'interprétation visuelle d'image. D'autre part, les données satellitaires peuvent être transformées directement en une carte thématique grâce à de nombreux processus de production qui seront présentés plus loin. Tous nécessitent l'interaction entre l'interprète compétent et l'informatique de pointe.

2.2 Interprétation visuelle d'image

2.2.1 Principes de base

L'interprétation visuelle d'image est souvent un moyen efficace et économique pour extraire de l'image satellite une information thématique utile. Il est néanmoins important de comprendre à la fois les procédures utilisées par un imagiste expérimenté exécutant une tâche d'interprétation, et les concepts sur lesquels reposent ces procédures.

Les principes de base en interprétation d'image satellitaire peuvent être énoncés comme suit:

(1) une image satellite de la Terre n'est rien de plus qu'une forme graphique de données fournissant une représentation imagée des motifs du paysage;

(2) ces motifs sont composés d'éléments, ou indicateurs, d'objets et de phénomènes qui reflètent les composantes physiques, biologiques et culturelles de la région;

(3) des conditions semblables dans des environnements semblables fournissent des formes similaires, et des conditions différentes fournissent des formes différentes;

(4) le type et la quantité d'informations qui peuvent être obtenues à partir d'une image satellite sont fonction des connaissances, de l'expérience, des capacités et de l'intérêt de l'imagiste, et de sa connaissance des limites des méthodes utilisées (Estes, 1980).

L'interprétation visuelle d'image n'a rien de magique. La compréhension des processus de création d'image, des éléments de forme, des objets, processus et phénomènes terrestres, et la "soif de connaissance" sont nécessaires au développement de l'art de l'interprétation des images-satellites pour en extraire des informations thématiques. Une image est seulement un outil et l'interprète doit extraire l'information en utilisant correctement cet outil. La capacité d'extraire des informations à partir des images-satellites s'accroît avec l'expérience— en d'autres termes, on apprend en pratiquant.

2.2.2 *Méthodes d'interprétation*

On peut définir l'interprétation visuelle d'image comme processus de détection, de délimitation, et d'identification des formes et/ou des attributs de l'image, et d'évaluation de leur signification. Les paramètres d'image qui permettent à l'interprète de localiser, délimiter, identifier et évaluer les objets à cartographier sont: la valeur, la couleur, la texture, la structure, la forme, la taille, l'ombre, la parallaxe, et les variations temporelles. Il est important de noter que certains facteurs sont en relation directe à la fois avec la perception de ces paramètres et avec l'interprétation résultante. Ces facteurs sont:

(1) la sensibilité du système imageur (chambres photographiques et films, dispositifs opto-électriques ou autres types de détecteurs);
(2) les caractéristiques de résolution du système imageur;
(3) l'exposition du film ou le traitement de données;
(4) la saison;
(5) l'heure;
(6) les effets atmosphériques;
(7) l'échelle de l'image;
(8) le mouvement de l'image;
(9) la parallaxe stéréoscopique;
(10) l'acuité visuelle et mentale de l'interprète;
(11) les techniques et les outils d'interprétation;
(12) les données auxiliaires.

Notons que les facteurs de 1 à 9 affectent plutôt la qualité de l'imagerie alors que 10, 11 et 12 reflètent davantage la capacité humaine à extraire l'information des images. A l'évidence, certaines combinaisons de ces facteurs permettront mieux que d'autres à l'interprète d'assurer des tâches variées d'interprétation d'image. C'est pourquoi un objectif essentiel durant l'analyse des images-satellites est de déterminer, au mieux de sa propre compétence, la combinaison optimale des paramètres nécessaires pour résoudre les problèmes spécifiques de cartographie thématique.

2.2.3 *Paramètres de l'image*

2.2.3.1 *Valeur*

La valeur dans une image-satellite correspond aux nuances de gris (c'est-à-dire la brillance) que peut prendre un objet, et elle résulte de la quantité d'énergie qu'il réfléchit et/ou émet. La valeur est fondamentale pour l'interprétation des images noir et blanc, et elle est un élément essentiel, utilisée avec d'autres outils de reconnaissance, pour identifier et interpréter les objets. Sur une image satellite, la valeur dépend d'un grand nombre de facteurs et celle d'un objet familier ne correspond que rarement à notre perception directe au sol. Par exemple, une étendue d'eau peut prendre toutes les valeurs de gris selon la position du soleil et le nombre des éléments de vagues réfléchissant l'énergie en direction du capteur. Quand l'imagiste a compris les facteurs qui déterminent la valeur, il (elle) considère les valeurs des objets à étudier comme les clés majeures de leur identité ou de leur genèse. Le pédologue se sert des variations de valeur pour classer les sols, le forestier pour différencier les feuillus des résineux, le géologue pour cartographier la lithologie et la structure des roches. Que ce soit sur une image satellite unique, où la forme des objets peut rarement être appréhendée, ou sur un couple stéréoscopique d'images satellites (cas de SPOT) où les objets intéressants peuvent n'avoir qu'une hauteur faible ou nulle, la valeur est particulièrement importante. L'imagiste peut mettre à profit les variations d'énergie émise ou réfléchie en utilisant des images prises dans les canaux spectraux qui donnent les meilleurs contrastes entre les objets à étudier (Colwell, 1961).

2.2.3.2 *Couleur*

Un objet possède une couleur lorsqu'il réfléchit des quantités différentes d'énergie dans des combinaisons particulières de longueurs d'onde. Par exemple, la végétation apparaît verte à l'oeil humain parce qu'en général les plantes renvoient un plus grand pourcentage d'énergie dans le vert que dans le rouge et le bleu. L'oeil humain peut distinguer environ 1000 fois plus de nuances de couleurs que de tons de gris, ce qui fait que la couleur permet de reconnaître et d'interpréter beaucoup plus d'informations sur la surface terrestre. Lorsque l'on s'intéresse aux roches, aux sols ou aux plantes, il peut y avoir un grand nombre de surfaces où les couleurs naturelles ont une grande importance; dans ce cas, la couleur peut considérablement faciliter l'interprétation par rapport au noir et blanc. De même, l'imagerie en fausses couleurs, qui combine des canaux spectraux autres que le bleu, le vert et le rouge, a montré son utilité pour certaines études sur l'état de santé ou la répartition des végétaux, l'humidité du sol, et un certain nombre d'autres thèmes.

2.2.3.3 Texture

En imagerie satellitaire, la texture naît de la répétition des variations de valeur ou de couleur dans un groupe d'objets trop petits pour être discernés individuellement. Il en résulte que la taille des objets produisant des effets de texture est liée à la résolution et à l'échelle des images. Par exemple, sur des photographies aériennes à haute résolution et à grande échelle, les arbres peuvent être discernés individuellement, contrairement à leurs feuilles, qui contribuent cependant à la texture de leur couronne. Sur des photographies aériennes de moindre résolution et à échelle plus petite, ainsi que sur la plupart des images-satellites, la couronne des arbres contribue à la texture de la forêt entière. Dans un intervalle d'échelles donné, la texture d'un groupe d'objets (commc une forêt), peut être assez perceptible pour servir de clé fiable pour identifier ces objets. La texture est un important facteur d'interprétation des images satellites. Par exemple, les formes d'érosion différentielle d'une zone renseignent sur la nature du sol et du sous-sol: une grande partie des ravines étant invisible sur la plupart des images satellites, l'érosion ne peut être remarquée que par sa texture. Dans d'autres cas, les surfaces peuvent être identiques en valeur et en couleur, tout en montrant des différences considérables au niveau de la texture (par ex. champs de lave plaqués sur plateau désertique).

2.2.3.4 Structure

Les spécialistes des sciences de la Terre attachent depuis toujours une grande importance à la structure, ou disposition spatiale des objets, pour expliquer leur origine ou leur fonction. Géographes et anthropologues étudient la structure des agglomérations et de leur répartition afin de comprendre les effets de diffusion et de migration, dans l'histoire de la culture. La disposition des affleurements fournit des indications sur la structure géologique, la lithologie et la texture du sol. De même les

relations variables entre les êtres vivants et leur environnement créent des formes particulières d'associations végétales. Les structures régionales qui autrefois ne pouvaient être étudiées qu'au prix de laborieuses observations au sol sont maintenant clairement et instantanément visibles sur les photographies aériennes et satellitaires. Les images captent souvent un grand nombre de structures significatives qui auraient pu être cachées à l'homme de terrain, ou mal interprétées par celui-ci. D'innombrables variations de structures classiques peuvent être vues et exploitées grâce à l'interprétation d'image. L'observateur entraîné, qui réalise des cartes thématiques, apprécie la signification de l'imagerie satellite essentiellement en fonction de sa propre compréhension des structures de la surface terrestre. Certaines de ces structures sont avant tout culturelles, d'autres naturelles. Il reste de toute façon peu de lieux sur Terre qui n'aient pas été affectés par l'activité humaine, et beaucoup de structures visibles par satellite résultent de l'interaction entre des facteurs naturels et culturels. Les structures dues aux peuplements historiques, à l'extraction minière et aux activités agricoles sont souvent visibles sur les images-satellites, soit de façon directe, soit à travers l'altération des formes causée par la végétation ou l'érosion. Les structures créées par les pratiques agricoles, les linéaments de failles, les réseaux d'écoulement et la végétation sont parmi les plus importants facteurs d'interprétation d'une image satellite. Ainsi que mentionné ci-dessus, des structures imbriquées peuvent se traduire, dans l'imagerie satellite, par des différences de texture. Dans beaucoup de cas, les structures régionales associées à d'autres paramètres picturaux permettent bien souvent d'interpréter complètement l'image.

2.2.3.5 Forme

Curieusement, la forme des objets ou des zones vue à la verticale est parfois difficile à interpréter. En effet, la vue de dessus d'un objet est si différente de l'habituelle vue de profil ou en oblique que bien souvent, les interprètes débutants ne réussissent pas à reconnaître sur une photographie aérienne verticale l'image du bâtiment dans lequel ils travaillent. La capacité de comprendre et d'utiliser une vue verticale doit donc être acquise comme tout langage mais elle se révèle alors comme un outil puissant. Pour l'interprète d'image satellite spécialisé en études d'urbanisme, la vue verticale d'une zone urbaine lui en apprend autant sur ses caractéristiques qu'une visite à pied ou en véhicule sur le terrain. De même, la vue verticale d'une forêt depuis l'espace peut révéler des informations sur son état, et la vue verticale d'une forme de terrain peut montrer des effets spectaculaires de processus tectoniques.

2.2.3.6 Taille

La taille d'un objet est l'une des clés d'interprétation les plus utiles. En mesurant la taille d'un objet inconnu sur une image satellite, l'interprète peut éliminer tout un groupe d'identifications possibles. Face à un objet inconnu, il est toujours conseillé de le mesurer, et lorsqu'il travaille sur des images satellites d'échelle variable, l'interprète devrait mesurer fréquemment les objets qui l'intéressent. Aussi l'interprète en imagerie spatiale doit toujours être conscient de l'échelle iconique lorsqu'il évalue la taille d'un objet, et bien sûr des objets majeurs peuvent n'occuper qu'une petite zone de l'image par suite de l'effet synoptique induit par les orbites spatiales.

2.2.3.7 Ombre

L'ombre est un phénomène familier, et dans la vie quotidienne, elle sert souvent à déduire la taille et la forme des objets dont elle est issue. Sur les photographies aériennes classiques, les ombres aident parfois l'interprète en lui fournissant des indications sur le profil des objets qui l'intéressent. Les ombres sont particulièrement utiles dans le cas d'objets très petits ou peu contrastés par rapport à leur environnement (US Department of Agriculture, 1966). Dans ces conditions, le fort gradient de valeur dû aux ombres peut permettre à l'interprète d'identifier des objets qui sont eux-mêmes à la limite de la reconnaissance. Si l'interprète n'est pas intéressé par une classe particulière, mais plutôt par le paysage dans son ensemble (ce qui est souvent le cas pour des images satellites), il doit renoncer à certains avantages de l'identification liée aux ombres, afin de voir le maximum de surface du terrain. Travailler en vue verticale oblige l'interprète à revoir sa conception du monde extérieur et à acquérir de nouvelles habitudes d'observation. En raison de la vue verticale et des très petites échelles, habituelles en imagerie spatiale, certains éléments apparents y prennent plus d'importance qu'au sol (par exemple les ombres qui rehaussent les traits topographiques).

2.2.3.8 Parallaxe

Sur un couple stéréoscopique de photos aériennes classiques, l'observateur voit les objets en trois dimensions et peut distinguer les objets proches des objets éloignés. En fait, il perçoit des angles de parallaxe. Dans le contexte naturel, ceux-ci sont déterminés par la distance entre les pupilles de l'observateur (ou distance oculaire). En photographie aérienne, la distance entre deux prises de vues successives (ou base de prise de vues) correspond à la distance oculaire. Elle est représentée sur les photos par la "base photographique" et elle est bien plus grande que la distance oculaire. C'est pourqoi l'impression de relief est très exagérée, et l'interprète expérimenté doit apprendre à la maîtriser. Cette exagération n'est pas toujours apparente pour un couple d'images-satellites en recouvrement, les angles de parallaxe étant relativement faibles. Bien entendu, des images qui ne se recouvrent pas ne peuvent pas être vues en 3 dimensions et la forme des objets n'apparaît ainsi que selon une vue en plan. Cependant la forme reste un facteur primordial d'interprétation, et est très utile dans l'identification des principales formes géo-morphologiques (par exemple coulées de lave, cônes volcaniques, dépôts alluviaux, champs de dunes, plages, baies, lacs, montagnes, etc.). Toutefois, avec l'amélioration des performances des capteurs (résolution, recouvrement, etc.) disponibles sur les satellites actuels et futurs, les observations stéréoscopiques et les mesures de parallaxe prennent de plus en plus d'importance pour les interprètes. Pour l'imagiste expéri-menté, l'intérêt de la forme en trois dimensions est de délimiter la catégorie à laquelle un objet inconnu doit appartenir; elle permet fréquem-ment d'aboutir à une identification certaine, et aide à comprendre la signification et la fonction de l'objet.

2.2.3.9 Phénomènes temporels

Très souvent, l'aspect temporel des phénomènes terrestres échappe à l'imagiste. Pour certaines applications de cartographie thématique, une somme considérable d'informations complé-mentaires sont fournies par la comparaison d'images prises à différentes périodes. Des analyses agricoles peuvent être plus complètes grâce à des images prises à divers stades du cycle de croissance. Le blé d'hiver a un cycle de croissance différent de celui des autres cultures.

La prairie a des stades de croissance différents de ceux des plantations. Parmi les autres phénomènes temporels pouvant être observés en imagerie satellitaire, on peut citer l'évolution de l'érosion des côtes, la sédimentation en profondeur, la répartition des plans d'eau, de la couverture neigeuse, etc.

2.2.4 Techniques d'interprétation et produits dérivés

Quand elles sont correctement appliquées, certaines techniques d'interprétation d'image peuvent améliorer la qualité et la quantité de l'information thématique extraite des images satellites (Colwell, 1987). Ces techniques nécessitent:

(1) des processus méthodiques;
(2) des recherches efficaces;
(3) la connaissance des facteurs qui commandent la formation de l'image;
(4) les connaissances de base et l'expérience de l'interprète;
(5) le concept de "convergence des preuves";
(6) le "système de la conférence";
(7) des informations sur des zones analogues;
(8) des données de référence;
(9) des équipements simples ou sophistiqués;
(10) des données de terrain.

L'imagiste doit savoir parfaitement comment l'image est formée, ce que les éléments d'image représentent, et quels processus et phénomènes terrestres y sont présents. Une approche systématique du problème est peut-être l'aspect le plus important de la télédétection. En général, elle implique la convergence des preuves empiriques issues de l'image, à travers des analyses régionales (aspects géographiques, physiographiques, géologiques, climatiques), ou locales (formes de terrain, de drainage, érosion, formes caractéristiques, végétation, phénomènes spectraux); l'exploitation de l'information exogène (rapports, cartes, données de terrain); la synthèse des résultats d'interprétation et le contrôle des zones-clés sur le terrain. D'autres gains et informations supplémentaires peuvent également être obtenus par une approche collective inter- et multi-disciplinaire.

Les résultats de ce type d'analyse systém-

atique peuvent être présentés sous la forme d'une carte thématique, soit en réalisant un ou plusieurs calques d'interprétations directement sur l'image satellite, soit en reportant l'information interprétée sur une carte de base existante en utilisant un procédé ad hoc de transfert de données, optique ou électronique.

2.3 Interprétation d'image assistée par ordinateur

2.3.1 Principes de base

De par leur faculté de traiter, évaluer et comparer, avec rapidité et précision, de grandes quantités de données numériques, les ordinateurs jouent un rôle de plus en plus important dans l'interprétation des images satellitaires multibandes utilisées en cartographie thématique. L'utilisation rationnelle des techniques d'interprétation assistée par ordinateur s'appuie sur quelques principes de base:

(1) les données du satellite sont souvent déjà sous forme numérique;
(2) il est généralement possible de corriger les erreurs dues aux capteurs ou aux systèmes de traitement;
(3) on dispose de diverses fonctions pour corriger l'éclairement;
(4) chaque élément de l'image (pixel) peut être analysé;
(5) on peut appliquer à ces données des traitements mathématiques ou statistiques sophistiqués;
(6) l'interprétation est objective et répétable;
(7) l'analyse numérique est plus fine que l'analyse visuelle;
(8) on peut traiter de grandes quantités de données;
(9) on peut facilement intégrer aux données numériques du satellite d'autres types de données (cartes, statistiques, autres capteurs, etc.);
(10) en sortie, on peut obtenir automatiquement des produits variés: cartes, tables, images, etc.

En général, l'interprétation d'image assistée par ordinateur est plus adaptée que l'interprétation visuelle au traitement de vastes surfaces qui requièrent des classes d'information détaillées.

Toutefois, cette technique présente un certain nombre d'inconvénients:

—investissement financier important;
—nécessité de former spécialement des programmeurs et des interprètes;
—coût plus élevé des données;
—moindre fiabilité dans des environnements complexes;
—résultats pas toujours compréhensibles pour l'utilisateur.

La décision de faire appel ou non aux techniques d'interprétation assistée par ordinateur d'images satellitaires en vue d'une cartographie thématique, dépend en grande partie de l'étendue de la zone étudiée, du degré de précision souhaité pour le résultat final, de la disponibilité des moyens informatiques (matériels et logiciels), et de la compétence de l'interprète.

2.3.2 *Prétraitement d'image*

2.3.2.1 *Corrections radiométriques*

Des lignes ou segments défectueux, ou des pertes de pixels provoquent l'apparition de rayures intermittentes. Lorsque ces problèmes se présentent dans des données numériques satellitaires, ils entraînent des erreurs de classification car les données fausses ne s'intègrent pas aux séries statistiques d'entraînement. Une technique utilisée pour insérer de nouvelles données est de remplacer la ligne défectueuse par la ligne précédente. Une seconde technique couramment employée est de substituer aux données fausses celles obtenues par interpolation des valeurs de brillance des pixels des lignes précédente et suivante. Bien que la différence entre ces deux techniques paraisse mince, la seconde semble fournir une valeur plus précise de la brillance correcte (Jensen, 1986).

Le striage radiométrique apparaît souvent sous forme de lignes systématiques claires ou sombres à travers l'image; elles proviennent des détecteurs du capteur; ceux-ci ont chacun des réponses légèrement différentes aux radiations reçues, d'où, pour une même intensité de radiation, une légère différence du voltage de sortie. Plus précisément, chaque détecteur possède différents gains et biais (Richards, 1985). Le striage radiométrique introduit des variations anormales dans la signature spectrale des divers éléments de la surface terrestre et

influencera la classification, soit par des variances élevées des données d'entraînement, soit par manque de données d'entraînement adéquates pour caractériser les signatures de pixels dans les lignes parasitées. Par ailleurs, si un des objectifs d'une phase d'analyse est d'accentuer le contraste de l'image, toute rayure fine sera également accentuée. Il peut même arriver que lorsqu'on fait subir ce traitement à des images très rayées, l'organisation spatiale des éléments du paysage s'en trouve détruite.

On dispose de plusieurs techniques pour normaliser les données et minimiser les effets du striage. Un procédé, qui fait référence à la correction d'histogramme, calcule la valeur moyenne de la brillance de chaque détecteur, dans chaque bande. On calcule un coefficient de normalisation en divisant, soit la moyenne des maximums, soit la moyenne des minimums, par la moyenne de chaque détecteur. Le coefficient de normalisation d'un détecteur donné est ensuite multiplié par la brillance de chaque pixel enregistré par ce détecteur, et le résultat constitue la nouvelle brillance du pixel.

On applique la correction d'ensoleillement à des scènes adjacentes lorsque, devant constituer une mosaïque numérique, elles ont été enregistrées dans des conditions d'ensoleillement différentes. Ce type de correction est aussi nécessaire quand on veut comparer les propriétés spectrales d'objets pour des scènes enregistrées dans des conditions d'ensoleillement différentes. La correction de la hauteur du soleil (angle du soleil) entraîne la multiplication de l'ensemble des valeurs de brillance de la scène par une constante, elle-même fonction de l'angle d'incidence. On notera que cette correction ne modifie pas les effets topographiques ni ne les corrige pour différents azimuts du soleil.

Enfin, l'atmosphère terrestre affecte les données enregistrées depuis l'espace de deux façons: diffusion et absorption. La diffusion atmosphérique est fonction de molécules (Rayleigh) et d'aérosols (Mie) présents dans l'atmosphère (McCartney, 1976). L'effet Rayleigh est facile à modéliser dans la mesure où il est largement invariant dans le temps à un endroit donné. La diffusion de Mie fait appel à des modèles complexes nécessitant des mesures d'aérosols, qui sont difficiles à obtenir. Quant à l'absorption atmosphérique, elle est largement liée à la présence de gaz tels que vapeur d'eau, dioxyde de carbone, et ozone (Asrar, 1989). Les

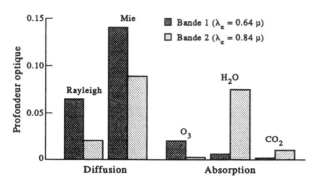

Fig. 2.1 Effets de diffusion et d'absorption atmosphérique sur les bandes 1 et 2 du radiomètre AVHRR (Advanced Very High Resolution Radiometer)—avec l'aimable autorisation de ST Systems Corporation.

modèles d'absorption s'appliquent en général directement, mais reposent sur des mesures qui sont également difficiles à obtenir. Diffusion et absorption atmosphériques ont une grande influence sur les mesures de caractéristiques spectrales des éléments du paysage terrestre (voir fig. 2.1). Si l'on doit comparer les mesures spectrales au sol à celles obtenues à bord du satellite, les effets atmosphériques devront être corrigés.

2.3.2.2 Corrections géométriques (généralités)
(Cet aspect est présenté de façon plus détaillée au Chapitre 3)

L'imagerie satellitaire recèle des erreurs géométriques systématiques et prévisibles, comme la distorsion transversale due à la rotation de la Terre, et la variation de longueur des lignes si le système d'observation comporte un dispositif de balayage par miroir. D'autres erreurs, variables et mesurables, incluent les distorsions dues aux variations de la vitesse du satellite, de son altitude et de son attitude (Williams, 1979).

Les distorsions transversales proviennent des mouvements combinés de la Terre (rotation) et du satellite (déplacement sur son orbite) au cours de l'enregistrement de l'image. L'importance de cette distorsion dans une scène est largement fonction de la latitude et de l'attitude du satellite. Chaque image est rectifiée à l'aide d'un algorithme qui translate chaque ligne vers la droite, selon un nombre de pixels calculé en fonction de la latitude terrestre estimée de la ligne; tous les éléments du terrain sont ainsi ramenés en position convenable sur l'ensemble de la scène.

Les systèmes qui utilisent un miroir oscillant génèrent des variations dans la longueur des lignes; ceci est du à l'inconstance de la vitesse de balayage du miroir. La longueur correcte d'une ligne est calculée en fonction du rythme de balayage, de l'échantillonnage du détecteur et de la largeur de la ligne de balayage.

Lorsqu'une image satellitaire recèle des erreurs géométriques non systématiques, les effets n'en sont pas connus à l'avance; pour corriger ce type d'erreurs, il faut donc d'abord en mesurer les effets. Cette correction peut être faite en mesurant les écarts apparents sur des points d'appui qui sont des objets identifiables sur l'image dont on connaît les coordonnées géographiques. Une fois ces écarts connus, on calcule les coordonnées du pixel correspondant dans l'image corrigée en utilisant des fonctions mathématiques de représentation cartographique. Ces coordonnées correspondent à des positions dans l'image non corrigée, qui coïncident rarement avec un pixel: c'est pourquoi il faut rééchantillonner l'image non corrigée pour déterminer la valeur de brillance que devra avoir le pixel dans l'image corrigée.

Il existe trois techniques communément utilisées de rééchantillonnage (cf. chapitre 3): interpolation au plus proche voisin, interpolation bilinéaire, ou bicubique. La première technique consiste à déterminer la valeur de la brillance du pixel dans l'image corrigée à partir de celle du pixel le plus proche du point homologue considéré dans l'image brute. La seconde technique conduit au même résultat, mais à partir de l'interpolation des valeurs de brillance des quatre pixels qui encadrent le point considéré. La troisième technique de rééchantillonnage par convolution cubique est d'ordre plus élevé et est souvent utilisée; la valeur de brillance du pixel dans l'image corrigée est interpolée à partir des valeurs de brillance d'un plus grand nombre de pixels les plus proches du point considéré.

Quand une image est enregistrée, des dégradations peuvent survenir qui réduisent la résolution apparente. Par exemple, une image peut être affectée de taches, résultant d'une optique déficiente ou d'un mouvement du capteur. Ces phénomènes sont modélisés selon une séquence de "filtrages" par le système imageur, chaque "filtre" représentant l'effet d'une seule cause de dégradation, telle que le vignettage optique. Dans le cas de l'accentuation spatiale, ces modèles peuvent être appliqués aussi bien dans le domaine fréquentiel, où une "fonction de transfert" d'un système imageur permet aux basses fréquences (les "tendances" des vastes surfaces) de passer préférentiellement aux hautes fréquences (les détails fins), que dans le domaine spatial, comme avec le procédé de fenêtre mobile, où la scène est "convoluée" avec la "fonction d'étalement de point" du système imageur.

La restauration d'image est le processus de correction de ces dégradations pour obtenir la meilleure finesse d'image possible. Les techniques de restauration impliquent généralement de modéliser la réponse du capteur, de calculer un filtre de correction, puis d'appliquer ce filtre à l'image dégradée. Des méthodes de correction telles que le "filtrage inverse" ou le "filtrage de Wiener" sont utilisées pour la restauration d'image (Andrews et Hunt, 1977). Ces filtres de correction peuvent être ou bien appliqués dans le domaine fréquentiel ou spatial, ou inclus dans le rééchantillonnage d'image au cours du processus dc correction géométrique.

2.3.3 Accentuation d'image

L'accentuation d'image a pour but d'améliorer la qualité de son interprétation visuelle, en amplifiant ses caractéristiques majeures, spectrales ou spatiales, tout en supprimant les caractéristiques mineures ou les redondances. A une image satellitaire de bonne qualité, on peut n'apporter qu'une "accentuation générale" minime. Des processus d'accentuation pourraient être appliqués pour améliorer l'imagerie en fonction de problèmes d'interprétation particuliers; malheureusement, l'accentuation d'une image pour l'interprétation d'un de ses éléments rend souvent plus difficile l'interprétation des autres éléments. Les caractéristiques du système oeil-cerveau de l'homme doivent être prises en compte en matière d'accentuation; ainsi, la plupart des gens peuvent différencier beaucoup plus de nuances de vert que de nuances de bleu; de même, le système optique humain opère une sorte d'accentuation des contours qu'il est difficile (et généralement inutile) de doubler par des moyens informatiques.

2.3.3.1 Accentuation de contraste

L'accentuation de contraste (ou étalement) est obtenue par amplification de la gamme des brillances (ou d'extraits choisis) d'une image.

Une accentuation générale peut être obtenue de la façon suivante: on détermine les valeurs maximale et minimale de brillance, puis on ajuste cette gamme pour qu'elle corresponde à la "portion droite" de la courbe de réponse du film qui sera utilisé pour enregistrer l'image. On obtient une accentuation plus spectaculaire en tronquant les extrémités minimale et maximale de la répartition, puis, une fois l'accentuation réalisée, en les réinsérant avec la valeur des seuils retenus. Si on utilise une fonction différente pour chaque bande du spectre, la balance des tonalités de la composition colorée peut être modifiée. Ce peut être parfois utilisé avec profit, mais ce serait inutile, par exemple, si les images doivent être mosaïquées.

Dans le cas d'une accentuation de contraste linéaire, chaque valeur de la brillance est multipliée par une constante, puis ajustée par addition d'une constante ou "biais".

Si l'on souhaite accentuer une partie seulement de la gamme des densités, une accentuation non-linéaire peut être faite. L'accentuation non-linéaire sera utilisée pour approximer une distribution uniforme, (ou une distribution en escalier), de telle sorte que chaque densité du film recueille un nombre approximativement égal de pixels. Les accentuations non-linéaires sont souvent très efficaces pour augmenter les contrastes pour des éléments spécifiques (généralement pour les densités prédominantes de la scène), mais parfois au détriment de l'interprétation d'autres éléments (essentiellement les petites surfaces ou les éléments peu nombreux) (Pinson et Lankford, 1981). La plupart des corrections non-linéaires font appel systématiquement à des fonctions dérivées des densités de probabilités telles que fonctions logarithmiques, Gaussiennes, ou sinusoïdales. D'autres fonctions d'accentuation dépendent des données et utilisent des histogrammes, qui doivent d'abord être créés avant que la fonction puisse être générée. Des exemples d'accentuation dépendant des données peuvent inclure la fonction de répartition en escalier, l'étalement selon la fonction de distribution des probabilités, l'étalement proportionnel à la distribution des fréquences, et l'égalisation d'histogramme.

2.3.3.2 Accentuation spatiale

Les accentuations spatiales adaptent la valeur de la brillance de chaque pixel par comparaison à celle des pixels proches. Par exemple, on renforce les limites en exagérant la différence entre un pixel et ses voisins; à l'inverse, une image lissée est obtenue en réduisant cette même différence. L'accentuation des limites est un filtrage des hautes fréquences, la correction de lissage est un filtrage des basses fréquences (Moik, 1980).

Une technique d'accentuation locale plus sophistiquée transforme (réellement) les données en un domaine de fréquences (transformation de Fourier par exemple); une fois la scène entièrement transformée, il est possible, par filtrage, de ne conserver que certaines fréquences. Les fréquences conservées sont alors retransformées dans leur format initial. On obtient pratiquement le même résultat dans le domaine spatial en utilisant une "fenêtre mobile" ou "moyenne mobile". Comme ce procédé est d'une mise en oeuvre et d'une utilisation plus simples, il reste le plus employé pour ce type de corrections.

L'accentuation des hautes fréquences est typiquement utilisé pour les images spatiales afin de rehausser les limites entre les différents éléments de la surface terrestre, ou pour faire ressortir des tendances géologiques significatives (Mayers *et al.*, 1988). Pour cette raison, on l'utilise souvent pour détecter des linéaments ou des réseaux de drainage. La fig. 2.2 montre un exemple d'application d'un filtre adaptatif appliqué à la bande 4 de Landsat Thematic Mapper.

Le filtrage des basses fréquences est employé pour lisser, généraliser ou "défocaliser" une image. Comme la généralisation est peu pratiquée par accentuation, une technique courante consiste à ignorer les différences d'un pixel avec son voisinage, à moins que ces différences soient suffisamment grandes pour indiquer une anomalie. On détermine un seuil au-delà duquel le lissage n'est pas appliqué, à moins qu'un pixel soit très différent de ses voisins, ce qui indique un "bruit" (par exemple, un manque dans la transmission des données). Lorsqu'un pixel est identifié comme un "bruit", il peut être remplacé par la moyenne des valeurs de ses voisins. On appelle cette opération suppression des bruits ou filtrage des bruits.

A la suite d'une classification, on peut effectuer un filtrage passe-bas pour généraliser

Original band 4

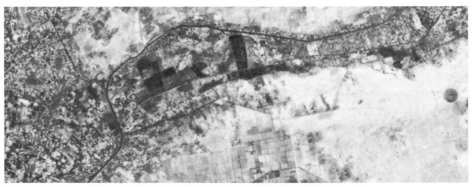

Adaptive boxcar filter

Fig. 2.2 Application d'un filtre passe-haut "adaptive boxcar" aux données de la bande 4 de Landsat Thematic Mapper. L'image non filtrée est en haut, l'image filtrée en bas (Mayers *et al.*, 1988).

et affiner l'image résultant de cette classification en réaffectant à chaque pixel la valeur de la classe qui domine dans son voisinage (Fosnight, 1989). Cette méthode est employée pour établir des cartes thématiques dont la définition est moins fine qu'un pixel: l'emploi d'un voisinage plus large permet de générer une image plus généralisée, avec une taille minimale également plus grande pour les unités cartographiques (voir fig. 2.3).

2.3.3.3 Combinaisons arithmétiques (ratio et différence)

Des combinaisons arithmétiques simples d'images satellitaires fournissent une méthode pour mettre en évidence les changements au

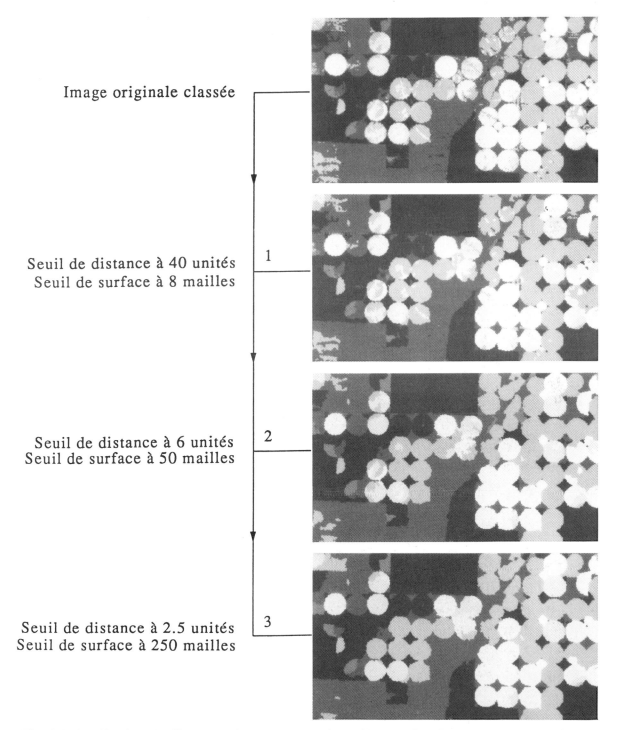

Image originale classée

Seuil de distance à 40 unités
Seuil de surface à 8 mailles — 1

Seuil de distance à 6 unités
Seuil de surface à 50 mailles — 2

Seuil de distance à 2.5 unités
Seuil de surface à 250 mailles — 3

Fig. 2.3 Application de filtres spatiaux basse-bas à une image classée sur une zone agricole irriguée (avec l'aimable autorisation de l'US Geological Survey, EROS Data Center).

cours du temps, pour réduire les effets topographiques et surtout d'éclairement, et pour combiner les données en un plus petit nombre de canaux(Schowengerdt, 1983).

Une image "ratio" est calculée en divisant la radiométrie de chaque pixel dans un canal par la valeur correspondante dans un autre canal ou groupe de canaux. De cette façon, toutes les constantes multiplicatives qui existent dans les 2 canaux sont annulées. Par exemple, la pente topographique et l'ensoleillement créent un effet d'éclairement supplémentaire qui peut, en grande partie, être éliminé en utilisant des méthodes de ratio.

Une autre façon de calculer un ratio consiste à établir une "image index" où, pixel par pixel, la valeur d'un canal est soustraite à celle d'un autre, puis divisée par la somme des deux. Cette technique, appliquée dans les bandes rouge et proche infra-rouge pour des données satellites multibandes, donne une excellente indication de la présence, ou de l'absence, de "végétation verte". Elle est souvent dénommée Image Index Normalisée de Différence de Végétation (INDV). La différence se fait en soustrayant une image de l'autre, laissant seulement les différences entre les images. Cette technique est tout à fait indiquée pour identifier les changements avec des données multitemporelles dans les bandes spectrales correspondantes. Du fait que seule l'information liée aux changements est conservée, cette méthode a ses limites, mais elle est tout à fait apte à traduire visuellement aussi bien les changements d'usage du sol que phénologiques (Gallo et Daughtry, 1987).

2.3.3.4 *Transformation spectrale (rotation)*

La transformation spectrale fournit une méthode pour réduire systématiquement les données hautement corrélées à un petit nombre de canaux non-redondants qui conservent l'essentiel de la variance contenue dans les données d'origine (voir fig. 2.4).

Une technique de rehaussement consiste à combiner des images multidates et multibandes et à concentrer l'information en une seule image couleur. Un avantage supplémentaire de la transformation spectrale est de n'avoir à traiter ensuite qu'un nombre réduit de canaux. La transformation spectrale est un outil puissant pour réduire l'information provenant de nombreuses images à un petit nombre d'images non-redondantes et non-corrélées. Un incon-

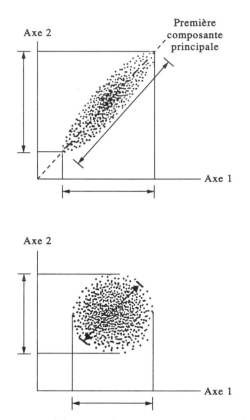

Fig. 2.4 Le schéma du haut montre une analyse en composantes principales de données très corrélées, où la projection sur le premier axe principal est plus longue que celle sur chacun des axes d'origine. Le schéma du bas montre une analyse en composantes principales de données non corrélées, où la projection sur tout axe nouveau sera égale à celle sur les axes d'origine (Jenson et Waltz, 1979).

vénient de cette méthode réside dans le fait que les couleurs obtenues sont grandement imprévisibles et difficiles à interpréter pour qui a l'habitude des images en couleurs naturelles ou en fausses couleurs. Un exemple de cette transformation est l'Analyse en Composantes Principales (ACP), qui rend les axes colinéaires aux vecteurs propres calculés, puis met à l'échelle et translate les données pour maximiser le contraste résultant, et ordonne les composantes de telle façon que la première composante contienne la plus grande partie de la variance totale de la scène (Fontanel *et al.*, 1975). La transformation "Canonical" diffère de l'ACP par le fait que des statistiques sont extraites de chaque classe définie par l'analyste. Les axes de rotation sont calculés de telle façon que les distances spectrales soient maximisées entre les classes, et minimisées à l'intérieur de celles-ci. Le but est d'accroître les distances spectrales entre les classes à interpréter, tout en minimisant la variabilité interne qui pourrait entraîner des confusions à l'intérieur d'une classe.

2.3.4 Classification des images

La classification assistée par ordinateur de l'imagerie satellite numérique consiste à regrouper un grand nombre de pixels individuels en un petit nombre plus commode de classes ou catégories thématiques. L'affectation d'un pixel inconnu à une classe thématique est basée sur la similarité des caractéristiques spectrales de ce pixel avec celles des pixels de catégorie connue. L'hypothèse principale est que chaque classe thématique a des caractéristiques spectrales uniques qui peuvent être extraites de l'imagerie satellite multibande. L'ensemble de la classification thématique peut permettre de décrire par exemple la végétation ou la couverture du sol, l'utilisation du sol, des zonages géologiques ou hydrologiques. A cause d'autres facteurs (pente, aspect, fréquence relative et densité des objets, etc.) la réponse spectrale pour une catégorie thématique donnée n'est pas un modèle spectral unique, mais variable. Quand les variations spectrales sont plus grandes entre les catégories thématiques qu'à l'intérieur de celles-ci, des classifications peuvent être réalisées avec succès (Lillesand et Kiefer, 1979).

2.3.4.1 Choix d'une classification

L'interprétation multibande nécessite deux étapes conceptuelles. La première est l'identification des modèles spectraux pour chaque catégorie thématique, la seconde est le choix d'une règle de classification qui affecte un pixel à une catégorie thématique en comparant les caractéristiques spectrales de celui-ci aux modèles spectraux définis au cours de la première étape (Schowengerdt, 1983). Diverses règles de classification (voir fig. 2.5) sont disponibles pour déterminer l'affectation de pixels inconnus à partir de leurs caractéristiques spectrales:

Distance minimale aux moyennes—Le pixel inconnu est affecté à la classe dont le modèle spectral moyen, déterminé à partir d'un échantillon de pixels connus, est le plus proche de celui du pixel inconnu.

Distance minimale au plus proche pixel d'une classe—Le pixel inconnu est affecté à la

Fig. 2.5 Exemple de règles de classification souvent utilisées pour assigner une valeur de pixel à une catégorie thématique (avec l'aimable autorisation de l'US Geological Survey, EROS Data Center).

même classe que celle du pixel connu le plus proche.

Parallélépipède—C'est un "rectangle multi-dimensionnel" défini par les limites spectrales inférieure et supérieure des pixels connus. Le pixel inconnu est affecté à la classe dont le parallélépipède contient le pixel inconnu.

Maximum de vraisemblance—Dans cette méthode, les pixels connus sont utilisés pour estimer les propriétés statistiques de chaque catégorie thématique. Chacune est décrite statistiquement par sa fonction multivariable de densité de probabilité. Cette fonction représente la probabilité que le modèle spectral d'une catégorie thématique tombe dans une région spectrale donnée. L'algorithme du maximum de vraisemblance classe un pixel inconnu dans la classe qui a la plus grande densité de probabilité dans la région spectrale du pixel inconnu. Cet algorithme est le plus couramment utilisé dans les processus interactifs de traitement d'images numériques.

2.3.4.2 Statistiques d'entraînement

Des jeux de données d'entraînement sont des descriptions statistiques des caractéristiques multibandes de catégories thématiques, utilisées pour "programmer" l'interprète en fonction des caractéristiques numériques de celles-ci. Les zones d'entraînement sont des échantillons d'une scène, choisis au hasard ou non. Elles sont utilisées pour extraire des statistiques d'entraînement et doivent refléter la variabilité spectrale de l'ensemble de la scène et contenir toutes les catégories thématiques de la classification. Plusieurs approches pour extraire des statistiques d'entraînement peuvent être utilisées, incluant classification dirigée, non dirigée, classifications hiérarchiques supervisées ou contrôlées.

L'établissement supervisé de statistiques d'entraînement consiste en la sélection a priori de zones d'entraînement de catégories thématiques connues et homogènes. Une approche non supervisée comprend un choix aléatoire de zones d'entraînement sans se préoccuper de leur composition thématique (Jensen, 1986). Dans les deux cas (supervisé ou non), la portion de scène échantillonnée dépend du nombre de catégories et de la variation spectrale dans la scène, mais est habituellement de 5 à 10%.

Le calcul de statistiques pratiques non supervisées utilise un algorithme de "nuées" qui répartit les pixels en groupes spectraux homogènes, indépendamment de leur composition thématique (Bryant, 1978). Le nombre maximal et la variabilité maximale des groupes spectraux ainsi que d'autres paramètres doivent être spécifiés avant d'appliquer l'algorithme de nuées. Des statistiques sont déterminées pour chaque nuée de pixels. Les statistiques d'entraînement dérivées de l'algorithme de nuées sont utilisées dans l'algorithme du maximum de vraisemblance pour classer la scène entière. A ce stade, chaque nuée est identifiée comme représentant une catégorie thématique grâce aux données de terrain ou issues de photographies. Idéalement, chaque thème sera représenté par au moins une nuée et chaque nuée représentera seulement un thème. Cette approche est utile quand il est difficile de définir des zones d'entraînement homogènes d'un point de vue spectral et thématique, ce qui est très fréquemment le cas dans la cartographie de zones inaccessibles et à condition de disposer d'une vaste réalité de terrain.

L'approche par nuées supervisées suppose la sélection d'un certain nombre de zones d'entraînement connues et homogènes pour chaque classe thématique, et utilise un algorithme de nuées pour calculer les statistiques d'entraînement en regroupant toutes les zones d'étude collectivement pour chaque catégorie thématique. Enfin l'approche par nuées dirigées utilise une sélection a priori de zones d'entraînement, comportant généralement plusieurs thèmes dans chaque zone; le calcul des statistiques d'entraînement applique le même algorithme de nuées, regroupant les zones d'entraînement par thème séparé ou collectivement. Chaque nuée est identifiée comme représentant une catégorie thématique grâce aux données d'appui, et les statistiques d'entraînement pour les nuées correspondant à plusieurs thèmes mélangés peuvent être éliminées (Bryant, 1978). Dans tous les cas, les statistiques d'entraînement sont évaluées en appliquant l'algorithme du maximum de vraisemblance à plusieurs zones d'entraînement et de test, et en comparant les résultats avec les données de photographies aériennes ou de terrain. Si les résultats de la classification préliminaire ne sont pas acceptables, de nouvelles statistiques d'entraînement doivent être calculées ou bien les objectifs de la classification doivent être réexaminés.

2.3.4.3 Classification

Avec la règle de classification selon le maximum de vraisemblance, chaque pixel est affecté à la classe spectrale qui a la fonction de densité de probabilité la plus grande pour les valeurs multibandes du pixel. Chaque pixel peut être affecté à une classe spectrale et une seule. Le classement par catégorie est obtenu en regroupant les résultats afin que chaque classe représente une catégorie thématique et une seule. La stratification de l'image est souvent utilisée pour améliorer les résultats de la classification au moyen de données auxiliaires et ceci peut être fait avant ou après la classification originale. La stratification peut être basée sur un ou plusieurs types d'informations comme des données topographiques ou environnementales, ou des limites administratives. De même, les données multibandes issues de différentes images peuvent aussi contenir différentes strates. Enfin, un lissage spatial est souvent utilisé pour améliorer l'apparence du produit cartographique final et réduire les erreurs de classification (Fosnight, 1988).

2.3.4.4 Précision de la classification

Une estimation non biaisée de la précision des résultats de la classification est souvent obtenue en déterminant par échantillonnage la proportion de pixels correctement classés. Une estimation de la précision globale de la classification d'une carte thématique issue de données satellites peut être obtenue aussi bien pour des thèmes particuliers de la carte.

Un simple échantillonnage aléatoire de points peut être utile pour obtenir des estimations de la précision globale. Un échantillonnage aléatoire stratifié, où chaque catégorie thématique peut être considérée comme une strate, est utile pour obtenir aussi bien des estimations de la précision globale que de chaque thème. Les pixels à échantillonner sont choisis au hasard, mais souvent on doit se limiter à des zones où figurent des données d'appui issues par exemple de photographies aériennes, ou à des zones de terrain accessibles. Les échantillons de points sont repérés sur les photographies aériennes par transfert depuis l'image satellite, ou en établissant un lien géométrique entre l'image satellite et une carte. Les points échantillonnés sont interprétés sur la photo ou sur la carte, l'interprétation est comparée aux résultats de la classification, et les pixels correctement classés sont décomptés. La proportion de pixels correctement classés est estimée en comparant le nombre de pixels échantillons correctement classés au nombre total de pixels échantillons. La précision de cette proportion estimée est déterminée en calculant son intervalle de confiance. Une évaluation des résultats de la classification peut aussi être résumée dans une *matrice de confusion* qui compare, pour chaque thème, les résultats obtenus avec ceux de la classification correcte. De cette façon, la nature et les types de confusions peuvent être identifiés et quantifiés.

2.3.4.5 *Produits de sortie des classifications*

Les données satellites sont géométriquement déformées et il est nécessaire de corriger ces déformations pour fournir les résultats de la classification sous une forme exploitable. Une référence géométrique est définie à l'aide de points d'appui identifiables à la fois sur l'image satellite et sur la carte. On écrit les équations de transformation pour ces points d'appui afin d'établir un lien entre l'image, la carte et les coordonnées géographiques. Les équations sont appliquées de manière à obtenir une géométrie correcte pour les calques et films issus de la classification, pour les points échantillons

sélectionnés, et pour la localisation, sur l'image, de données issues des cartes. La production de calques thématiques suppose d'abord de déterminer la partie des résultats de la classification correspondant à la carte, et ensuite de corriger la position des pixels afin de les positionner correctement sur la carte. Des calques cartographiques montrant les résultats de la classification peuvent être produits au moyen de nombreux matériels tels que imprimantes en ligne, traceurs à plat et phototraceurs.

Des statistiques de surface peuvent être obtenues simplement en comptant le nombre de pixels affectés à chaque thème et en mutipliant ce résultat par un facteur ad hoc pour obtenir la surface du terrain (en acres, hectares, milles carrés, km^2, etc.). Souvent on souhaite décomposer ces résultats suivant les zonages fonciers ou administratifs. Une transformation géométrique peut permettre de déterminer la position des limites correspondantes sur l'image et de les tracer sur la carte. Les calculs de surface peuvent alors être effectués. Une grille d'échantillonnage peut aussi servir de base pour choisir des échantillons pour estimer les attributs des zones thématiques. Par exemple, on peut calculer la somme, pondérée ou non, des pixels affectés à chaque thème dans chaque maille de la grille. Cette information est particulièrement utile si les mailles échantillons doivent être choisies avec des probabilités variables. Les coordonnées géographiques des mailles choisies peuvent être déterminées en utilisant la transformation géométrique appropriée.

2.4 Synthèse et conclusions

Divers processus de production pour extraire l'information thématique relative à la surface de la Terre à partir de données satellites sont décrits dans cette partie du manuel. Deux méthodes générales sont présentées: l'interprétation visuelle de l'image et l'analyse d'image assistée par ordinateur, permettant la production cartographique. Le lecteur doit être averti, cependant, que ces méthodes peuvent être appliquées suivant des combinaisons variées et à différents niveaux. Par exemple, un premier niveau peut être la production d'images accentuées radiométriquement et corrigées géométriquement, pouvant faire l'objet d'une petite interprétation complémentaire permettant de satisfaire un grand nombre d'usagers: c'est alors une "iconocarte"; à un second niveau, l'image brute ou accentuée est interprétée visuellement pour des applications spécifiques ; à un troisième niveau on peut créer une image index basée sur un ratio ou d'autres combinaisons mathématiques de canaux spectraux (par exemple l'Index Normalisé de Différence de Végétation) avec peu ou pas d'interprétation visuelle complémentaire; à un quatrième niveau on peut produire une carte thématique à partir d'une classification non dirigée qui montre simplement les différences spectrales des éléments du paysage, mais où ces éléments ne sont pas identifiés; un cinquième niveau pourrait être celui des classifications par nuées, montrant les différences spectrales dans les éléments du paysage, avec interprétation complète et classement de ces éléments dans des thèmes explicites, tout en fournissant une estimation de la précision de ces catégories thématiques. Quel que soit le niveau de combinaison des méthodes utilisées, il y a presque toujours nécessité d'interaction effective homme–machine. De plus, la quantité et la qualité de l'information thématique qui peut être extraite des données satellites augmentent quand on peut combiner différents types de données satellites et quand celles-ci sont combinées avec d'autres types de données, graphiques ou tableaux (Nyquist, 1987; Lauer, 1986)—ce qui est le sujet du chapitre suivant.

Références

ANDREWS, H. C. and HUNT, B. R. 1977. *Digital Image Restoration.* Prentice-Hall, Englewood Cliffs, New Jersey.

ASRAR, G. 1989. *Theory and Applications of Optical Remote Sensing,* Chapter 9, The atmospheric effect on remote sensing and its corrections. Kaufman, Y. J. (Ed.). John Wiley & Sons, Inc., New York.

AVERY, T. E. 1987. *Interpretation of Aerial Photographs,* Burgess Publishing Company, Minnesota.

BRYANT, J. 1978. *Applications of Clustering in Multi-image Data Analysis.* College Station, Texas, Department of Mathematics, Texas A & M University, Report No. 18.

COLWELL, R. N. 1961. Some practical applications of multiband spectral reconnaissance. *American Scientist,* Vol. 49, No. 1, March 1961, pp. 9–36.

COLWELL, R. N. 1965. The extraction of data from aerial photographs by human and mechanical means. *Photogrammetria,* Vol. 20, pp. 211–228.

COLWELL, R. N. 1987. Remote sensing—Past, present and future. *Proceedings, Study Week on Remote Sensing and Its Impact on Developing Countries,* Vatican City, Italy, 16–21 June 1986, Pontifical Academy of Sciences, pp. 3–141.

ESTES, J. E. 1980. Attributes of a well-trained remote sensing technologist. *Proceedings, Conference of Remote Sensing Educators* (CORSE-78), Stanford University, California, 26–30 June 1978; NASA Scientific and Technical Information Office, Conference Publication 2102, 1980, pp. 103–118.

FONTANEL, A., BLANCHET, C. and LALLEMAND C. 1975. Enhancement of Landsat imagery by combination of multispectral classification and principal component analysis. *NASA Earth Resources Surv. Symp.* July 1975, Houston, Texas. NASA-TMX-58168, pp. 991–1012.

FOSNIGHT, E. A. 1988. Applications of spatial post-classification models. *International Symposium on Remote Sensing of Environment, 21st,* Ann Arbor, Michigan, October 1987. Ann Arbor, Environmental Research Institute of Michigan, pp. 469–485.

GALLO, K. P. and DAUGHTRY, C. S. T. 1987. Differences in vegetation indicies for simulated Landsat-5 MSS and TM, NOAA-9 AVHRR, and SPOT-1 sensor systems. *Remote Sensing of Environment,* Vol. 23, pp. 439–452. Elsevier Science Publishing Company, Inc., New York.

JENSEN, J. R. 1986. *Introductory Digital Image Processing: A Remote Sensing Perspective.* Prentice Hall, New Jersey.

JENSON, S. K. and WALTZ, F. A. 1979. Principal components analysis and conconical analysis in remote sensing. *Proceedings American Society of Photogrammetry/American Congress of Surveying and Mapping Annual Meeting,* Washington, D.C., 18–23 March 1979.

LAUER, D. T. 1986. Applications of Landsat data and the data base approach. *Photogrammetric Engineering and Remote Sensing,* Vol. 52, No. 8, pp. 1193–1199.

LILLESAND, T. M. and KIEFER, R. W. 1979. *Remote Sensing and Image Interpretation.* John Wiley & Sons, Inc., New York.

MAYERS, M., WOOD, L. and HOOD, J. 1988. Adaptive spatial filtering. *Proceedings American Congress on Surveying and Mapping (ASCM), American Society for Photogrammetry and Remote Sensing (ASPRS),* Fall Convention, Virginia Beach, Virginia, September 1988. Falls Church, Virginia, ASPRS, pp. 99–105.

McCARTNEY, E. J. 1976. *Optics of the Atmosphere: Scattering by Molecules and Particles.* John Wiley & Sons, Inc., New York.

Moik, J. G. 1980. *Digital Processing of Remotely Sensed Images.* NASA Scientific and Technical Information Branch. NASA-SP-431. US Government Printing Office, Washington, D.C.

Nyquist, M. O. 1987. The Integration of remotely sensed data into a geographic information system—rediscovered!?? *Proceedings 21st International Symposium on Remote Sensing of the Environment,* Ann Arbor, Michigan, 26–30 October 1987, pp. 487–493.

Pinson, L. J. and Lankford, J. P. 1981. *Research on Image Enhancement Algorithms.* Tullahoma, Tennessee, Technical Report RG-CR-81–3. University of Tennessee Space Institute.

Richards, M. E. 1985. An evaluation of a new statistical approach to traditional linear destriping. *Proceedings American Society of Photogrammetry Annual Meeting, 51st,* Washington, D.C. March 1985. Falls Church, Virginia, American Society of Photogrammetry, vol. 2, pp. 557–575.

Schowengerdt, R. A. 1983. *Techniques for Image Processing and Classification in Remote Sensing.* Academic Press, New York.

Simon, K. W. 1975. Digital image reconstruction and resampling for geometric manipulation. *Proceedings International Symposium of Machine Processing of Remotely Sensed Data, 1st.* West Lafayette, Indiana, 1975. West Lafayette, Indiana, Purdue University, pp. 3A1–3A11.

US Department of Agriculture. 1966. *Foresters Guide to Aerial Photo Interpretation.* US Forest Service, Agricultural Handbook 308, Washington, D.C.

Williams, J. M. 1979. *Geometric Correction of Satellite Imagery.* Farnborough, Hants, United Kingdom. Technical Report 79121, Royal Aircraft Establishment.

Chapitre 3

MÉTHODES DE COMBINAISON DES INFORMATIONS D'ORIGINE SATELLITAIRE ET DES INFORMATIONS DE SOURCES CLASSIQUES

Auteurs
Sten Folving* et Jean Denègre†

*Institute for Remote Sensing Applications, Joint Research Centre, Ispra, Italy
†Conseil National de l'Information Géographique, Paris, France

Traducteurs
Jean-Pierre Delmas (*IGN*) et Jean Denègre (*CNIG*)

TABLE DES MATIERES

3.1 Introduction

L'apparition, depuis 1972, des données d'observation de la Terre d'origine satellitaire a apporté des innovations considérables dans le contenu, l'expression et la perception des informations géographiques en général. En même temps elle a créé pour le cartographe des problèmes également nouveaux, dans la mesure où est apparue l'obligation de combiner les données satellitaires avec les données de sources classiques qui apparaissent souvent comme complémentaires. En effet, les données satellitaires apportent rarement la totalité des informations souhaitées et, de toute façon, elles nécessitent, pour être exploitables par l'usager, d'être *combinées* à un minimum d'éléments de référence, topographiques ou thématiques, permettant au lecteur de se repérer et d'analyser le contenu de la "spatiocarte" ainsi obtenue.

Si l'on essaie de classer, d'une façon très générale, les différents cas où cette combinaison intervient, on doit d'abord considérer les formes sous lesquelles les données satellitaires sont fournies. En allant de la plus élémentaire à la plus élaborée, on peut distinguer trois niveaux, déjà décrits précédemment (chapitre 2):

(i)   Un niveau "image", où il n'y a aucune interprétation des données.

(ii)   Un niveau "image classée" où l'information satellitaire interprétée est fournie sous forme de pixels affectés à des classes en nombre limité (par ex. 10 ou 15).

(iii)   Un niveau "carte de trait" où l'information satellitaire interprétée est fournie sous forme de vecteurs ou de zonages de contenu (polygones thématiques).

Dans ce dernier cas, le problème cartographique de la combinaison avec des données issues d'autres sources (terrain, photos aériennes, etc.) ne présente pas de différence particulière avec les problèmes classiques de compilation de documents de trait, bien connus des cartographes.

Le problème cartographique ne se pose donc de façon nouvelle que dans les cas (i) et (ii), lorsque l'information satellitaire conserve son caractère d'"image" sous forme de pixels, que ceux-ci soient interprétés (cas ii) ou non (cas i). Ce problème est inédit dans la mesure où l'image satellitaire est d'une richesse telle que toute incrustation de surcharges cartographiques se fait au détriment des détails apportés par l'image: ce problème existe déjà pour les photocartes, mais se trouve ici amplifié par la petitesse de l'échelle, qui accentue le divorce entre détails réels et symboles artificiels, entre positions vraies et déplacements conventionnels, entre image et carte. Il s'y ajoute, pour des images satellitaires très agrandies, la structure maillée en pixels de grande taille, qui ne se combinent pas aisément à des éléments topographiques en mode vecteur.

Si l'on considère maintenant les deux familles principales de cartes où l'imagerie satellitaire joue un rôle, à savoir les cartes générales (ou topographiques) et les cartes thématiques, on se trouve finalement conduit à considérer trois types de documents où cette imagerie se trouve combinée à des données exogènes (voir fig. 3.1):

A.   *Cartes générales*, constituées par un fond d'image satellitaire non *interprétée*, avec des surcharges topographiques (objets, toponymes).

B.   *Cartes thématiques*, constituées par une classification (ou une photo-interprétation) d'image satellitaire, avec des surcharges *topographiques* allégées (objets, toponymes) pour un repérage minimal.

C.   *Cartes thématiques* constituées par une classification (ou une photo-interprétation) d'images satellitaire, avec comme fond de carte l'image satellitaire elle-même, non interprétée, servant de repérage d'ensemble, en général monochrome, et complétée éventuellement de quelques éléments topographiques.

A chacun de ces trois types de documents, correspondent des problèmes différents et des méthodes spécifiques de combinaison des informations en présence. Pour décrire ces méthodes en fonction des problèmes posés, les aspects suivants seront examinés:

1.   Calage géométrique des informations satellitaires et des informations classiques.
2.   Assemblage et découpage de l'image le long des limites de la carte.
3.   Traitement de l'image devant servir de fond cartographique.
4.   Incrustation d'informations classiques dans l'image satellitaire (interprétée ou non).

1 et 2 concernent les types A, B et C.

3 concerne les types A et C.

4 concerne les types A et B.

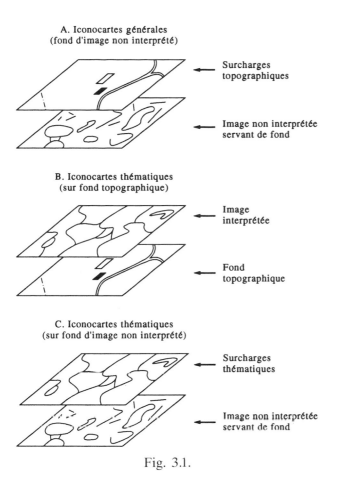

A. Iconocartes générales
(fond d'image non interprété)

← Surcharges topographiques

← Image non interprétée servant de fond

B. Iconocartes thématiques
(sur fond topographique)

← Image interprétée

← Fond topographique

C. Iconocartes thématiques
(sur fond d'image non interprété)

← Surcharges thématiques

← Image non interprétée servant de fond

Fig. 3.1.

3.2 Corrections géométriques

Les images satellitaires, par rapport aux cartes, présentent des déformations géométriques d'origines diverses: effets systématiques dus au dispositif d'enregistrement et à la rotation de la terre, défauts d'attitude du satellite, etc. (voir chapitre 2).

Sauf exception, ce sont les informations "classiques" qui fournissent la référence géométrique sous forme de cartes existantes ou de points d'appui issus d'autres sources.

On modélise généralement les déformations par une fonction polynomiale (coordonnées lignes-colonnes dans l'image en fonction des coordonnées cartographiques, par ex. UTM) dont les coefficients sont déterminés en introduisant dans les calculs les coordonnées d'une dizaine de points d'appui identifiés dans les deux systèmes, image et référence.

3.2.1 *Coordonnées* X–Y *des points d'appui*

Les coordonnées-image ("iconiques") des points d'appui sont relevées par pointé sur écran imageur, les coordonnées cartographiques correspondantes par pointé sur table à numériser, ou simplement par lecture sur la carte. Le succès de cette modélisation dépend évidemment de la bonne identification des points d'appui, de la précision des pointés et de la qualité de la carte utilisés dans ce type de correction.

3.2.1.1 *Calage d'image à carte*

Les points d'appui, données en coordonnées cartographiques et en ligne/colonne dans l'image, sont utilisés pour modéliser la déformation au moyen d'une grille. Par exemple, le cas de 4 points d'appui peut être modélisé par un polynôme de la forme:

$$x = a_0 + a_1 x' + a_2 y' + a_3 x'y' + a_4 (x')^2 + a_5 (y')^2$$
$$y = b_0 + b_1 x' + b_2 y' + b_3 x'y' + b_4 (x')^2 + a_5 (y')^2$$

où x et y sont les coordonnées dans l'image déformée et x' et y' sont celles de l'image corrigée. Si les termes de puissance 2 sont nuls, on a les cas suivants (voir fig. 3.2):

Une approximation géométrique par morceaux peut être obtenue au moyen d'un réseau continu de points d'appui disposés en quadrilatères, avec différentes approximations polynomiales définies pour chaque quadrilatère (voir fig. 3.3).

Si les termes de puissance 2 doivent être conservés, une transformation par moindre carrés doit être calculée, ce qui conduira à des

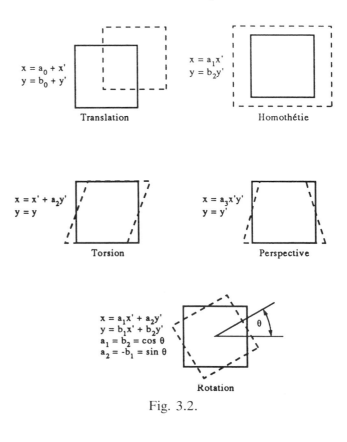

$x = a_0 + x'$
$y = b_0 + y'$
Translation

$x = a_1 x'$
$y = b_2 y'$
Homothétie

$x = x' + a_2 y'$
$y = y$
Torsion

$x = a_3 x'y'$
$y = y'$
Perspective

$x = a_1 x' + a_2 y'$
$y = b_1 x' + b_2 y'$
$a_1 = b_2 = \cos \theta$
$a_2 = -b_1 = \sin \theta$
Rotation

Fig. 3.2.

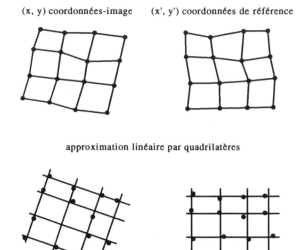

(x, y) coordonnées-image   (x', y') coordonnées de référence

approximation linéaire par quadrilatères

approximation polynomiale globale par moindres carrés

Fig. 3.3.

résidus. Une simple transformation affine suffira dans la plupart des cas ci-dessus, mais il est recommandé de calculer plusieurs fois la transformation par itération afin de valider les points d'appui, opération le plus souvent nécessaire avant de traiter toute l'image.

Remarque: Si un logiciel standard de traitement d'image est disponible, les coordonnées cartographiques UTM peuvent être translatées en coordonnées iconiques comme suit:

$$\text{ligne} = (N_0 - N)/\text{pixels} - 1$$
$$\text{colonne} = (E - E_0)/\text{pixels} - 1$$

où N_0 et E_0 sont les coordonnées UTM du coin supérieur gauche de l'iconocarte corrigée. La taille des pixels doit être dans les mêmes unités que les coordonnées cartographiques utilisées.

3.2.1.2   Calage d'image à image

Le calage précis d'image à image est très souvent aussi nécessaire que le calage d'image à carte. Normalement l'opération doit simplement pointer au voisinage des points d'appui sur l'image à caler, les points sur l'image de référence étant déjà définis. Un processus de corrélation peut être utilisé pour optimiser la recherche de la correspondance correcte entre les points d'appui homologués; les pixels ayant obtenu le meilleur taux de corrélation spatiale sont ensuite utilisés pour calculer la transformation polynomiale comme pour le calage d'image à carte.

3.2.2   Interpolation des informations radiométriques

Pour chaque colonne et ligne (i,j) de l'image traitée, il faut déterminer les coordonnées correspondantes (c,l) dans l'image d'origine, non corrigée (voir fig. 3.4).

La radiométrie du pixel (i,j) est obtenue en recherchant la position (c,l) de son homologue dans l'image originale, via la fonction de déformation, et en calculant une valeur radiométrique (niveau de gris) à partir des pixels occupant cette position. Comme le polynôme "pointe" rarement exactement sur un pixel de l'image originale, un certain nombre de choix et d'approximations doivent être faits afin de définir la valeur de la radiométrie du nouveau pixel. On utilise souvent son plus proche voisin, mais une interpolation du premier degré peut être également employée (voir fig. 3.5).

D'autres méthodes plus complexes, comme par exemple la convolution cubique, peuvent également être employées. Dans de nombreux cas, cette solution sera bien adaptée aux projets

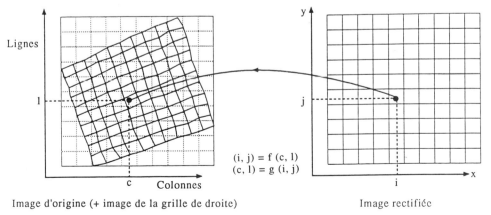

$$(i, j) = f(c, l)$$
$$(c, l) = g(i, j)$$

Image d'origine (+ image de la grille de droite)   Image rectifiée

Fig. 3.4.

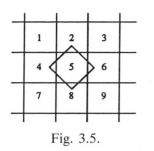

Fig. 3.5.

cartographiques, car elle a tendance à produire un résultat légèrement lissé; cependant, on ne la recommandera pas pour les traitements d'images numériques destinées à l'extraction et la compilation d'informations (voir fig. 3.6).

3.3 Mosaïquage

Si l'image ne couvre pas entièrement la zone à cartographier, plusieurs images sont nécessaires; dans le cas d'images contiguës appartenant au même segment enregistré le même jour, les images brutes sont rigoureusement identiques

dans leur partie commune et on peut les assembler avant tout autre traitement.

Dans tous les autres cas, deux scènes contiguës n'ont jamais exactement la même radiométrie dans leur zone de recouvrement. Pour chaque canal, il faut alors définir une règle de transformation de la radiométrie, qui ramènera la radiométrie d'une image au niveau de celle de l'image choisie comme référence dans leur partie commune. Cette transformation est alors appliquée à l'ensemble de l'image (voir le paragraphe 3.4).

Malgré tout, les radiométries de la zone commune sont légèrement différentes et, un assemblage direct d'images brutes laissera voir une ligne de coupe rectiligne et brutale.

3.3.1 Cas simple

Si l'on dispose d'un système interactif permettant de visualiser l'image, la ligne de coupe, choisie dans la partie commune aux deux

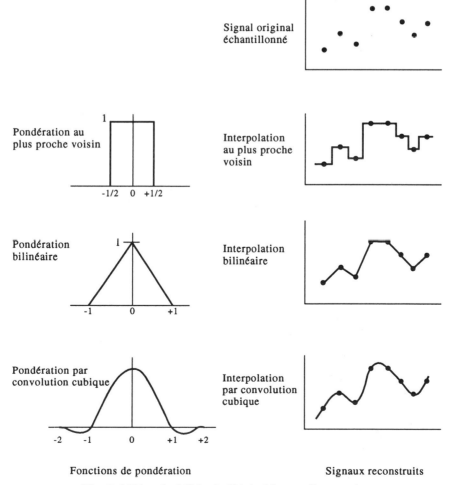

Signal original échantillonné

Pondération au plus proche voisin

Interpolation au plus proche voisin

Pondération bilinéaire

Interpolation bilinéaire

Pondération par convolution cubique

Interpolation par convolution cubique

Fonctions de pondération   Signaux reconstruits

Fig. 3.6 D'après Niblack, Digital Image Processing.

images, sera tracée interactivement à l'écran. Dans ce cas, on s'attachera à ce que la ligne de coupe suive des discontinuités naturelles, comme par exemple des haies ou des lignes du paysage. Une fois la ligne de coupe choisie, le mosaïquage devient une simple procédure automatique ou programmable d'entrée-sortie. Cette méthode toutefois requiert un recouvrement important entre les deux images (voir fig. 3.7).

3.3.2 Mosaïquage complexe

Quand le recouvrement latéral entre les images est faible ou que les orbites ne sont pas parallèles et inadaptées au découpage Nord–Sud des cartes—ce qui est souvent le cas pour les cartes aux (très) petites échelles—une méthode particulière peut être employée. Chaque image est traitée séparément en géométrie comme si elle couvrait toute la zone. Un masque dichotomique correspondant est créé: un "0" signifie l'absence d'image sur la zone correspondante de la carte et un "1" signifie l'existence de données-images. Les images corrigées sont additionnées ainsi que les masques. La carte finale est alors obtenue en divisant, pixel par pixel, la somme des images par la somme des masques (voir fig. 3.8).

Auparavant il faut bien entendu faire les corrections radiométriques quelle que soit l'utilisation, et même alors les très petites différences de radiométrie seront facilement perceptibles; il faudra par conséquent utiliser une procédure de lissage avant d'effectuer le mosaïquage définitif. La véritable ligne de coupe peut être estompée par une procédure calculant

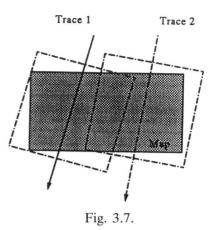

Fig. 3.7.

une sorte de moyenne mobile le long du trait de coupe. Ces valeurs sont ensuite utilisées à la place des valeurs originales de la zone de contact des deux images.

Une fois la carte entièrement couverte par l'assemblage des images, ce dernier peut être découpé selon les limites géographiques de la carte, soit, en général, en suivant des méridiens et des parallèles. C'est également l'occasion de produire les informations géométriques en marge de la carte, c'est à dire le cadre, le réseau des croisillons et amorces du carroyage, par ex. toutes les 10′, et ceux du quatrillage, par ex., tous les 10 km.

3.4 Traitement radiométrique

De nombreux facteurs ont une influence sur la qualité des informations de l'imagerie obtenue par télédétection. Les conditions d'enregistrement sont toujours différentes d'un jeu d'informations à un autre. L'ensemble de ces

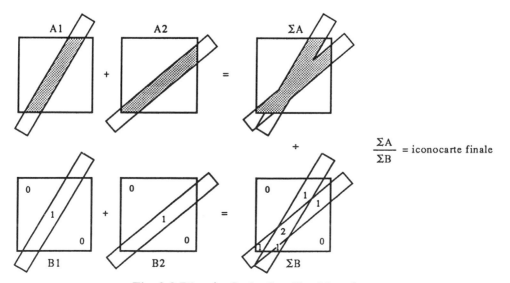

Fig. 3.8 D'après Carle, Satellite Mapping.

différents paramètres a été décrit au chapitre 2. L'éclairement dépend de la relation pixel-soleil: hauteur et azimut du soleil. La situation topographique du pixel influe grandement sur les relations géométriques entre le rayonnement reçu par le pixel et l'énergie réfléchie reçue par le capteur. La composition de l'atmosphère est décisive quant à la modulation du signal transmettant les informations du terrain.

(Les rayures de l'image—provenant des différences de sensibilité des capteurs et/ou d'un défaut de calibration—doivent être éliminées soit par égalisation d'histogramme, soit par moyenne des lignes adjacentes; à moins que chaque capteur puisse être calibré séparément, ainsi qu'il est décrit dans les paragraphes suivants.)

3.4.1 Traitement primaire

Si les données de télédétection sont destinées à confectionner un fond image supportant d'autres informations cartographiques, il est important que le paysage ou la surface du terrain soit restitué avec la plus grande fidélité. Ce qui revient à dire que les proportions spectrales de l'image doivent être aussi proches que possible de la réalité. Il est alors souvent pertinent de recalculer les données numériques normalement fournies en valeurs ayant une signification physique par exemple en watts par unité de surface et en pourcentage de réflectance. La radiance en mW/cm^2 peut être calculée par la formule:

$$P_r = \frac{D_n}{D_{\max}}(L_{\max} - L_{\min}) + L_{\min},$$

où P_r est la radiance du pixel en question, D_n est la valeur numérique du pixel, Dmax est la valeur maximale du système de comptage utilisé (256 pour les données codées sur 8 bits); $L_{\max}$ et $L_{\min}$ sont des paramètres du scanneur qui donnent, d'une part la valeur du rayonnement qui sature le capteur, d'autre part le plus petit rayonnement détectable. En multipliant P_r par $\pi(=3,1416)$ et en divisant par la valeur du rayonnement reçu, E_0—qui est la valeur maximale du rayonnement atmosphérique pour la bande spectrale considérée—multipliée par le sinus de la hauteur du soleil:

$$R_g = \frac{\pi P_r}{E_0 \sin a},$$

on obtient la valeur de la réflectance R_g du terrain.

3.4.2 Correction atmosphérique

Il est fortement recommandé d'éviter d'utiliser des images n'ayant pas été enregistrées dans des conditions atmosphériques quasi idéales. L'atmosphère doit être aussi claire que possible. Si c'est le cas, les données peuvent être utilisées directement, le traitement primaire décrit précédemment devant être suffisant. Bien entendu, la présence de nuages doit être évitée (il faut aussi se rappeler que leurs ombres masquent une partie du terrain).

Si des conditions atmosphériques optimales ne peuvent être trouvées parmi les scènes existantes, il convient d'apporter des corrections atmosphériques. Mais faire de telles corrections sans mesures simultanées est plutôt compliqué. Une méthode couramment employée est la suivante:

Considérons la méthode de prétraitement décrite en 3.4.1: c'est visiblement une transformation linéaire. En utilisant la compensation et la pente d'une fonction donnée de calibration du capteur, on obtient la radiance spectrale du canal i, $P_r(i)$, à partir de $D_n(i)$ par la relation:

$$P_r(i) = C_a(i) + C_b(i) \times D_n(i)$$

où C_a et C_b sont l'ordonnée à l'origine et la pente de la fonction de calibration du canal i.

La réflectance du terrain est:

$$R_g = \frac{1}{A_0(i)}[\pi P_r(i)/E_0(i) - A_a(i)].$$

A_0 et A_a sont des fonctions atmosphériques qui relient la réflectance maximale de l'atmosphère R_a à la réflectance du terrain R_g:

$$R_a = A_0 + A_a R_g.$$

Le plus souvent, la fonction atmosphérique devra être estimée. Une manière de procéder est d'estimer A_0 à partir des zones d'ombre ou en utilisant la réflectance de l'eau telle que provenant de canaux individuels. La transmission atmosphérique, A_a, peut être reliée à la visibilité générale qui peut être obtenue auprès des stations météorologiques.

3.4.3 Filtrage

L'objet de la carte et le type de renseignements cartographiques qui viendront en superposition de l'image satellitaire déterminent la façon dont les données satellites devront être présentées. Il peut être nécessaire d'accentuer certains détails, ou parfois, il faut réduire la variation dynamique des données satellites. En filtrant les données, on pourra obtenir le résultat correspondant à l'objectif de la carte.

De nombreux types de filtres peuvent être employés, certains requérant beaucoup de temps d'ordinateur; les deux principaux—lissage et accentuation des limites—sont décrits ci-dessous.

3.4.3.1 Lissage

Un filtre de lissage (ou filtre passe-bas) peut être considéré comme un algorithme qui calcule une sorte de moyenne mobile des données à deux dimensions. Il en résulte souvent une image atténuée car les éléments à forts contrastes sont supprimés. Le nombre de cellules ainsi que leurs poids respectifs déterminent le degré d'atténuation du résultat.

Exemple de filtres passe-bas:

$$\frac{1}{9} \times \begin{array}{ccc} 1 & 1 & 1 \\ 1 & 1 & 1 \\ 1 & 1 & 1 \end{array} = \begin{array}{ccc} \frac{1}{9} & \frac{1}{9} & \frac{1}{9} \\ \frac{1}{9} & \frac{1}{9} & \frac{1}{9} \\ \frac{1}{9} & \frac{1}{9} & \frac{1}{9} \end{array}; \quad \frac{1}{25} \times \begin{array}{ccccc} 1 & 1 & 1 & 1 & 1 \\ 1 & 2 & 2 & 2 & 1 \\ 1 & 2 & 5 & 2 & 1 \\ 1 & 2 & 2 & 2 & 1 \\ 1 & 1 & 1 & 1 & 1 \end{array}.$$

[Le filtre est situé dans le coin supérieur gauche, les poids des cellules sont multipliés par les valeurs des pixels correspondants de l'image, additionnés et placés dans la nouvelle image filtrée à la position de la cellule centrale du filtre. Le filtre se déplace alors d'un ou plusieurs pas (pixels)—le calcul est alors répété pour la nouvelle position, et ainsi de suite.]

3.4.3.2 Accentuation des limites

Le filtrage passe-haut ou accentuation des limites est employé à la fois pour faire ressortir, d'une façon générale, les éléments linéaires du paysage et pour rehausser les lignes dans des directions spécifiques.

Le résultat du filtrage est additionné à l'image originale, ou bien un filtre à haute intensité est utilisé. Dans un filtrage à haute intensité, l'image originale est multipliée par une constante K.

Filtrage haute intensité:

$K \times$ original+filtre passe-bas=
$(K+1) \times$ original−filtre passe-haut.

Exemples de filtres passe-haut:

$$\frac{1}{9} \times \begin{array}{ccc} -1 & -1 & -1 \\ -1 & 8 & -1 \\ -1 & -1 & -1 \end{array} = \begin{array}{ccc} \frac{-1}{9} & \frac{-1}{9} & \frac{-1}{9} \\ \frac{-1}{9} & \frac{8}{9} & \frac{-1}{9} \\ \frac{-1}{9} & \frac{-1}{9} & \frac{-1}{9} \end{array}; \quad \frac{1}{25} \times \begin{array}{ccccc} -1 & -1 & -1 & -1 & -1 \\ -1 & -1 & -1 & -1 & -1 \\ -1 & -1 & 24 & -1 & -1 \\ -1 & -1 & -1 & -1 & -1 \\ -1 & -1 & -1 & -1 & -1 \end{array}.$$

Exemples de filtres directionnels:

$$\begin{array}{ccc} -1 & 0 & 1 \\ -1 & 0 & 1 \\ -1 & 0 & 1 \end{array}; \quad \begin{array}{ccc} -1 & -2 & -1 \\ 0 & 0 & 0 \\ 0 & 2 & 1 \end{array}.$$

3.4.3.3 Inversion topographique

Les variations d'éclairement dues au relief peuvent souvent provoquer des erreurs d'interprétation pour les utilisateurs/lecteurs de cartes habitués aux estompages cartographiques des versants faisant face au lecteur. Comme la plupart des images satellitaires prises dans l'hémisphère nord et destinées à servir de fond de cartes sont enregistrées de jour (éclairement réel venant du sud au lieu d'un éclairement conventionnel venant du nord-ouest), on sera confronté à ce problème dans les zones à relief très marqué.

Au moyen des modèles numériques de terrain (MNT), la variation d'éclairement des images satellitaires peut être corrigée. Il est éventuellement possible d'inverser la variation naturelle d'éclairement et d'obtenir ainsi une topographie "cartographique", avec accentuation du fond de carte par estompage.

Si l'on ne dispose d'aucun MNT et qu'on veuille éliminer les ombres et les variations d'éclairement du terrain, on peut calculer des ratios en divisant les canaux deux à deux (can.1/ can.2; can.2/can.3, etc.). L'éclairement est ainsi "pseudo-normalisé", ce qui est, à l'occasion, meilleur que le "mauvais" estompage du relief; toutefois, comme les images en ratios ont très souvent tendance à être bruitées, il faut les filtrer pour obtenir un résultat esthétique et efficace.

3.5 Traitement de l'image pour servir de fond de carte

Dans les cas où l'image satellitaire (non interprétée) sert de fond de carte (cas A et C cités dans l'introduction), il y a lieu de traiter spécialement l'ensemble de la radiométrie pour permettre une représentation lisible et expressive des informations cartographiques ajoutées. En particulier, dans le cas C (cartes thématiques sur fond d'image), la carte doit associer aussi harmonieusement que possible des zones de pixels "naturels", des aplats colorés et d'éventuels contours ou plus généralement des traits en superposition. Le choix de leur couleur doit permettre de distinguer parfaitement les différentes surcharges entre elles, et de les distinguer d'avec le fond.

3.5.1 Cas général

Si le fond est monochrome, le problème principal consiste à choisir le canal le plus expressif (ou le plus contrasté) compatible avec une bonne lisibilité des surcharges. L'impression en noir ou en bistre est la plus fréquente.

Si le fond est en couleur, deux options principales sont généralement retenues:

  (i) La représentation en infra-rouge couleur ("fausses couleurs").
  (ii) la représentation en couleurs "pseudo-naturelles".

(i) L'infra-rouge couleur (analogue à celui des photos aériennes) est bien connu: il consiste à utiliser trois canaux de l'image multispectrale et à les représenter:

(a) soit en synthèse *soustractive* (impression sur papier):
  —Le canal vert, ex. SPOT 0,50–0,59 μm imprimé en jaune
  —Le canal rouge, ex. SPOT 0,61–0,68 μm imprimé en magenta
  —Le canal infra-rouge, ex. SPOT 0,79–0,89 μm imprimé en cyan avec une densité d'impression *inversement proportionnelle* à la radiométrie du pixel dans chaque canal: les tachèles d'eau ayant une radiométrie nulle en infra-rouge, les pixels correspondants seront imprimés en aplat cyan (100%).

(b) Soit en synthèse *additive* (affichage sur écran couleur):
  —Le canal vert sera affiché en bleu-violet (complémentaire du jaune)
  —Le canal rouge sera affiché en vert (complémentaire du magenta)
  —Le canal infra-rouge sera affiché en rouge (complémentaire du cyan) avec une intensité *proportionnelle* à la radiométrie du pixel dans chaque canal.

(ii) La représentation en couleurs pseudo-naturelles consiste à essayer de retrouver les couleurs conventionnelles des cartes (i.e. végétation en vert, eau en bleu, etc.); pour cela, on prend généralement les règles suivantes (en synthèse additive):

La couleur verte s'obtient à partir du canal vert.

La couleur rouge s'obtient à partir du canal rouge.

La couleur bleue s'obtient artificiellement (pour Landsat MSS et pour SPOT, qui n'ont pas de canal bleu) par combinaison linéaire des trois canaux (vert, rouge, infra-rouge) (voir paragraphe ci-dessous).

3.5.2 Cas particulier de la couleur bleue (en couleurs pseudo-naturelles)

La combinaison linéaire précédente est à calculer différemment suivant les thèmes à représenter en *bleu* (eau, sol nu, végétation), ce qui amène à faire une classification sommaire de l'image dans l'espace des radiométries (X=Rouge–Vert, et Y=Infra-rouge–Rouge).

Pour chaque quadrant, le bleu est défini par la combinaison linéaire:

$$\text{Bleu}\,(X,Y)=A_i X+B_i Y+C_i$$
$$(i=1 \text{ à } 8 \text{ suivant le quadrant})$$

et les coefficients $A_i B_i C_i$ sont déterminés interactivement, tout en assurant la continuité du bleu sur les frontières entre quadrants.

Toutefois, ce traitement donne des résultats

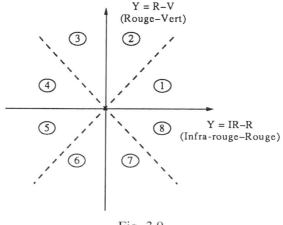

Fig. 3.9.

souvent peu satisfaisants pour l'eau. Il convient donc d'isoler ce thème et de lui appliquer un traitement particulier.

3.5.3   *Cas particulier des surfaces d'eau*

Comme la combinaison linéaire mentionnée ci-dessus ne donne que rarement des résultats satisfaisants pour l'eau, et comme la radiométrie des surfaces d'eau dépend de plusieurs facteurs tels que la profondeur, la turbidité et les concentrations d'algues, une classification peut être faite. Cette classification nécessite des zones d'entraînement pour les différents types de surfaces d'eau. Le résultat de cette classification permettra d'établir un masque grâce auquel on ajustera la teinte bleue. En employant l'"analyse en composantes principales", un masque peut aussi être obtenu, souvent suffisamment puissant pour délimiter les surfaces d'eau.

Les masques obtenus soit par classification soit par analyse en composantes principales permettent de procéder à un étalement dynamique, spécifique à chaque canal, et qui est déterminant pour l'application du bleu dans la représentation du fond hydrographique.

3.5.4   *Transformation des couleurs (accentuations particulières)*

Ainsi qu'il est mentionné ci-dessus l'analyse en composantes principales peut être utilisée pour transformer les diverses couleurs. C'est une méthode—disons plutôt un ensemble de méthodes—très puissante pour analyser des jeux de données multibandes et multidates.

L'analyse en composantes principales, utilisée pour fournir des fonds d'image à diverses spatiocartes, peut être employée par ailleurs pour réduire la "dimensionnalité" de l'image multibande c'est à dire pour en extraire un maximum de variances sous la forme d'un nombre réduit d'images monocanal. Souvent, plus de 70 à 80% de variances d'un fichier de données multibandes peut être représenté par une seule image monocanal, la première composante principale; les secondes et troisième composantes principales contiendront respectivement 15 à 20 % et 3 à 10% de la variance. En utilisant directement ces composantes principales pour la production de cartes, on s'assure qu'un maximum de la variance est présenté dans la carte, mais par ailleurs, la représentation colorée n'aura rien à voir avec les couleurs naturelles, ce qui rendra le travail de l'utilisateur plus délicat et demandera au cartographe plus d'efforts dans la rédaction d'une légende bien adaptée.

Souvent, de nombreux fichiers de données multibandes sont très corrélés. L'analyse en composantes principales peut être utilisée pour réaliser un étalement décorrélé des données. Dans un premier temps, les données originales prétraitées sont transformées, ou subissent une rotation; ensuite, les différentes composantes sont étalées indépendamment, enfin les composantes principales étalées sont retransformées dans l'espace d'origine (voir fig. 3.10).

1. Les canaux *y* et *x* sont fortement corrélés. Une analyse en composantes principales transforme les données dans un nouveau système de coordonnées, où la première coordonnée contient la majeure partie de la variance, la seconde coordonnée principale—perpendiculaire à la première—contient une moindre partie de la variance.
2. Les données après la transformation en composantes principales.
3. Les données sont étalées séparément le long de l'axe de la principale composante pour lui faire occuper la majeure partie du champ disponible.

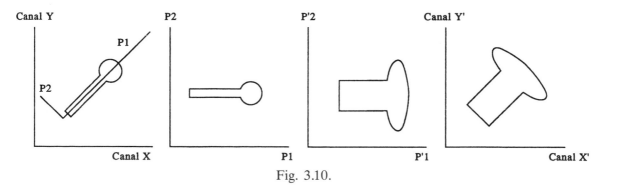
Fig. 3.10.

4. Après avoir été retransformées, les données seront décorrélées et contiendront un champ de valeurs dynamiques plus étendu.

La méthode de l'étalement de décorrélation peut être utilisée quel que soit le nombre de canaux; quand on n'utilise que 3 canaux, par exemple les 3 premières bandes de Landsat TM (bleu, vert et rouge), une transformation TSI peut être mise en oeuvre. En transformant les 3 composantes colorées RVB en Teinte, Saturation et Intensité (TSI), il devient possible d'étaler (et décorréler) les composantes de saturation et d'intensité de l'image. Les données TSI étalées peuvent être alors retransformées en composantes RVB.

Une autre possibilité consiste à substituer la composante d'intensité par des données issues d'images différentes de celle utilisée pour la première transformation et d'inclure ainsi des données auxiliaires dans la présentation/ impression des couleurs RVB des images. Cette méthode sera souvent employée quand on désirera inclure du noir et blanc, de satellite monocanal—ou photographie aérienne numérisée—dans des fichiers de données multibandes de plus faible résolution.

Il est possible d'utiliser les trois premières bandes de Landsat TM pour les données RVB et d'employer une transformation TSI. Comme les données TM ont une résolution géométrique de 30 m, il peut être profitable d'y inclure les données du canal panchromatique à résolution 10 m de SPOT. Par un ajustement adapté des niveaux de gris des données SPOT aux intensités obtenues par la transformation TSI des données TM, les données SPOT peuvent se substituer aux intensités TM dans la retransformation dans l'espace des couleurs RVB.

Cette procédure impose un calage extrêmement précis des deux fichiers de données.

3.5.5 *Sortie graphique du fond d'image*

Après les traitements géométriques et radiométriques décrits ci-dessus, le fond d'image est prêt à être restitué sous forme graphique, de façon à pouvoir servir de support au complètement topographique et/ou l'addition d'informations exogènes.

Dans le cas où une impression offset est requise, le calcul d'une synthèse soustractive (pour déterminer les pourcentages imprimants d'une trichromie cyan, magenta et jaune) doit être effectué à partir de l'image numérique obtenue généralement sur écran par synthèse additive.

On a évidemment intérêt, dans ce cas, à sortir directement sur film monochrome (à l'aide d'un phototraceur) les planches-mères correspondantes, tramées et prêtes à la reproduction pour l'impression.

L'autre solution consiste à sortir une image couleur sur film (phototraceur couleur), ou un tracé couleur sur papier (traceur électrostatique ou à jet d'encre), et à appliquer à ce document les techniques classiques de séparation des couleurs et d'impression.

3.6 Ajout d'informations classiques à l'image satellite

Compléter ou enrichir l'imagerie de télédétection avec des informations cartographiques "normales" est devenue une partie très importante de l'exploitation de la télédétection. Il reste toujours difficile d'interpréter les données multibandes de satellites, en particulier parce que les usagers s'attendent de plus en plus à ce que les données satellites leur fournissent "toute" l'information, voire même l'expliquent. Le fait un peu frustrant que les données de télédétection contiennent plus d'information que n'en apporte une interprétation rapide, a conduit à développer également l'approche combinée de la télédétection et des informations géographiques classiques, non seulement à des fins de recherche, mais aussi à des fins d'application.

L'ajout d'informations cartographiques classiques vise à aider l'utilisateur dans sa lecture de l'imagerie satellitaire, en lui donnant des points de repère ou des références indispensables à l'interprétation. Les informations de type cartographique peuvent provenir de différentes sources:

—soit de cartes existantes,
—soit d'une interprétation visuelle partielle des images satellites (par exemple routes, rivières, agglomérations),
—soit de systèmes d'informations géographiques (SIG) existants.

Quelle que soit leur origine, les informations doivent être représentées sous une forme compatible avec les caractéristiques de l'image (voir chapitre 4). Plusieurs critères interviennent dans cette compatibilité:

(i) Richesse de l'information cartographique ajoutée à l'image (densité, lisibilité, etc.).
(ii) Fiabilité géométrique de l'information ajoutée.
(iii) Fiabilité thématique de l'information ajoutée.
(iv) Actualité de l'information ajoutée.
(v) Esthétique de la représentation cartographique.

Ces différents critères ont chacun une part directe dans la *qualité* du produit fini. En termes de méthodes d'élaboration, le choix des sources et la sélection des informations dépendent directement des objectifs (ou des thèmes) de la carte à établir, et des disponibilités en matière de sources existantes. On ne peut donc pas donner de règles générales en la matière. En termes de techniques de fabrication, deux aspects sont traités ici:

(i) La rédaction proprement dite (sémiologie graphique): c'est l'objet du chapitre 4.
(ii) Le tracé et l'élaboration des planches-mères en vue de l'impression finale: c'est l'objet des paragraphes ci-après.

3.6.1 Les deux filières analogique et numérique

Pour élaborer le tracé des surcharges cartographiques et les planches-mères de l'édition finale, deux filières peuvent être suivies: la première utilise les techniques photo-mécaniques classiques (filière analogique), la seconde exploite le caractère numérique de l'image satellite en utilisant les informations classiques également sous forme numérique, afin de les combiner selon une procédure entièrement informatique.

La première filière fait appel aux méthodes de rédaction classiques, et elle est donc bien connue. On ne fera pas ici de commentaire particulier, excepté en ce qui concerne la fabrication des planches-mères. Etant donné que l'image satellite multibande est le plus souvent imprimée en trichromie ou en quadrichromie (avec les couleurs habituelles), il y a avantage à adopter la même séparation des couleurs pour les surcharges cartographiques:

ceci permet de les combiner directement aux films de l'image, et donc de conserver le même nombre de planches-mères à l'impression.

Il doit être noté que la filière analogique, entièrement indépendante du traitement d'image (excepté pour la phase finale que l'on vient de citer), permet d'utiliser l'intégralité des ressources de la rédaction classique, en matière de sémiologie, de types de lettres, de reproductions d'extraits de cartes, etc. Cette situation contraste encore avec la filière numérique, où la rédaction cartographique est soumise aux limites imposées par le mode maillé (résolution de la trame élémentaire). Toutefois cette différence tend à disparaître avec l'amélioration continelle des performances de l'infographie en mode maillé, et de la filière numérique dans son ensemble.

La filière numérique débouche, d'autre part, sur l'exploitation généralisée des systèmes d'information géographique (SIG), et elle représente, à l'évidence, la convergence la plus porteuse d'avenir des 3 secteurs technologiques concernés: le traitement d'image, l'infographie, et les banques de données. Elle fait l'objet du paragraphe 3.6.2.

3.6.2 Méthodes informatiques (avec SIG)

Dans ce contexte, il faut insister sur le fait que l'intégration de la Télédétection aux Systèmes d'Information Géographique (SIG) n'a pas essentiellement de rapport avec la cartographie traditionnelle. La carte numérique, dans l'association Télédétection/SIG, est avant tout un outil orienté vers l'écran et combinant analyse et synthèse des données. Les résultats sont rarement exploités en production cartographique (c'est pratiquement la même chose pour l'application technique des SIG servant principalement à gérer des réseaux de distribution). S'ils sont copiés et diffusés, ce sera presque toujours sous forme numérique. Bien entendu, le contenu de l'écran peut être transcrit à l'aide d'un dispositif d'impression comme par exemple un traceur à jet d'encre colorée ; mais on peut difficilement parler de production de cartes avec cette filière. Dans le paragraphe suivant, nous ne traiterons pas d'éditions numériques au sens cartographique tradi-tionnel—celles-ci sont considérées comme des méthodes analogiques, telles que la surimpres-sion.

3.7 Rôle des systèmes d'information géographique (SIG)

3.7.1 Définition et potentialités des SIG

La notion de SIG est apparue au début des années 1980 et résulte de l'extension des systèmes de gestion de bases de données (SGBD) à tous les types de données géographiques. Parmi celles-ci, la télédétection satellitaire constitue un apport original, qui accentue la spécificité des SIG par rapport aux systèmes d'informations au sens large.

Une définition américaine a été publiée par le "Federal Interagency Coordinating Committee on Digital Cartography" (FICCDC, 1988)

> "System of computer hardware, software, and procedures designed to support the capture, management, manipulation, analysis, modeling, and display of spatially referenced data for solving complex planning and management problems."

En France, la Société Française de Photogrammétric ct dc Télédétection (SFPT) a proposé (octobre 1989) la définition suivante:

> "Système informatique permettant, à partir de diverses sources, de rassembler ct organiser, de gérer, d'analyser et de combiner, d'élaborer et de présenter des informations localisées géographiquement, contribuant notamment à la gestion de l'espace."

En conséquence, les caractéristiques suivantes peuvent être retenues:

—la gestion de données localisées multiples (dont la TD),
—la capacité de produire des documents cartographiques,
—l'aptitude à des traitements complexes en vue de la gestion du territoire, et donc afin d'aider à la prise de décision.

A titre d'exemples d'applications, citons quelques opérations de gestion courante de l'espace:

—délivrance de permis de construire (habitation, usine, etc.),
—imposition foncière,
—choix d'un tracé routier, autoroutier ou ferroviaire,
—définition de zones constructibles ou inconstructibles (réglementation de l'occupation des sols),
—aménagement agricole (drainage, irrigation, etc.),
—aménagement urbain ou industriel ou touristique, etc.,
—planification régionale,
—suivi de l'évolution de l'environnement (désertification, déforestation, changement climatique, etc.),
—prévention contre les risques naturels ou artificiels majeurs (inondations, glissements de terrain, pollutions, etc.),
—intervention en cas de catastrophes,
—aide à la circulation automobile informatisée, etc.

En aucune façon, un système d'information géographique n'implique ou ne dépend de l'usage des données de télédétection. Si l'on définit un SIG comme outil d'aide à la planification et au développement via des décisions juridiques, administratives ou économiques, il consiste en une base de données contenant des informations à référence géographique, concernant les infrastructures et les aspects économiques, ainsi que l'environnement et le paysage; il est employé pour le recueil systématique de données, pour leur mise à jour ct leur traitement, et pour la représentation de celles-ci et des résultats dérivés sous diverses formes.

Cependant, les relations mutuelles entre SIG, et systèmes de télédétection sont évidentes: la télédétection constitue une source de données géographiques potentiellement énorme, dont une très faible partie, seulement, a été jusqu'ici exploitée et intégrée; réciproquement les SIG, élaborés à partir de cartes existantes ou de données de terrain, peuvent apporter une aide considérable à l'interprétation des images de télédétection, par l'intermédiaire des classifications assistées par ordinateur.

3.7.2 Architecture générale et composants

Il découle des définitions précédentes l'organisation globale représentée ci-dessous (d'après FICCDC, 1988) (voir fig. 3.11).

Le système de gestion de bases de données (SGBD) est un ensemble de logiciels assurant le stockage et l'utilisation de toutes les données du système. Le SGBD d'un SIG doit permettre en particulier les opérations liées à la topologie des données (recherche sur localisation, sur

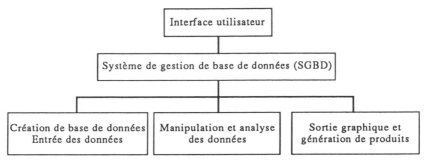

Fig. 3.11.

adjacence, sur inclusion, etc.) et à leur transcription (cartographique, etc.).

L'interface utilisateurs est constituée par l'ensemble des outils grâce auxquels les utilisateurs communiquent avec le système, c'est-à-dire avec l'ensemble des bases de données auxquelles il a accès, et avec l'ensemble des modules d'applications disponibles. Ces outils consistent en une panoplie de logiciels, tels que "menus" de fonctions, notices d'écrans, affichages graphiques, etc.

Compte tenu des définitions précédentes et des caractéristiques de l'information géographique, on peut assigner un certain nombre de fonctions générales à un SIG, qui précisent leurs principaux domaines d'utilisation.

Un SIG doit être a priori capable d'assurer les fonctions générales suivantes, définies du point de vue de l'utilisateur (non limitatives, et citées dans un ordre quelconque):

—**donner** une information géographique d'ensemble concernant un territoire donné (tous objets et thèmes confondus) (ex. inventaire de tous les objets ou thèmes présents sur une commune);
—**extraire** tous les objets ou thèmes répondant à une question donnée (quelles sont toutes les routes de largeur 4 m dans tel département?);
—**croiser** les informations disponibles pour rechercher des corrélations entre différents thèmes ou objets, selon des algorithmes prédéfinis;
—**permettre de combiner** des données spécifiques apportées par l'utilisateur avec celles du SIG;
—**pouvoir fournir** des réponses sous des formes appropriées aux problèmes traités: fichiers numériques, textes, cartes (sur écran ou sur papier), diagrammes, images, ou même sous forme vocale;

—**permettre une mise à jour à tout moment**, ainsi que l'introduction de nouvelles données ou de nouveaux thèmes.

Au sein d'un SIG, toutes les informations sont ou doivent être rendues accessibles sous une forme homogène qui permet les manipulations et les traitements, ainsi que l'expression cartographique finale des résultats.

3.7.3 Intégration de la télédétection et des SIG

Les systèmes "intégrés" gérant à la fois des données géographiques classiques (le plus souvent en mode vecteur) et des images de télédétection sont encore rares, ou seulement en développement, comme l'attestent de nombreuses communications (Boursier 1986; Huguet 1987, etc.). Or, seule cette intégration permet de traiter simultanément, au sein d'un système unique (voir fig. 3.12).

Il en résulte que l'articulation générale d'un tel système peut se présenter sous la forme suivante (voir fig. 3.13).

Ce type de schéma, où les images sont stockées, après traitement, sous forme rectifiée autorisant la superposition directe avec d'autres données géographiques, correspond à un système dit géocodé (cas des systèmes SEP France, ou LDIAS du Centre Canadien de Télédétection, ou IBIS du Jet Propulsion Laboratory du California Institute of Technology). Une autre approche peut consister à ne pas stocker a priori les images rectifiées mais à exécuter les traitements de mise en superposition (corrections géométriques) seulement lors des requêtes de l'utilisateur, et d'une façon d'ailleurs transparente pour lui (cas du système d'informations climatologiques du Centre scientifique d'IBM-France, avec des

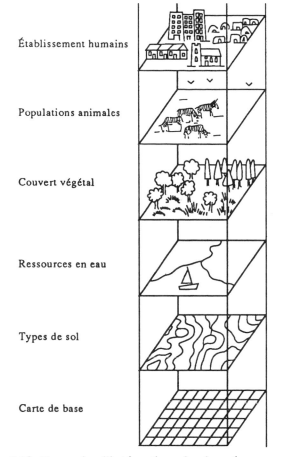

Établissement humains

Populations animales

Couvert végétal

Ressources en eau

Types de sol

Carte de base

Fig. 3.12 Exemple d'intégration de données au sein d'un SIG (d'après GRID, United Nations, Geneva).

images NOAA et Meteosat, gérées à l'aide du SGBD. SQL/DS).

Il est facile de voir que les informations de télédétection peuvent être intégrées de deux manières: soit comme des données iconiques prétraitées, soit comme des classifications (en général). Le fait de choisir l'une ou l'autre option dépend du but recherché (et du logiciel du SIG employé).

3.7.4 Aspects technologiques des SIG

Dans la plupart des SIG à but administratif (appelé souvent "Systèmes d'informations territoriaux" quand l'information est purement administrative), les données seront traitées le plus souvent en mode vecteur, alors que les systèmes destinés à la recherche et au développement sont orientés à la fois vers le mode vecteur et le mode maillé.

Un SIG orienté-vecteur traite les données comme des points dans un système de coordonnées rectangulaires ou géographiques. Le fichier est habituellement stocké comme une longue chaîne de coordonnées, associées à des informations sur les points et/ou les zones adjacentes.

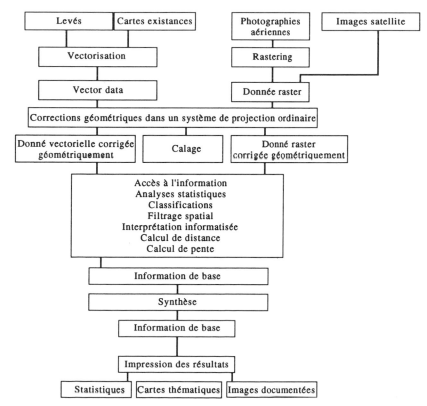

Fig. 3.13 Système hybride vecteur–raster (d'après Hugeut, 1987).

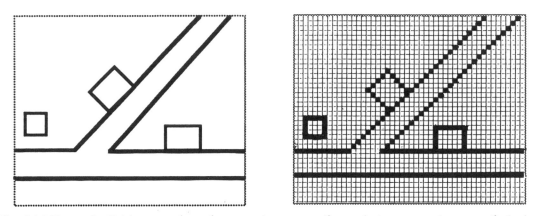

Fig. 3.14 Exemple d'objets représentés en mode vecteur (à gauche) et en mode raster (à droite).

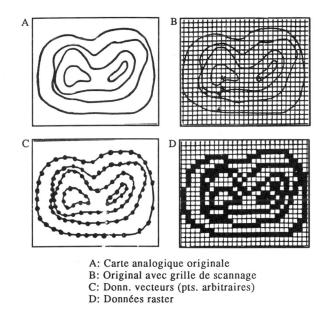

A: Carte analogique originale
B: Original avec grille de scannage
C: Donn. vecteurs (pts. arbitraires)
D: Données raster

Fig. 3.15 Conversion raster/vecteur.

En principe, un SIG orienté-mailles stocke les données sous forme d'une grande surface continue composée de points portant les données. Donc, mais seulement en principe, une base de données maillées requiert plus d'espace de stockage qu'en mode vecteur (voir fig. 3.14).

Un système orienté-mailles aura son origine dans le coin supérieur gauche, l'axe des Y sera à l'inverse d'un système de coordonnées normal tandis que l'axe des X sera comme le système de données vecteur qui a son origine dans le coin inférieur gauche. La base de données maillées (sur le dessin) requerra au moins 30×30 données stockées -sauf compression spéciale; dans la base de données vecteur, on aura besoin (1) de deux couples de coordonnées X,Y pour chaque ligne droite et (2) suivant la résolution voulue, d'un nombre suffisant de couples X,Y pour décrire les lignes courbes.

Bien sûr les lignes brisées et les courbes peuvent perdre en résolution (rendu cartographique) quand elles sont représentées en mode maillé.

Ce n'est évidemment pas un problème d'intégrer des données satellitaires dans un SIG travaillant en mode maillé, car elles sont déjà dans ce mode. Si la base de données est en mode vecteur, le problème principal pour l'intégration sera de transformer les données satellitaires en mode vecteur ce qui est impossible pour une scène dans son entier. Une image demi-teinte entière ne peut être vectorisée; une classification (ou d'autres produits dérivés) doit être envisagée au préalable (voir fig. 3.15).

La transformation de données vectorielles en mode maillé implique une transformation dans un nouveau système de coordonnées et un calcul pour déterminer quels seront les points qui devront être projetés dans le plan maillé, chaque position devant être affectée d'une certaine valeur.

Quand des données en mode maillé sont à convertir en mode vecteur, les choses deviennent plus compliquées. Le nombre nécessaire de noeuds (coordonnées X,Y) doivent être trouvés et transformés dans le nouveau système de coordonnées; ceci requiert souvent la squelettisation des détails linéaires pour déterminer les noeuds vectoriels optimaux. L'image maillée originale est restituée en image vectorielle simplement en reliant les noeuds par des arcs.

Une fois formatées dans la structure appropriée au système de traitement en question, les données satellitaires et les données cartographiques classiques peuvent être superposées, analysées et synthétisées selon l'objectif poursuivi.

De nombreuses procédures employées pour l'analyse des données dans un SIG sont les mêmes que celles mises en oeuvre pour l'analyse d'une image; ce sont des procédures d'analyse qui travaillent principalement sur des points, par exemple des opérations arithmétiques, booléennes, de Fourier, ou de classification.

Les polygones et les opérations relatives au voisinage sont particulièrement intéressantes dans l'analyse de type SIG. Ces opérations concernent:

—Analyse d'identifiants, de surfaces et de périmètres—employée pour les reclassifications et redésignations de zones, les calculs de surface et de périmètre, souvent aussi les classifications de silhouette et de forme— des classifications hiérarchiques combinées de surfaces et de formes.
—Corrélation de polygones, définition de nouveaux polygones, fusion et combinaison, masquage.
—Analyse de distance et de voisinage. Calcul de courbes de niveau et interpolations.
—Analyse de surface et de la topographie.
—Optimisation des directions d'écoulement, sélection d'itinéraires et désignation de sites particuliers; extraction de surfaces, de points et de couloirs potentiels.
—Synthèses.

3.7.5 Sortie graphique de l'iconocarte numérique

Une fois les données géographiques combinées à l'image au moyen des procédures précédentes, et visualisées sur écran, il reste à réaliser les sorties graphiques si l'on souhaite conserver le résultat sur support permanent.

Différents cas se présentent: pour une simple conservation sur papier, une sortie sur reprographe (photographique) couleur ou sur un traceur électrostatique ou à jet d'encre couleur

est suffisante. Ce processus convient pour un rendu cartographique moyen et pour un nombre d'exemplaires limités (de l'unité à quelques dizaines). En revanche, si l'on souhaite une qualité cartographique supérieure et surtout un nombre d'exemplaires important, il est préférable de sortir un document sur film, reproductible, et si possible en planches de couleur séparée.

On a évoqué au paragraphe 3.5.5 les opérations nécessaires à la sortie graphique de l'image seule: les opérations sont exactement les mêmes pour l'image combinée, sous réserve d'avoir prévu, pour les surcharges cartographiques, les mêmes types de séparation des couleurs (en synthèse soustractive, et en pourcentages imprimants).

Références

BOURSIER, P. 1986. The integration of cartographic and image data into geographic information systems. *International Electronic Image Week*, IGN, France.

CARLE, C. 1982. *Satellite Mapping, Geometric Correction of Remote Sensing Images.* Meddelelser No. 12. Institut for Landmaaling of Fotogrammetri, Danmarks Tekniske Hoejskole.

CASTLEMAN, K. R. 1979. *Digital Image Processing.* Prentice-Hall, New Jersey.

FICCDC. 1988. *A Process for Evaluating Geographic Information Systems.* Federal Interagency Coordinating Committee on Digital Cartography, USGS, Reston, USA.

HUGUET P. 1987. *Traitement et analyse thématique d'images; de cartes et de données associées.* Actes du Forum Fi3G-Lyon Conseil National de l'Information Géographique, Paris, France.

NIBLACK, W. 1985. *An Introduction to Digital Image Processing.* Strandberg, Denmark.

SCHOWENGERDT, R. A. 1983. *Techniques for Image Processing and Classification in Remote Sensing.* Academic Press, New York.

STUCKI, P. 1979. *Advances in Digital Image Processing.* Plenum Press, New York.

Chapitre 4
CONCEPTION ET SÉMIOLOGIE DE LA REPRÉSENTATION CARTOGRAPHIQUE

Auteurs
Janos Lerner* et Jean Denègre†

*Université R. Eötvös, Départment de cartographie, Kun Bela ter 2, Budapest, Hongrie
†Conseil National de l'Information Géographique (CNIG), 136 bis rue de Grenelle, 75700 Paris, France

TABLE DES MATIÈRES

4.1 Introduction

La présentation des données obtenues par télédétection et traitées se fait généralement sous une forme qui s'apparente à une carte. Comment faire pour obtenir ce résultat? C'est la question à laquelle il fallut répondre durant ces quelque trente années dc cartographie spatiale. La Terre, avec son relief, sa surface, sa couleur, ses ombres et son atmosphère, était la donnée initiale. Comment faire le tri, comment s'y retrouver? Ces questions se sont posées sitôt la publication des premières photographies prises par Gemini et Apollo dans les années 60. Quand le problème fut résolu, les images Landsat apparurent, puis celles de SPOT et d'autres encore; avec à chaque fois de nouvelles possibilités et toujours les mêmes questions et des réponses différentes.

Les premières photographies montrèrent une vue inconnue et inhabituelle de la surface terrestre et tous les détails furent expliqués sous une forme bien connue: la carte. Les images nouvelles devinrent courantes, et apparurent sur des couvertures de livres, d'atlas et de calendriers comme étant la nouvelle et unique représentation "cartographique". Après l'apparition des images en fausses couleurs, elles devinrent rapidement non seulement courantes mais même familières. Vint donc l'époque où il n'était plus nécessaire d'expliquer chaque détail; les cartes au trait pouvaient être mises de côté, l'image étant surchargée de symboles et de lettres. Landsat apporta les conditions géométriques nécessaires à la production de photocartes "classiques". La capacité stéréoscopique de SPOT offrit la possibilité d'y ajouter des données altimétriques.

En télédétection, il existe deux étapes d'édition. L'une précède la procédure de classification et l'autre présente les résultats de l'interprétation. La cartographie spatiale s'appuie sur cette dernière. Si le résultat est un croquis ou une carte conventionnelle, nous nous trouvons en face d'une curieuse contradiction: un procédé moderne, de conception entièrement nouvelle, se traduit par une des plus anciennes formes de représentation.

Les techniques de l'informatique s'efforcent de résoudre cette contradiction. Les techniques de sortie graphique et de restitution transforment le pixel d'origine (valeur spectrale) en une information (couleur) cartographiable. Les cartes traduisent les informations par des symboles. Les images satellites fournissent des données sous forme de pixels et de densités. Comment réaliser le compromis? Répondre à cette question est l'objectif de la cartographie spatiale.

4.2 Extraction d'informations pour la cartographie spatiale

4.2.1 Données iconiques

Dans le chapitre 3, une description détaillée des informations contenues dans une image a été donnée. Les techniques d'interprétation analogique ou numérique fournissent bon nombre de données utiles à la représentation

cartographique des résultats. Mais les images "brutes" (c'est-à-dire: photographies en couleurs naturelles, en fausses couleurs ou en noir et blanc, images monocanal, images radar, etc.) sont aussi utilisées à des fins cartographiques. Des images prétraitées, accentuées, des compositions colorées sont quelquefois assimilées à des "cartes" bien qu'elles ne donnent aucune information classifiée. Pour les projets cartographiques, l'accentuation des données radiomètriques (voir les éléments d'image au chapitre 2) est primordiale.

4.2.2 Données thématiques

Ce sont les informations essentielles à représenter. On entend par données thématiques: les informations scientifiques déjà connues avant l'analyse d'image (informations de contexte); les catégories d'objets résultant des opérations de classification; les données permettant de se repérer (c'est-à-dire toponymes et coordonnées).

Les résultats de l'interprétation visuelle analogique (voir chapitre 2) apparaissent sous forme graphique (c'est-à-dire sous forme de symboles, de traits ou de surfaces colorées); l'interprétation numérique donne une carte en mode maillé, à base de pixels, chacun étant codé par une couleur. De nombreuses sciences utilisent les images spatiales à des fins de cartographie thématique, les applications existantes couvrant presque une centaine de disciplines différentes. L'occupation ou couverture du sol et les sciences de la Terre sont les premières applications concernées. Le *Rapport international sur la cartographie thématique derivée des images satellitaires* publié par J. Denègre, Elsevier, 1988) donne un panorama de ces applications.

Les méthodes de représentation des données thématiques sont classiques. La conception des produits de cartographie satellitaire se caractérise principalement par le choix de la méthode de représentation adéquate. Ces produits sont avant tout de nature cartographique—critère décisif quant au choix des données à y faire figurer.

4.2.3 Représentation cartographique

Les cartes thématiques disposent d'un système de représentation cartographique bien connu et très répandu. Pour traduire la réalité, les cartes utilisent des méthodes de classification, de simplification, et de symbolisation.

Classer signifie créer des catégories, qui regroupent artificiellement des éléments caractéristiques. Simplifier ou généraliser consiste à sélectionner, réduire, lisser et combiner des données significatives quant au but cartographique à atteindre. Symboliser signifie traduire la réalité sous forme abstraite, l'information géographique devenant un point, un trait ou une surface, caractérisés par une taille, une forme, un motif et une couleur.

Du point de vue cartographique, les méthodes de représentation employées par les cartes thématiques peuvent être distinguées par leur aspect graphique (points, lignes, surfaces), par leur contenu (en qualité, mais aussi en quantité), par ce qu'elles peuvent montrer (direction, répartition, fréquence), par les types de surface auxquels elles font référence (point, ligne, surface continue ou discontinue, etc.). Les informations qualitatives sont généralement représentées par des variables de forme ou de couleur, les informations quantitatives plutôt par des variables de taille, de texture ou de densité. Ces règles ont été développées conceptuellement et pratiquement par J. Bertin (1967).

4.3 Familles de produits spatiocartographiques

Pour plus de clarté, on peut distinguer deux grandes familles de produits de cartographie thématique, dérivés des images satellites:

—*Les cartes au trait*, où toute l'information est tracée et représentée avec les variables graphiques habituelles, et où toute trace des images initiales a disparu (sauf, dans certains cas, la forme des pixels qui reste visible).

—*Les iconocartes* (cartes à fonds d'images), où le fond de carte utilise tout ou partie des images satellites initiales, et où l'information thématique est représentée sous des formes diverses.

4.3.1 *Cartes au trait*

La typologie des cartes thématiques au trait, lorsqu'elles dérivent d'images satellites, n'est pas fondamentalement différente de celle des cartes thématiques classiques, sauf peut-être pour ce qui concerne les cartes de type maillé. Les types de documents les plus courants sont:

—De simples cartes d'étude avec des symboles linéaires (par exemple des lignes tectoniques).

—Des cartes d'étude ayant une référence cartographique (c'est-à-dire avec des renseignements topographiques comme les routes, le réseau hydrographique, les agglomérations—généralement en teinte monochrome), les informations thématiques étant représentées par des points et des lignes de couleur (voir par ex. chapitre 5, applications 8, 15, 20, etc.).

—Des cartes maillées (chaque maille correspondant à un pixel de l'image d'origine), sorties sur imprimante d'ordinateur, généralement en noir et blanc, où les caractères correspondent aux catégories de surfaces.

—Des cartes maillées restituées sur traceur, définies par des mailles colorées, les couleurs représentant les mêmes thèmes que ceux des images sur imprimante (voir par ex. chapitre 5, applications 5, 10, 13, etc.).

—Des cartes polygonales (choroplèthes) avec des zonages thématiques en couleur, qui ne se distinguent en rien des cartes classiques de même type (voir par ex. chapitre 5, applications 6, 7, 9, etc.).

4.3.2 *Iconocartes (cartes sur fond d'images)*

Quand on veut créer une iconocarte, on a le choix entre les trois types de combinaison des données cartographiques déjà mentionnés au chapitre 3 dans lequel les processus techniques étaient résumés.

Des documents graphiques couleur peuvent être tracés en combinaison avec des cartes de référence (montrant les frontières, les routes, le carroyage géographique, les toponymes, voir par ex. chapitre 5, applications 8, 12, 17, etc.), en teinte monochrome et dans certains cas en plusieurs couleurs. Ces cartes de référence peuvent être établies par procédé analogique, numérique, ou obtenues directement à partir d'une carte topographique. Dans ce dernier cas, une rectification géométrique est nécessaire pour adapter l'image à la carte, c'est-à-dire que les coordonnées de l'image sont transformées dans la projection de la carte (voir chapitre 3.1).

La cartographie possède un système classique, bien défini, de représentation graphique constitué de symboles, de points, de lignes (isolées ou en trames), d'ombres et de couleurs avec leurs différentes combinaisons. Les images présentent une structure particulière faite de pixels noirs, gris, blancs ou colorés. Pour l'utilisateur, les pixels en tant qu'élément de l'image n'ont pas une importance primordiale; il est intéressé principalement par l'aspect "photographique" de l'image c'est-à-dire une certaine répartition "esthétique" des pixels. Il ne veut pas la voir sous son aspect de mosaïque mais identifiés et interprétés. Les iconocartes doivent associer des données issues de la carte traditionnelle et des données iconiques. Cette association peut être envisagée sous l'aspect du contenu de l'image (voir chapitre 3) ou sous l'aspect de la conception cartographique.

4.4   Présentation générale des iconocartes

Une iconocarte est la combinaison d'éléments graphiques (carte) et d'éléments iconiques (image). A quoi une iconocarte doit-elle ressembler? Il n'y a pas de règle générale définissant un "standard" d'iconocarte comme on le verra au paragraphe 5. Il existe en revanche des formats standards associant images en mode maillé et données cartographiques en mode vecteur (par ex. GIS-GEOSPOT et SDTS aux USA ou EDIGEOen France). Mais il existe plusieurs possibilités, c'est-à-dire plusieurs façons de combiner les éléments graphiques et les informations iconiques (intensités, couleurs); tout dépend de ce que l'on compte conserver de l'image pour atteindre à un objectif scientifique donné (par exemple géologie, occupation du sol). En considérant le rapport "fond d'image/ contenu", il existe trois sortes d'iconocartes:

—iconocartes à images accentuées,
—iconocartes à éléments graphiques accentués,
—iconocartes à thèmes accentués.

4.4.1   Iconocartes à image accentuée

Dans ce type de carte, le fond photographique prend le pas sur les éléments graphiques (voir par ex. chapitre 5, applications 2 et 3). L'image, qu'elle soit en noir et blanc, en fausses couleurs, en couleurs naturelles, ou mosaïquée, subit un traitement d'amélioration qui lui donne la meilleure qualité possible. Dans la plupart des cas, le contenu graphique apparaît en teinte

monochrome sous la forme de symboles, de points, de traits ou de trames. Quand des thèmes de surface sont interprétés, ceux-ci sont limités par un trait et sont renseignés par des lettres ou des codes. Quand les thèmes sont des éléments linéaires (par exemple lignes tectoniques, réseaux hydrographiques), ils peuvent être représentés par différents types de trait et/ou par différentes couleurs. Si on utilise des symboles de différentes couleurs, on choisit généralement une teinte monochrome pour le fond.

Avantages:
—représentation proche de la réalité,
—possibilité d'interprétation ultérieure,
—possibilité de contrôler des interprétations antérieures,
—aspect esthétique.
Inconvénients:
—lecture difficile pour un non-spécialiste du fait de l'absence d'une légende standard,
—si les thèmes surfaciques ne correspondent pas à des zones naturelles, il se produit des "interférences" du fait de la non-super-position des classes d'objets.

4.4.2   Iconocartes à éléments graphiques accentués

Nous parlerons d'iconocartes avec éléments graphiques renforcés quand les surcharges graphiques ont plus d'importance que le fond d'image. L'image est alors toujours mono-chrome, constituant véritablement un "fond" ou une "base" pour les informations thématiques. Le fond d'image apparaît généralement en gris ou en brun clair, dans un ton neutre. Dans certains cas, le fond d'image subit un traitement particulier pour accentuer un thème spécifique (par exemple des éléments linéaires pour un thème géologique, couvert végétal ou forestier pour l'occupation du sol). Les informations thématiques apparaissent sous forme de surfaces ou de symboles colorés. Le relief peut être représenté par des courbes de niveau. Les routes et les agglomérations sont également indiquées par des signes conventionnels particuliers. Les surfaces d'eau libre sont renforcées par une teinte bleue. En général, on choisit des teintes claires pour que le fond reste lisible. Dans certains cas, une technique de "zonage" est utilisée pour faire ressortir les informations thématiques seulement sur certaines parties du fond monochrome ou coloré de l'iconocarte (voir par ex. applications 10, 14, etc.).

Avantages:
—possibilité d'informations détaillées,
—pas de "blanc" dans la carte (comme dans les cartes classiques),
—aspect esthétique parfois remarquable.

Inconvénients:
—l'image n'est plus interprétable car les surcharges thématiques occultent l'information iconique qui par ailleurs manque de contraste,
—dans certains cas, l'information iconique va jusqu'à disparaître du fait de l'emploi d'aplats colorés appliqués principalement à l'hydrographie et aux agglomérations urbaines.

4.4.3 Iconocartes à thèmes accentués

Le problème général des iconocartes est de trouver un équilibre entre le fond d'image et les surcharges (graphiques) thématiques. En 4.4.1

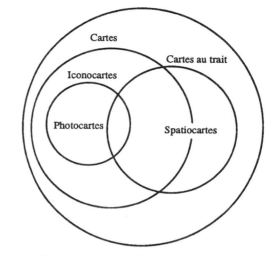

Fig. 4.1 Cartes au trait et iconocartes (CNIG, 1989).

les informations iconiques, et en 4.4.2 les surcharges thématiques prennent, tour à tour, le pas les unes sur les autres. Comment établir une iconocarte surmontant les inconvénients ci-dessus? Un processus conduisant à un tel produit est illustré par la fig. 4.2. En fonction des catégories des thèmes de surface et les

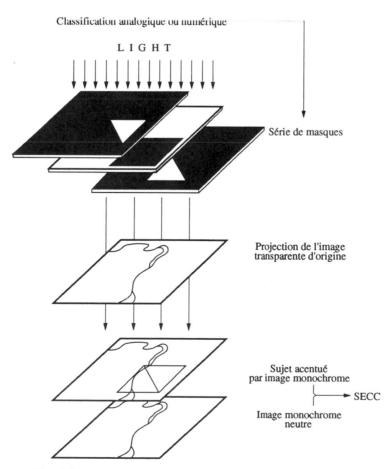

Fig. 4.2 Processus d'accentuation des zones thématiques.

couleurs choisies pour la facture finale, on construit une série de masques. Par l'intermédiaire d'un procédé classique de pré-masquage, on obtient des planches tramées par couleurs séparées en utilisant comme une image monocanal originale déjà tramée. Une méthode numérique pour accentuer les thèmes consiste à imposer au pixel interprété une nuance de couleur associée à la radiométrie de l'une des bandes du spectre. Les informations graphiques apparaissent en teinte monochrome et peuvent être combinées avec l'image en couleur.

Avantages:
—le pixel original (point) et les rapports spectraux (contrastes) ne sont pas perdus et l'image peut toujours être interprétée,
—les thèmes surfaciques sont bien définis et peuvent faire l'objet d'une légende,
—les thèmes surfaciques sont bien distingués même s'ils ne correspondent pas à des zones naturelles.

Inconvénients:
—la réalisation d'une iconocarte à thèmes renforcés est plus longue et plus onéreuse que celle des autres iconocartes,
—les meilleurs résultats sont obtenus pour les cartes d'occupation du sol. Quelquefois, la diversité des types de couverture du sol entraîne des variations de contraste à l'intérieur d'un thème et peut perturber le résultat,
—quelquefois, des thèmes surfaciques colorés détruisent l'harmonie de la carte, rendant la lecture difficile. Pour avoir une image mieux interprétable, l'image classifiée peut être combinée (fusionnée) avec une image monochrome "neutre" issue de la même bande spectrale, ou d'une autre, ceci dépendant du thème de la classification.

4.4.4 *Un cas particulier: la toponymie des iconocartes*

Une toponymie correcte est un problème général en cartographie. Comment positionner les noms géographiques au bon endroit, sans perturber le contenu de la carte, tout en la rendant lisible? Comment utiliser les différentes polices de caractères pour donner un sens aux toponymes? En cartographie classique, le problème est plus ou moins résolu, car on peut trouver (ou créer) des vides pour disposer les noms, utiliser différentes polices de caractères et

différentes couleurs pour distinguer par exemple les noms des villes de ceux des cours d'eau. Mais comment résoudre le problème dans le cas d'iconocartes?

Les surfaces "vides"—c'est-à-dire des espaces de couleur ou de teinte uniforme—sont rares (grandes étendues d'eau, déserts, forêts, terrains enneigés par exemple). Créer un vide pour placer un nom n'est pas la meilleure solution car on perd ainsi en moyenne trois fois plus d'informations contenues dans l'image que si l'on se contentait simplement d'ajouter les lettres. Comment améliorer la lisibilité des noms sur une iconocarte? Il faut abandonner certaines conventions et règles cartographiques. On peut (on doit!) employer différentes tonalités ou couleurs pour des noms appartenant à un même thème (lettres de couleur blanche ou jaune sur un fond sombre, ou de couleur noire ou bistre sur les parties claires par exemple), mais on doit conserver comme convention de n'utiliser qu'un type de caractères pour un même thème. Les figs 4.3a–d, montrent quelques exemples typiques de méthodes employées en matière de toponymie sur les iconocartes.

En fait, le problème est pratiquement le même que pour les photocartes réalisées classiquement à partir de photographies aériennes, avec toutefois deux différences fréquemment rencontrées: l'emploi de la couleur est plus généralisée en imagerie satellite qu'en imagerie aérienne, et surtout, le caractère très souvent numérique des images satellites permet d'exploiter toutes les ressources de l'informatique pour optimiser, en particulier, la représentation des toponymes. Ainsi l'emploi des couleurs pour ceux-ci peut être généralisé à l'infini (ce qui n'est pas le cas en cartographie classique), sans augmenter pour autant le nombre de planches-mères à imprimer: il suffit pour cela de combiner, au moment de l'établissement final de planches-mères, l'ensemble des écritures avec les fichiers de données "iconiques", "thématiques" et "graphiques". On peut ainsi obtenir l'ensemble de toutes les informations sur les 4 films dc quadrichromie (noir, jaune, cyan et magenta) (cf. chapitre 3).

4.5 Typologie des iconocartes

D'une manière générale, et compte tenu de tous les chapitres précédents, 5 critères paraissent pouvoir caractériser une iconocarte:

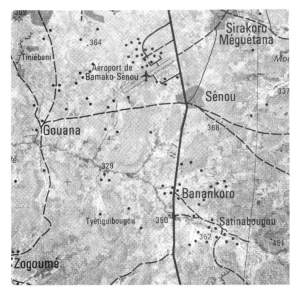

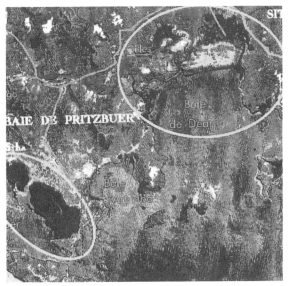

Composition photomécanique: simple surcharge en noir sur le fond d'image

Composition numérique en mode raster: surcharges en négatif couleur (c à d texte en jaune sur les parties sombres de l'image, en blanc sur les parties rouges, etc.) voir Chap. V, applic. 15

Composition photomécanique en noir avec un léger détourage des lettres (pour une meilleure lecture)

Composition numérique en mode raster: lettres "pixélisées" en négatif en arabe, avec un léger ombrage (lumière du N-O)

Fig. 4.3 Exemples de méthodes de rédaction des toponymes sur des iconocartes satellitaires.

Critères métriques
1. Valeur du traitement géométrique de l'image servant de fond.
2. Cohérence du système de représentation cartographique, découpage, format, habillage, légende.

Critères cartographiques
3. Richesse de l'information cartographique ajoutée à l'image.
4. Mise en place et fiabilité de l'information cartographique ajoutée à l'image.

5. qualité de l'esthétique cartographique de l'ensemble rédaction-édition.

Pour chaque critère, la qualité du produit peut être caractérisée, comme pour les cartes classiques, par un niveau défini comme suit:

Iconocarte régulière
Iconocarte d'étude
(Classe intermédiaire se présentant souvent: le produit n'a pas toutes les caractéristiques d'une carte ou bien, s'il les a toutes, celles-ci

n'atteignent pas la qualité requise pour une carte régulière).

Document "image" (pratiquement sans redressement géométrique et/ou sans habillage).

Ces définitions sont explicitées dans le chapitre suivant, qui décrit, pour chaque niveau des cinq critères ci-dessus, la typologie correspondante des cartes sur fonds d'images (voir fig. 4.4 à la fin).

4.5.1 *Typologie des iconocartes régulières*

Au sens de la cartographie contemporaine, toute carte régulière suppose une correspondance rigoureuse entre les positions représentées et les positions réelles dans l'espace.

Une carte régulière sur fond d'image doit donc comporter les caractéristiques générales suivantes:

4.5.1.1 *Critère no. 1. Traitement géométrique de l'image servant de fond*

L'image est corrigée géométriquement, de façon à satisfaire à une précision planimétrique caractérisée par:

Un écart-type de 0,3 mm sur le document édité (soit statistiquement: *90% des points à mieux que 0,5 mm* de leur position exacte, norme utilisée fréquemment à l'étranger; ou encore: *99% des points à mieux que 0,8 mm* de leur position exacte, norme utilisée généralement en France sous le nom de "tolérance").

Lorsque les données sont définies uniquement sous forme numérique, ces valeurs doivent évidemment être données en unités de terrain. On en déduit alors l'échelle maximale de représentation à respecter pour obtenir une carte régulière.

Selon le capteur ou le vecteur utilisé, les données à traiter et les processus à mettre en oeuvre peuvent varier sensiblement. Dans certains cas (définis suivant les paramètres précédents) on peut obtenir une géométrie correcte par redressement photographique ou correction géométrique sur points d'appui (cas des photos redressées ou des images SPOT prétraitées au niveau 2). Dans d'autres cas, notamment pour les zones à relief accidenté, une correction avec modèle numérique de terrain est nécessaire (cas des orthophoto-graphies ou des images SPOT prétraitées au niveau 3).

4.5.1.2 *Critère no. 2. Système de représentation*

—Pour les cartes de série, le découpage doit être normalisé de préférence, ou au moins régulier, et appuyé sur les axes rectangulaires ou géographiques. Le format doit être en principe standardisé pour les cartes de base.

—Pour les productions hors série, découpage et format peuvent être adaptés à la configuration générale de la zone cartographiée; l'ensemble doit être cohérent.

Système de représentation plane. Pour les cartes régulières, la représentation des éléments du système de coordonnées doit être complète: amorces périphériques et angles de coupure chiffrés, croisillons. Il est utile de faire figurer, aux moyennes et petites échelles, le quadrillage géographique.

Lorsque la représentation complète du système de coordonnées n'est pas nécessaire à l'utilisateur (carte touristique par exemple), elle peut être allégée. L'information fournie devra néanmoins permettre de définir sans ambiguïté les coordonnées planes de tout point représenté.

Habillage—Légende. Un certain nombre d'indications de base doivent figurer en marge, qu'elles soient générales (titre, échelle, auteurs, date d'édition, etc.) ou spécifiques:

—Date d'acquisition, origine des images, échelle de saisie s'il s'agit de photographies, et éventuellement traitement appliqué.

—Légende permettant d'interpréter l'image: "caissons" ou extraits d'image montrant les aspects types de la végétation, de l'hydro-graphie, etc.

—Légende complète des signes conven-tionnels ajoutés à l'image.

4.5.1.3 *Critère no. 3. Richesse de l'information cartographique (ajoutée à l'image)*

Les éléments représentés, soit par leur image, soit par un signe additionnel, sont ceux répertoriés dans les spécifications générales de la cartographie nationale ou dans un document particulier, par exemple un appel d'offres (cahier des clauses techniques particulières).

	CRITERES D'EVALUATION	VAL.	CARACTÉRISTIQUES	IMAGES			CARTES D'ETUDES		CARTES REGULIERES	
QUALITE METRIQUES	No.1 TRAITEMENT GEOMETRIQUE DE L'IMAGE	3 2 1	Corrections de toutes les déformations (capteur-relief) Mise en projection Corrections partielles Pas de projection orthogonale Pas de correction Mise à échelle moyenne	1,5 à 1	2 à 1	3 à 2	3 à 2	3 à 2	3	3
QUALITE METRIQUES	No.2 SYSTEME DE REPRESENTATION HABILLAGE	3 2 1	DÉCOUPAGE / PROJECTION / LÉGENDE Système international / Représentée / Complète Système local / Allégée / Réduite Aucun / Non représentée / Inexistante	1	2 à 1	3 à 1	3 à 2	3 à 2	3	3
QUALITES CARTOGRAPHIQUES	No.3 RICHESSE DE L'INFORMATION AJOUTEE	3 2 1	Surcharge d'informations fiables et complètes Surcharge d'informations approximatives et généralisées Pas de surcharge	1	1	1	2	2	3 et 2,5	3 et 2,5
QUALITES CARTOGRAPHIQUES	No.4 MISE EN PLACE DE L'INFORMATION AJOUTEE	3 2 1	Données et processus cartographiques réguliées conformes aux normes Données et processus cartographiques adaptés aux conditions locales Données et processus cartographiques facultatifs	1	1	1	3 à 2	3 à 2	3	3
QUALITES CARTOGRAPHIQUES	No.5 ESTHETIQUE CARTOGRAPHIQUE	3 2 1	Image: raccords entre images et feuilles assurés. Rédaction du trait aux normes réguliées Image: raccords non retouchés Rédaction du trait et toponymie de facture simple Image brute	1	2	2,5 à 2	3 à 2	3 à 2	3 à 2,5	3 à 2,5
	TOTAL			5 à 5,5	6 à 8	2 à 11,5	10 à 14	10 à 14	14 á 15	14 á 15
	DOCUMENTS SUR FONDS D'IMAGES			Image brute	Assemblage d'images	Mosaique d'images	Carte d'etude topogr.	Carte d'etude themat.	Carte régul. topogr.	Carte régul. themat.

Fig. 4.4 Typologie provisoire de produits iconocartographiques basée sur cinq critères.

Information topographique (planimétrie)

—Les éléments du paysage dont l'image est explicite peuvent ne pas être surchargés par un signe.

—Les détails importants justiciables de l'échelle adoptée, qui ne sont pas visibles sur l'image, ou visibles de façon trop discontinue, ou d'identification malaisée, sont représentés par un signe combiné à l'image.

—Les écritures (toponymes, désignations) respectent au mieux les critères de densité, localisation, classification, des cartes classiques de même échelle.

—De façon générale, la densité et le mode d'expression graphique des éléments combinés à l'image doivent être harmonieusement équilibrés pour que l'ensemble reste clair et lisible et que l'information apportée par l'image soit globalement conservée.

Information topographique (altimétrie). L'altimétrie est représentée par des courbes de niveau et/ou des points cotés selon le paysage. L'équidistance des courbes, la densité des points sont celles des normes générales de référence ou de l'appel d'offres.

Information thématique. L'information thématique ajoutée à l'image provient, soit d'une interprétation de l'image elle-même (photo-interprétation, classification assistée par ordinateur, etc.), soit d'autres sources d'information (cartes existantes, observations sur le terrain, etc.).

—La richesse de cette information (nombre de thèmes) doit être compatible avec une bonne lisibilité de la carte sur fond d'image.

—Le thème représenté a priorité sur la représentation planimétrique additionnelle, qui doit toutefois rester suffisante pour la localisation des phénomènes.

4.5.1.4   Critère no. 4. Mise en place et fiabilité de l'information cartographique (ajoutée à l'image)

Fiabilité de la position géométrique. La qualité géométrique de l'image étant évaluée par le critère no. 1, la valeur du renseignement ajouté, au sens qualitatif, étant estimée par le critère no. 3, on examine la valeur cartographique, au sens géométrique, de ce renseignement. On doit donc évaluer ici l'ensemble des données et des processus techniques utilisés pour cartographier ces éléments, et la précision de leur mise en place au vu des normes internationales ou des spécifications particulières (précision planimétrique et altimétrique).

Cette mise en place doit offrir la même qualité géométrique que l'image (voir critère no. 1).

Actualité de l'information. La date des données ajoutées à l'image doit être en rapport avec la date de celle-ci. En particulier, si ces données proviennent d'une carte existante, celle-ci a été préalablement actualisée selon les normes habituelles. En tout état de cause, l'origine, la date d'édition et/ou de mise à jour des données cartographiques ajoutées à l'image sont clairement mentionnées.

Fiabilité de l'information thématique. On doit indiquer l'origine de l'information thématique et son processus d'élaboration (photo-interprétation, classification assistée par ordinateur, données externes, etc.), ainsi que la nature et la valeur du critère d'évaluation des résultats, par exemple sous forme d'un intervalle de confiance ou d'une matrice de confusion.

4.5.1.5   Critère no. 5. Esthétique cartographique

Image servant de fond. Sa date d'enregistrement doit être choisie dans une période favorable (état atmosphérique, végétation, etc.); elle doit avoir reçu les traitements nécessaires pour que sa qualité générale (définition, densité moyenne, contraste, etc.) soit conservée, voire améliorée.

—Si une coupure se compose de plusieurs images de même date, leurs lignes de raccord doivent être pratiquement invisibles. La densité ou la radiométrie de part et d'autre des raccords doivent être équilibrées.

—Entre coupures adjacentes, le fond d'image ne doit pas présenter d'écart sensible.

—Si des images de dates différentes doivent être utilisées (ou si certains aspects du paysage sont sensiblement différents), des écarts sont tolérés entre elles (densité, couleurs, etc.).

Ensemble rédaction—édition. L'ensemble de ces phases doit conserver la qualité de l'image et offrir un niveau de qualité satisfaisant pour les informations ajoutées à l'image, les indications en marge et l'habillage.

Les écritures doivent avoir la qualité cartographique habituelle, quelle que soit leur origine: typographie, photocomposition, photo-traceur incrémental. Celles qui seront combinées à l'image devront être rendues lisibles par tout procédé (couleur, alourdi, réserve en blanc, etc.).

Cette qualité d'ensemble est homogène dans l'ensemble de la production.

4.5.2   Typologie des iconocartes d'étude

Certaines des spécifications requises pour la carte régulière ne sont pas respectées.

4.5.2.1   Critère no. 1. Traitement géométrique de l'image

Le niveau de qualité de l'ensemble données-traitements n'est pas suffisant pour considérer l'image résultante comme une projection ortho-gonale du terrain. Par exemple, correction d'image SPOT au niveau 2 en zone à fort relief.

4.5.2.2   Critère no. 2. Système de représentation

Aucune spécification de système cartographique officiel n'est retenue. Découpage et format sont adaptés à la forme du chantier. Le système de projection peut être représenté de façon succincte: angles de coupure, amorces et croisillons dans le seul quadrillage rectangulaire. La légende est réduite, voire incomplète.

4.5.2.3   Critère no. 3. Richesse de l'information cartographique ajoutée

La surcharge des informations est très allégée. Toutefois, il ne faut pas confondre une information allégée et une information de faible densité provoquée par le caractère semi-désertique de la région cartographiée.

4.5.2.4   Critère no. 4. Mise en place et fiabilité de l'information cartographique ajoutée

Sa qualité n'est pas garantie au sens carto-graphique, c'est-à-dire qu'elle n'est pas appuyée sur une information précise (complètement sur le terrain, compilation de carte ancienne sans mise à jour). De même la fiabilité de l'informa-tion thématique représentée n'est pas caractér-isée à l'aide des critères d'évaluation habituels.

4.5.2.5   Critère no. 5. Esthétique cartographique

La qualité d'ensemble est bonne, mais peut varier légèrement entre coupures adjacentes (processus différents, etc.). A l'intérieur d'une coupure, les raccords d'image peuvent être visibles, mais de façon discontinue.

4.5.3   Typologie des documents appelés "images"

Aucune des spécifications définies précédem-ment n'est exigée. Seuls l'origine des données de base et les processus sont indiqués.

Références

Bertin, J. 1983. *Semiology of Graphics.* University of Wisconsin Press, USA.

ICA-ACI Denegre, J. *et al.* 1988. *International Report on Thematic Mapping from Satellite Imagery.* Elsevier, U.K.

CNIG. 1989. National Council for Geographic Information. *Typology and Terminology of Image Maps.* Paris.

Galtier, B. 1992. Le rôle de la spatiocarte en tant que produit cartographique. *Proceedings 17th Conference ISPRS.*

Keyword index

Index de mots-clé